PEARSON EDEXCEL INTERNATIONAL AS/A LEVEL

PHYSICS

Student Book 1

Miles Hudson

Published by Pearson Education Limited, 80 Strand, London, WC2R 0RL.

www.pearsonglobalschools.com

Copies of official specifications for all Pearson Edexcel qualifications may be found on the website: https://qualifications.pearson.com

Text © Pearson Education Limited 2018
Designed by Tech-Set Ltd, Gateshead, UK
Edited by Kate Blackham and Jane Read
Typeset by Tech-Set Ltd, Gateshead, UK
Original illustrations © Pearson Education Limited 2018
Cover design by Pearson Education Limited 2018
Picture research by Aptara, Inc
Cover photo © RUSSELL CROMAN/SCIENCE PHOTO LIBRARY
Inside front cover photo: Dmitry Lobanov

The right of Miles Hudson to be identified as author of this work has been asserted by him in accordance with the Copyright, Designs and Patents Act 1988.

First published 2018

25 24
15 14

British Library Cataloguing in Publication Data
A catalogue record for this book is available from the British Library
ISBN 978 1 2922 4487 7

Printed by Neografia in Slovakia

Acknowledgements

Logos
Logo on page 168 from LIFEPAK® 1000 DEFIBRILLATOR brochure, http://www.physio-control.com/uploadedFiles/Physio85/Contents/Emergency_Medical_Care/Products/Brochures/LP1000_Brochure 20w 20Rechargable 20Battery_3303851_C.pdf, copyright © 2012 Physio-Control, Inc.

Tables
Table on page 84 from Mountaineering: The Freedom of the Hills, 6th Edition, published by Mountaineers Books, Seattle and Quiller Publishing, Shrewsbury 148, copyright © 1997. Illustration and table reprinted with permission of the publisher; Table on page 84 adapted from 'The 2013/14 catalogue of DMM International', www.dmmwales.com. Reproduced with permission from DMM International Ltd; Table on page 144 from 'Electrical Power Annual' report, December 2013, The United States Energy Information Association Source: U. S. Energy Information Administration (Dec 2013).

Text
Extract on page 54 from manufacture hockey goalkeeping equipment, 8 July 2014, http://www.obo.co.nz/the-o-lab, OBO. Reproduced by permission; Extract on page 70 from 'What is a Plimsoll line?', http://oceanservice.noaa.gov/facts/plimsoll-line.html. Source: NOAA's National Ocean Service; Extract on page 96 from USGC FAQs, http://www.usgs.gov/faq. Source: U.S. Geological Survey Department of the Interior/USGS; Extract on page 114 adapted from 'Tuning the Marimba Bar and Resonator' http://lafavre.us/tuning-marimba, copyright © 2007 Jeffrey La Favre; Extract on page 126 from 'Forensic Glass Comparison: Background Information Used in Data Interpretation', publication number 09-04 of the Laboratory Division of the Federal Bureau of Investigation, Vol.11, No 2 (Maureen Bottrell),April 2009, Source: Federal Bureau of Investigation, Quantico, Virginia, USA; Extract on page 168 from LIFEPAK® 1000 DEFIBRILLATOR brochure, pp. 56–57, http://www.physiocontrol.com /uploadedFiles/Physio85/Contents/Emergency_Medical_Care/Products/Brochures/LP1000_Brochure%20w%20Rechargable%20Battery_3303851_C.pdf, copyright © 2012 Physio-Control, Inc. ; Extract on page 190 from 'Rover Team Working to Diagnose Electrical Issue', 2 November 2013, http://www.jpl.nasa.gov/news/news.php?feature=3958, Source: NASA.

Every effort has been made to contact copyright holders of material reproduced in this book. Any omissions will be rectified in subsequent printings if notice is given to the publishers.

For Photo and Figure Acknowedgements please see page 214

ABOUT THIS BOOK

This book is written for students following the Pearson Edexcel International Advanced Subsidiary (IAS) Physics specification. This book covers the full IAS course and the first year of the International A Level (IAL) course.

The book contains full coverage of IAS units (or exam papers) 1 and 2. Each unit in the specification has two topic areas. The topics in this book, and their contents, fully match the specification. You can refer to the Assessment Overview on page X for further information. Students can prepare for the written Practical Skills Paper (unit 3) by using the IAL Physics Lab Book (see page viii of this book).

Each Topic is divided into chapters and sections to break the content down into manageable chunks. Each section features a mix of learning and activities.

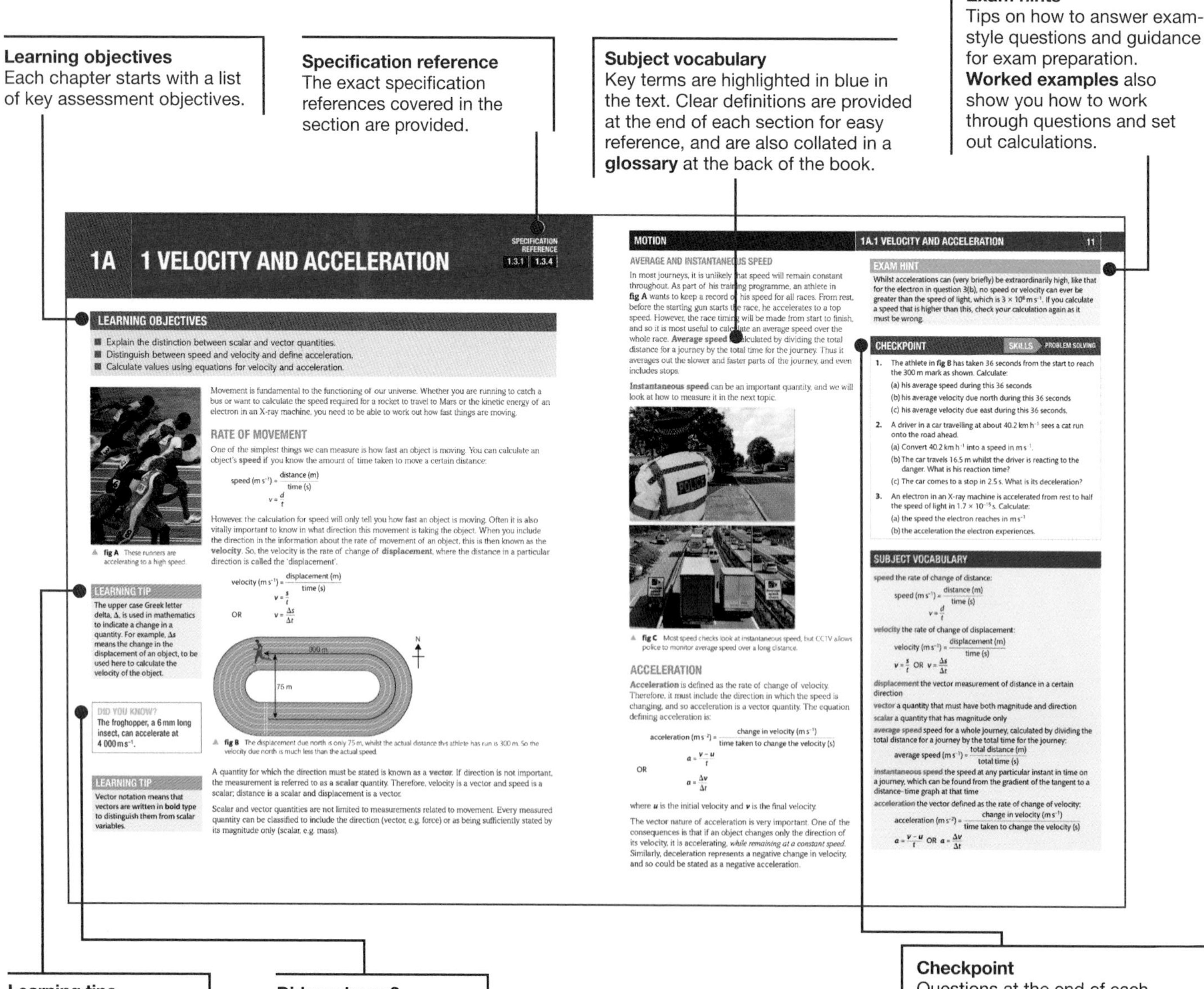

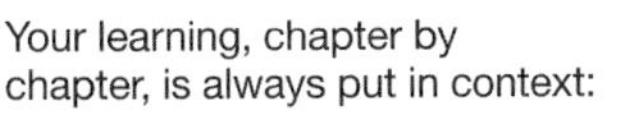

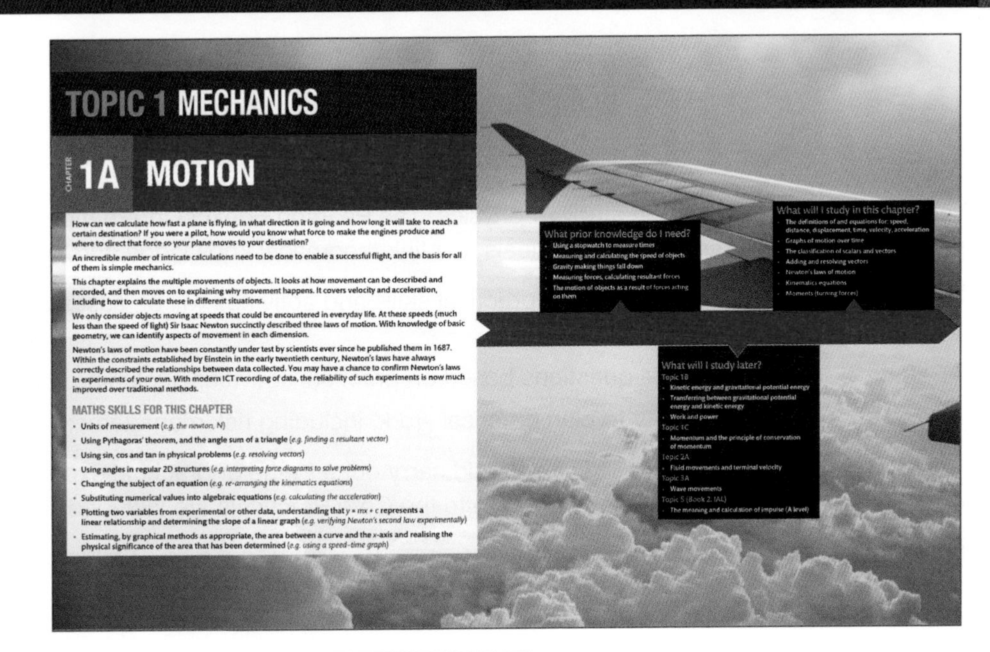

Your learning, chapter by chapter, is always put in context:

- Links to other areas of Physics include previous knowledge that is built on in the topic, and future learning that you will cover later in your course.
- A checklist details maths knowledge required. If you need to practise these skills, you can use the **Maths Skills** reference at the back of the book as a starting point.

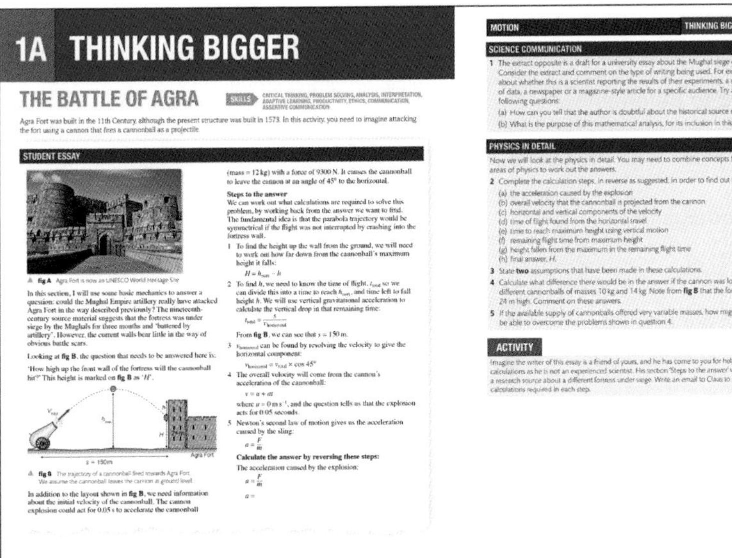

Thinking Bigger
At the end of each topic, there is an opportunity to read and work with real-life research and writing about science. The activities help you to read real-life material that's relevant to your course, analyse how scientists write, think critically and consider how different aspects of your learning piece together.

Skills
These sections will help you develop transferable skills, which are highly valued in further study and the workplace.

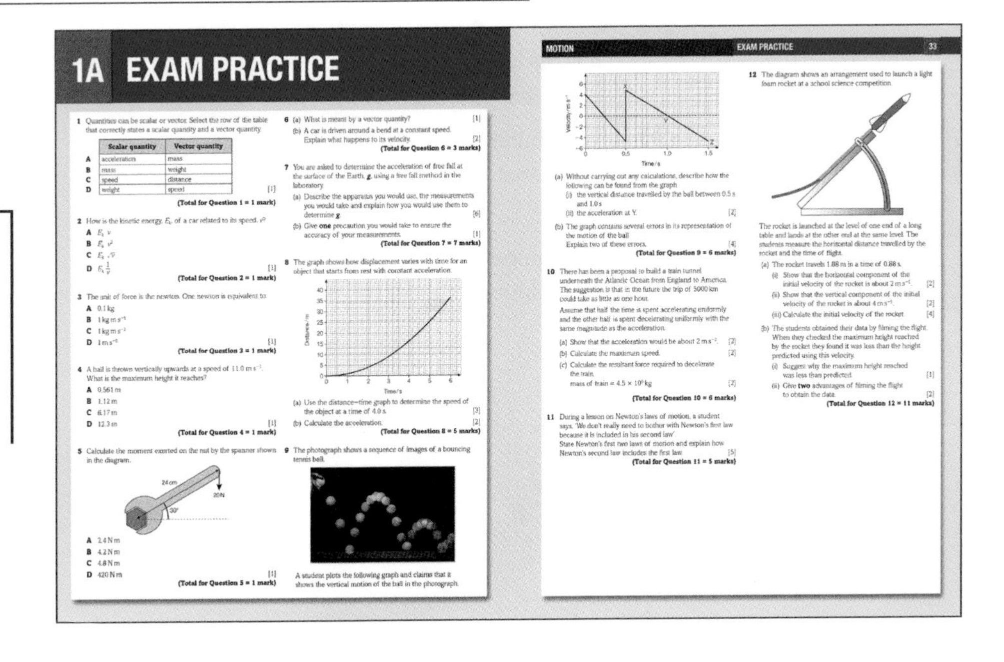

Exam Practice
Exam-style questions at the end of each chapter are tailored to the Pearson Edexcel specification to allow for practice and development of exam writing technique. They also allow for practice responding to the command words used in the exams (see the **command words glossary** at the back of this book).

PRACTICAL SKILLS

Practical work is central to the study of physics. The Pearson Edexcel International Advanced Subsidiary (IAS) Physics specification includes eight Core Practicals that link theoretical knowledge and understanding to practical scenarios.

Your knowledge and understanding of practical skills and activities will be assessed in all examination papers for the IAS Level Physics qualification.

- Papers 1 and 2 will include questions based on practical activities, including novel scenarios.
- Paper 3 will test your ability to plan practical work, including risk management and selection of apparatus.

In order to develop practical skills, you should carry out a range of practical experiments related to the topics covered in your course. Further suggestions in addition to the Core Practicals are included below.

STUDENT BOOK TOPIC	IAS CORE PRACTICALS	
TOPIC 1 MECHANICS	**CP1**	Determine the acceleration of a freely-falling object
TOPIC 2 MATERIALS	**CP2**	Use a falling-ball method to determine the viscosity of a liquid
	CP3	Determine the Young modulus of a material
TOPIC 3 WAVES AND THE PARTICLE NATURE OF LIGHT	**CP4**	Determine the speed of sound in air using a two-beam oscilloscope, signal generator, speaker and microphone
	CP5	Investigate the effects of length, tension and mass per unit length on the frequency of a vibrating string or wire
	CP6	Determine the wavelength of light from a laser or other light source using a diffraction grating
TOPIC 4 ELECTRIC CIRCUITS	**CP7**	Determine the electrical resistivity of a material
	CP8	Determine the e.m.f. and internal resistance of an electrical cell

UNIT 1 (TOPICS 1 AND 2) MECHANICS AND MATERIALS

Possible further practicals include:

- Strobe photography or the use of a video camera to analyse projectile motion
- Determine the centre of gravity of an irregular rod
- Investigate the conservation of momentum using light gates and air track
- Hooke's law and the Young modulus experiments for a variety of materials

UNIT 2 (TOPICS 3 AND 4) WAVES AND ELECTRICITY

Possible further practicals include:

- Estimating power output of an electric motor
- Using a digital voltmeter to investigate the output of a potential divider and investigating current/voltage graphs for a filament bulb, thermistor and diode
- Determining the refractive index of solids and liquids, demonstrating progressive and stationary waves on a slinky

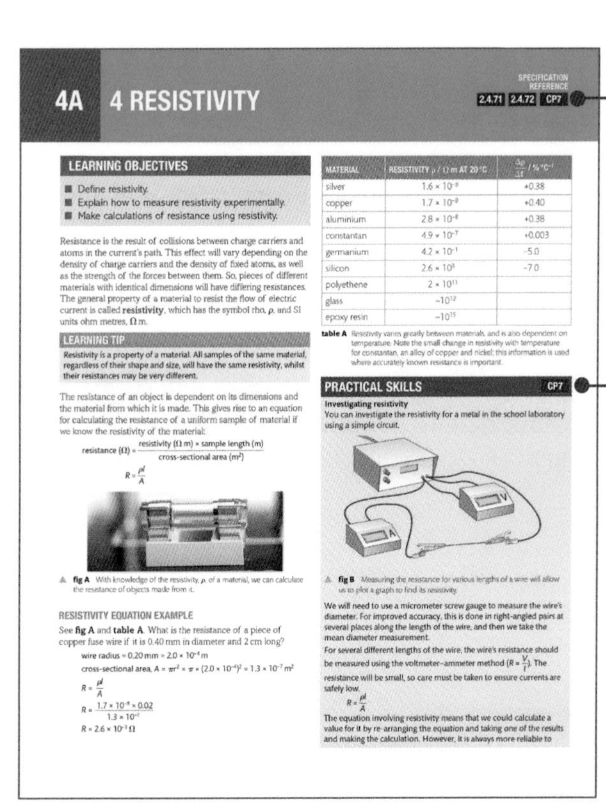

4A 4 RESISTIVITY

SPECIFICATION REFERENCE 2.4.71 2.4.72 CP7

LEARNING OBJECTIVES

- Define resistivity.
- Explain how to measure resistivity experimentally.
- Make calculations of resistance using resistivity.

Resistance is the result of collisions between charge carriers and atoms in the current's path. This effect will vary depending on the density of charge carriers and the density of fixed atoms, as well as the strength of the forces between them. So, pieces of different materials with identical dimensions will have differing resistances. The general property of a material to resist the flow of electric current is called **resistivity**, which has the symbol rho, ρ, and SI units ohm metres, Ω m.

LEARNING TIP

Resistivity is a property of a material. All samples of the same material, regardless of their shape and size, will have the same resistivity, whilst their resistances may be very different.

The resistance of an object is dependent on its dimensions and the material from which it is made. This gives rise to an equation for calculating the resistance of a uniform sample of material if we know the resistivity of the material:

$$\text{resistance } (\Omega) = \frac{\text{resistivity } (\Omega \text{ m}) \times \text{sample length (m)}}{\text{cross-sectional area } (\text{m}^2)}$$

$$R = \frac{\rho l}{A}$$

fig A With knowledge of the resistivity, ρ, of a material, we can calculate the resistance of objects made from it.

RESISTIVITY EQUATION EXAMPLE

See **fig A** and **table A**. What is the resistance of a piece of copper fuse wire if it is 0.40 mm in diameter and 2 cm long?

wire radius = 0.20 mm = 2.0×10^{-4} m

cross-sectional area, $A = \pi r^2 = \pi \times (2.0 \times 10^{-4})^2 = 1.3 \times 10^{-7}$ m^2

$$R = \frac{\rho l}{A}$$

$$R = \frac{1.7 \times 10^{-8} \times 0.02}{1.3 \times 10^{-7}}$$

$$R = 2.6 \times 10^{-3}\ \Omega$$

MATERIAL	RESISTIVITY ρ / Ω m AT 20 °C	$\frac{\Delta\rho}{\Delta T}$ / % °C^{-1}
silver	1.6×10^{-8}	+0.38
copper	1.7×10^{-8}	+0.40
aluminium	2.8×10^{-8}	+0.38
constantan	4.9×10^{-7}	+0.003
germanium	4.2×10^{-1}	−5.0
silicon	2.6×10^{3}	−7.0
polyethene	2×10^{11}	
glass	~10^{12}	
epoxy resin	~10^{15}	

table A Resistivity varies greatly between materials, and is also dependent on temperature. Note the small change in resistivity with temperature for constantan, an alloy of copper and nickel; this information is used where accurately known resistance is important.

PRACTICAL SKILLS CP7

Investigating resistivity

You can investigate the resistivity for a metal in the school laboratory using a simple circuit.

fig B Measuring the resistance for various lengths of a wire will allow us to plot a graph to find its resistivity.

We will need to use a micrometer screw gauge to measure the wire's diameter. For improved accuracy, this is done in right-angled pairs at several places along the length of the wire, and then we take the mean diameter measurement.

For several different lengths of the wire, the wire's resistance should be measured using the voltmeter–ammeter method ($R = \frac{V}{I}$). The resistance will be small, so care must be taken to ensure currents are safely low.

$$R = \frac{\rho l}{A}$$

The equation involving resistivity means that we could calculate a value for it by re-arranging the equation and taking one of the results and making the calculation. However, it is always more reliable to

In the **Student Book**, the Core Practical specification and Lab Book references are supplied in the relevant sections.

Practical Skills

Practical skills boxes explain techniques used in the Core Practicals, and also detail useful skills and knowledge gained in other related investigations.

CORE PRACTICAL 1:
DETERMINE THE ACCELERATION OF A FREELY-FALLING OBJECT

SPECIFICATION REFERENCE 1.3.11

Procedure

1 Drop the object from rest and record the time taken, t, for:
  (a) the sphere to fall through the trap door
  (b) the dowel to pass through the light gate.
2 Repeat step **1** twice more and calculate the mean value of t for each method.
3 Measure and record the height, h, fallen by the object.
4 Vary the height and repeat steps **1–3**. You should take readings at at least six different heights.
5 Use half the range in your readings for t as the uncertainty in t. Calculate the percentage uncertainty in t.
6 For method **(b)**, you should measure the length of the dowel.

Learning tips

- Make sure that points plotted on a graph take up more than half of the available space on each scale. You do not always need to include the origin.
- Keep scales simple: one large square as 5, 10 or 20 is ideal. A scale where one large square represents 3 or 7 units (or similar) is very difficult to plot and can often lead to errors.
- Always consider whether the graph line should go through the origin.
- Straight lines should be drawn with the aid of a rule long enough to cover the full length of the line.
- Since the object is falling at constant acceleration, use the appropriate kinematics equation:
  (a) $s = ut + \frac{1}{2}at^2$ where $u = 0$, $a = g$, and $s = h$
  This can be rearranged to: $t^2 = \frac{2h}{g}$
  Comparison with $y = mx + c$ shows that plotting t^2 against h should give a straight line passing through the origin with gradient $\frac{2}{g}$.
  (b) $v^2 = u^2 + 2as$ where $u = 0$, $a = g$, and s is h
  Therefore: $v^2 = 2gh$
  Comparison with $y = mx + c$ shows that plotting v^2 against h should give a straight line passing through the origin with gradient $2g$.

Objectives

- To measure the acceleration due to gravity, g, of an object falling freely and to consider the following alternative methods:
  (a) object falling through a trap door
  (b) object falling through a light gate

Equipment

- metre rule or tape measure with millimetre resolution

For **(a)**:
- steel sphere
- electronic timer
- electromagnet to retain steel sphere
- trap door switch
- clamp and stand
- low voltage power supply

For **(b)**:
- falling object, such as a dowel with 2 cm diameter, 10 cm long
- means to guide dowel through light gate
- light gate and datalogger

Safety

- Make sure the stand cannot topple over by clamping it securely.
- Keep hands and face away from the falling objects.
- Turn off the electromagnet between 'drops' so that it doesn't overheat and cause burns.

6

CORE PRACTICAL 1:
DETERMINE THE ACCELERATION OF A FREELY-FALLING OBJECT

SPECIFICATION REFERENCE 1.3.11

5 The percentage uncertainty (%U) in t^2 is twice that in t. Use this to draw error bars onto your graph for method **(a)** – in the y-direction only. You can use a typical mid-range value to calculate uncertainties – you do not need to work out a separate error bar for each value. Draw a new line of best fit and use this to calculate the %U in your value for g.

6 Calculate the percentage difference (%D) between your value and the accepted value of g (9.81 m s^{-2}) and comment on the accuracy of method **(a)**.

Questions

1 Describe an advantage of using light gates in this experiment.

2 Discuss the effect of air resistance on your value for g.

3 Explain why the graph should be a straight line.

10

This Student Book is accompanied by a **Lab Book**, which includes instructions and writing frames for the Core Practicals for students to record their results and reflect on their work.

Practical skills checklists, practice questions and answers are also provided.

The Lab Book records can be used as preparation and revision for the Practical Skills Paper.

ASSESSMENT OVERVIEW

The following tables give an overview of the assessment for Pearson Edexcel International Advanced Subsidiary course in Physics. You should study this information closely to help ensure that you are fully prepared for this course and know exactly what to expect in each part of the examination. More information about this qualification, and about the question types in the different papers, can be found on page 200 of this book.

PAPER / UNIT 1	PERCENTAGE OF IAS	PERCENTAGE OF IAL	MARK	TIME	AVAILABILITY
MECHANICS AND MATERIALS Written exam paper Paper code WPH11/01 Externally set and marked by Pearson Edexcel Single tier of entry	40%	20%	80	1 hour 30 minutes	January, June and October First assessment : January 2019

PAPER / UNIT 2	PERCENTAGE	PERCENTAGE OF IAL	MARK	TIME	AVAILABILITY
WAVES AND ELECTRICITY Written exam paper Paper code WPH12/01 Externally set and marked by Pearson Edexcel Single tier of entry	40%	20%	80	1 hour 30 minutes	January, June and October First assessment June 2019

PAPER / UNIT 3	PERCENTAGE	PERCENTAGE OF IAL	MARK	TIME	AVAILABILITY
PRACTICAL SKILLS IN PHYSICS 1 Written examination Paper code WPH13/01 Externally set and marked by Pearson Edexcel Single tier of entry	20%	10%	50	1 hour 20 minutes	January, June and October First assessment : June 2019

ASSESSMENT OBJECTIVES AND WEIGHTINGS

ASSESSMENT OBJECTIVE	DESCRIPTION	% IN IAS	% IN IA2	% IN IAL
A01	Demonstrate knowledge and understanding of science	34–36	29–31	32–34
A02	(a) Application of knowledge and understanding of science in familiar and unfamiliar contexts.	34–36	33–36	34–36
	(b) Analysis and evaluation of scientific information to make judgments and reach conclusions.	9–11	14–16	11–14
A03	Experimental skills in science, including analysis and evaluation of data and methods	20	20	20

RELATIONSHIP OF ASSESSMENT OBJECTIVES TO UNITS

UNIT NUMBER	ASSESSMENT OBJECTIVE			
	A01	A02 (a)	A02 (b)	A03
UNIT 1	17–18	17–18	4.5–5.5	0.0
UNIT 2	17–18	17–18	4.5–5.5	0.0
UNIT 3	0.0	0.0	0.0	20
TOTAL FOR INTERNATIONAL ADVANCED SUBSIDIARY	**34–36**	**34–36**	**9–11**	**20**

WORKING AS A PHYSICIST

Throughout your study of physics, you will develop knowledge and understanding of what it means to work scientifically. You will develop confidence in key scientific skills, such as handling and controlling quantities and units and making estimates. You will also learn about the ways in which the scientific community functions and how society as a whole uses scientific ideas.

At the end of each chapter in this book, there is a section called Thinking Bigger. These sections are based broadly on the content of the chapter just completed, but they will also draw on your previous learning from earlier in the course or from your previous studies and point towards future learning and less familiar contexts. The Thinking Bigger sections will also help you to develop transferable skills. By working through these sections, you will:

- read real-life scientific writing in a variety of contexts and aimed at different audiences
- develop an understanding of how the professional scientific community functions
- learn to think critically about the nature of what you have read
- understand the issues, problems and challenges that may be raised
- gain practice in communicating information and ideas in an appropriate scientific way
- apply your knowledge and understanding to unfamiliar contexts.

You will also gain scientific skills through the hands-on practical work that forms an essential part of your course. As well as understanding the experimental methods of the practicals, it is important that you develop the skills necessary to plan experiments and analyse and evaluate data. Not only are these very important scientific skills, but they will be assessed in your examinations.

MATHS SKILLS FOR PHYSICISTS

- **Recognise and make use of appropriate units in calculations** (*e.g. knowing the difference between base and derived units*)
- **Estimate results** (*e.g. estimating the speed of waves on the sea*)
- **Make order of magnitude calculations** (*e.g. estimating approximately what an answer should be before you start calculating, including using standard form*)
- **Use algebra to rearrange and solve equations** (*e.g. finding the landing point of a projectile*)
- **Recognise the importance of the straight line graph as an analysis tool for the verification and development of physical laws by experimentation** (*e.g. choosing appropriate variables to plot to generate a straight line graph with experimental data*)
- **Determine the slope and intercept of a linear graph** (*e.g. finding acceleration from a velocity–time graph*)
- **Calculate the area under the line on a graph** (*e.g. finding the energy stored in a stretched wire*)
- **Use geometry and trigonometry** (*e.g. finding components of vectors*)

What prior knowledge do I need?

- Understanding and knowledge of physical facts, terminology, concepts, principles and practical techniques
- Applying the concepts and principles of physics, including the applications of physics, to different contexts
- Appreciating the practical nature of physics and developing experimental and investigative skills based on the use of correct and safe laboratory techniques
- Communicating scientific methods, conclusions and arguments using technical and mathematical language
- Consideration of the implications, including benefits and risks, of scientific and technological developments
- Understanding how society uses scientific knowledge to make decisions about the implementation of technological developments
- Understanding of how scientific ideas change over time, and the systems in place to validate these changes

What will I study in this section?

- The difference between base and derived quantities and their SI units
- How to estimate values for physical quantities and use these estimates to solve problems

What will I study later?

- Knowledge and understanding of further physical facts and terminology, deeper concepts, principles and more complex practical techniques
- Practical skills and techniques for some key physics experiments
- How to communicate information and ideas in appropriate ways using appropriate terminology
- The implications of science and their associated benefits and risks
- The role of the scientific community in validating new knowledge and ensuring integrity
- The ways in which society uses science to inform decision making

1 STANDARD UNITS IN PHYSICS

LEARNING OBJECTIVES

- Understand the distinction between base and derived quantities.
- Understand the idea of a fixed system of units, and explain the SI system.

BASE AND DERIVED QUANTITIES

▲ **fig A** The international standard kilogram, officially known as the International Prototype Kilogram, is made from a mixture of platinum and iridium and is held at the Bureau International des Poids et Mesures in Paris. All other masses are defined by comparing with this metal cylinder.

Some measurements we make are of fundamental qualities of things in the universe. For example, the length of a pencil is a fundamental property of the object. Compare this with the pencil's speed if you drop it. To give a value to the speed, we have to consider a distance moved, and the rate of motion over that distance – we also need to measure time and then do a calculation. You can see that there is a fundamental difference between the types of quantity that are length and speed. We call the length a base unit, whilst the speed is a derived unit. At present, the international scientific community uses seven base units, and from these all other units are derived. Some derived units have their own names. For example, the derived unit of force should be $kg\,m\,s^{-2}$, but this has been named the newton (N). Other derived units do not get their own name, and we just list the base units that went together in deriving the quantity. For example, speed is measured in $m\,s^{-1}$.

BASIC QUANTITY	UNIT NAME	UNIT SYMBOL
mass	kilogram	kg
time	second	s
length	metre	m
electric current	ampere	A
temperature	kelvin	K
amount of substance	mole	mol
light intensity	candela	cd

table A The base units.

The choice of which quantities are the base ones is somewhat a matter of choice. The scientists who meet to decide on the standard unit system have chosen these seven. You might think that electric current is not a fundamental property, as it is the rate of movement of charge. So it could be derived from measuring charge and time. However, scientists had to pick what was fundamental and they chose current. This means that electric charge is a derived quantity found by multiplying current passing for a given time.

SI UNITS

For each of the base units, a meeting is held every four or six years of the General Conference on Weights and Measures, under the authority of the Bureau International des Poids et Mesures in Paris. At this meeting, they either alter the definition, or agree to continue with the current definition. As we learn more and more about the universe, these definitions are gradually moving towards the fundamental constants of nature.

▲ **fig B** A standard metre, made to be exactly the length that light could travel in 1/299 792 458 of a second.

The current definition of each of the seven base units is listed below:

- The kilogram is the unit of mass; it is equal to the mass of the International Prototype Kilogram, as in **fig A**.
- The second is the duration of 9 192 631 770 periods of the radiation corresponding to the transition between the two hyperfine levels of the ground state of the caesium-133 atom.
- The metre is the length of the path travelled by light in vacuum
- during a time interval of $\frac{1}{299\,792\,458}$ of a second (see **fig B**).
- The ampere is that constant current which, if maintained in two straight parallel conductors of infinite length, of negligible circular cross-section, and placed 1 m apart in vacuum, would produce between these conductors a force equal to 2×10^{-7} newton per metre of length.
- The kelvin, unit of thermodynamic temperature, is the fraction $\frac{1}{273.16}$ of the thermodynamic temperature of the triple point of water.
- The mole is the amount of substance of a system which contains as many elementary entities as there are atoms in 0.012 kg of carbon-12. (When the mole is used, the elementary entities must be specified and may be atoms, molecules, ions, electrons, other particles or specified groups of such particles.)

EXAM HINT

Table A has the complete list of SI base units. You will not be asked questions about the candela in the exam.

LEARNING TIP

'Metrology' is the study of the science of measurement, and 'metrics' refers to ways of standardising measuring techniques.

DERIVED UNITS

In **table B** you will see many of the derived units that we will study in this book, but this is only a list of those that have their own name.

DERIVED QUANTITY	UNIT NAME	UNIT SYMBOL	BASE UNITS EQUIVALENT
force	newton	N	$kg\,m\,s^{-2}$
energy (work)	joule	J	$kg\,m^2\,s^{-2}$
power	watt	W	$kg\,m^2\,s^{-3}$
frequency	hertz	Hz	s^{-1}
charge	coulomb	C	A s
voltage	volt	V	$kg\,m^2\,s^{-3}\,A^{-1}$
resistance	ohm	Ω	$kg\,m^2\,s^{-3}\,A^{-2}$

table B Some well known derived units.

POWER PREFIXES

Sometimes the values we have to work with for some quantities mean that the numbers involved are extremely large or small. For example, the average distance from the Earth to the sun, measured in metres, is 150 000 000 000 m. Scientists have made an easier system for writing such large values by adding a prefix to the unit which tells us that it has been multiplied by a very large or very small amount. In the Earth orbit example, the distance is equivalent to 150 billion metres, and the prefix giga- means multiply by a billion. So the Earth–sun distance becomes 150 gigametres, or 150 Gm.

FACTOR	NAME	SYMBOL	FACTOR	NAME	SYMBOL
10^1	deca-	da	10^{-1}	deci-	d
10^2	hecto-	h	10^{-2}	centi-	c
10^3	kilo-	k	10^{-3}	milli-	m
10^6	mega-	M	10^{-6}	micro-	μ
10^9	giga-	G	10^{-9}	nano-	n
10^{12}	tera-	T	10^{-12}	pico-	p
10^{15}	peta-	P	10^{-15}	femto-	f
10^{18}	exa-	E	10^{-18}	atto-	a
10^{21}	zetta-	Z	10^{-21}	zepto-	z
10^{24}	yotta-	Y	10^{-24}	yocto-	y

table C Prefixes used with SI units.

CHECKPOINT

SKILLS PROBLEM SOLVING

1. Refer to **table B** and answer the following questions:
   (a) Pick any quantity that you have studied before and explain how its base unit equivalent is shown.
   (b) All of the derived quantity units are named after scientists. Compare their names and abbreviations. What do you notice?
2. Write the following in standard form:
   (a) 9.2 GW
   (b) 43 mm
   (c) 6400 km
   (d) 44 ns.
3. Write the following using an appropriate prefix and unit symbol:
   (a) 3 600 000 joules
   (b) 31 536 000 seconds
   (c) 10 millionths of an ampere
   (d) 105 000 hertz.

2 ESTIMATION

LEARNING OBJECTIVES

- Estimate values for physical quantities.
- Use your estimates to solve problems.

ORDER OF MAGNITUDE

In physics, it can be very helpful to be able to make approximate estimates of values to within an order of magnitude. This means that the power of ten of your estimate is the same as the true value. For example, you are the same height as the ceiling in your classroom, if we consider the order of magnitude. The ceiling may be twice your height, but it would need to be ten times bigger to reach the next order of magnitude.

This is made clearer if we express all values in standard form and then compare the power of ten. You are likely to be a thousand times taller than an ant, so we would say you are three orders of magnitude larger.

typical ant height: 1.7 mm = 1.7×10^{-3} m

typical human height: 1.7 m = 1.7×10^{0} m

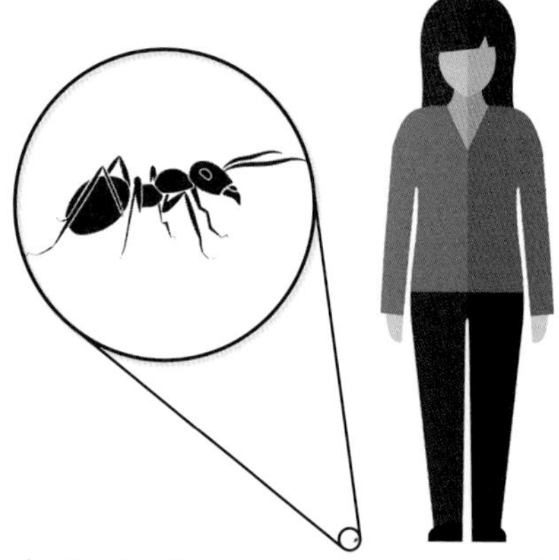

fig A We are three orders of magnitude taller than an ant.

In many situations, physicists are not interested in specific answers, as circumstances can vary slightly and then the specific answer is incorrect. An order of magnitude answer will always be correct, unless you change the initial conditions by more than an order of magnitude. So a physicist could easily answer the question 'What is the fastest speed of a car?' because we don't really want to know the exact true value. To give an exact answer would depend on knowing the model of car, and the weather and road conditions, and this answer would only be correct for that car on that day. By estimating important quantities, like a typical mass for cars, we can get an approximate – order of magnitude – answer. The reason for doing so would be that it allows us to develop ideas as possible or impossible. Also it helps us focus on developing the ideas along lines that will eventually be feasible when we get to developing a specific solution. This reduces time and money wasted by following ideas that are impossible. It also helps us quickly notice any miscalculations in an answer to a question. If we used an equation to calculate the answer to the fastest speed of a particular car in particular conditions, and the answer came out as 300 000 000 metres per second (the speed of light), we would immediately know that the answer is wrong, and re-check the calculation.

ORDER OF MAGNITUDE SCALE	TYPICAL OBJECT
1×10^{13} m	size of the solar system
1×10^{11} m	size of Earth's orbit around the sun
1×10^{8} m	size of Moon's orbit around Earth
1×10^{4} m	diameter of Manchester
1×10^{0} m	human height
1×10^{-3} m	ant height
1×10^{-5} m	biological cell diameter
1×10^{-8} m	wavelength of ultraviolet light
1×10^{-10} m	diameter of an atom
1×10^{-14} m	diameter of an atomic nucleus

table A Examples of object scales changing with powers of ten.

FERMI QUESTIONS

fig B Enrico Fermi was one of the developers of both nuclear reactors and nuclear bombs, along with other work on particle physics, quantum physics and statistical mechanics. He was awarded the 1938 Nobel Prize for Physics for the discovery of new radioactive elements and induced radioactivity.

Enrico Fermi was an Italian physicist who lived from 1901 to 1954. He was a pioneer of estimation. What have become known as Fermi questions are seemingly specific questions, to which only an order of magnitude answer is expected. It is common for the question to appear very difficult, as we do not have enough information to work out the answer. One of Fermi's most interesting thought experiments was a consideration of whether or not alien life exists. Over a lunch with other scientists in 1950, Fermi surprised the group by asking 'Where is everybody?' referring to extraterrestrials. There seems to be no evidence of the existence of alien life. That is still as true today as it was in 1950. However, when Fermi made an estimation of what would be necessary for an extraterrestrial civilisation to travel to visit us, his estimate came out at a much shorter amount of time than the age of our galaxy.

Conditions and likelihood for a visit by extraterrestrials

- *A planet that will support life – the galaxy holds about 100 billion stars, so there is a high probability that some other solar systems will have an Earth-like planet.*
- *Time to develop life – many of the stars in the galaxy are much older than the sun, so alien life developing at the same rate as our own should have been established as long as a billion years before ours.*
- *Time to develop interstellar travel – even if humans have to live our entire history again in order to develop spaceships that can travel to other stars, that means we will reach other stars in less than a million years of human existence.*
- *Time to explore the whole galaxy – from the nearest star to exploration of the whole galaxy is, by extrapolation, only a matter of a few million years.*

▲ **fig C** The Fermi Paradox: even the most conservative estimates of the requirements of exploring the galaxy mean that aliens should reach Earth within ten million years of their life beginning. If they existed, they would be here.

You need to work out what steps are needed to make an estimation. First, think about what steps you would take to reach an answer, if you could have any information you wanted. Then, when the necessary data is not all available, make an estimate for the missing numbers. Making sensible assumptions is the key to solving Fermi questions.

WORKED EXAMPLE

Probably the most famous example of a Fermi question was this challenge to a class:

'How many piano tuners are there in Chicago?'

The only piece of information he provided was that the population of Chicago was 3 million.

Step 1: How many pianos in Chicago?

If each household is 4 people, then there are:

$$\frac{3\,000\,000}{4} = 750\,000 \text{ households}$$

If one household in ten owns a piano, then there are:

$$\frac{750\,000}{10} = 75\,000 \text{ pianos}$$

Step 2: How many pianos per piano tuner?

Assume each piano needs tuning once a year. Further assume a piano tuner works 200 days a year, and can service 4 pianos a day. Each tuner can service: $200 \times 4 = 800$ pianos.

Step 3: How many tuners?

Each piano tuner works with 800 pianos, and there are 75 000 pianos in total. So there are: $\frac{75\,000}{800} = 94$ piano tuners.

Your answer to Fermi would be 'There are 100 piano tuners in Chicago'. This is not expected to be the exactly correct answer, but it will be correct to order of magnitude. We would not expect to find that Chicago has only 10 piano tuners, and it would be very surprising if there were 1000.

CHECKPOINT

SKILLS ADAPTIVE LEARNING

1. Give an order of magnitude estimate for the following quantities:
   (a) the height of a giraffe
   (b) the mass of an apple
   (c) the reaction time of a human
   (d) the diameter of a planet
   (e) the temperature in this room.
2. ▶ Answer the following Fermi questions, showing all the steps and the assumptions and estimates you make.
   (a) How many tennis balls would fit into a soccer stadium?
   (b) How many atoms are there in your body?
   (c) How many drops of water are there in a swimming pool?
   (d) In your lifetime, how much money will you make in total?
   (e) How many Fermi questions could Enrico Fermi have answered whilst flying from Rome to New York?

TOPIC 1 MECHANICS

CHAPTER 1A MOTION

How can we calculate how fast a plane is flying, in what direction it is going and how long it will take to reach a certain destination? If you were a pilot, how would you know what force to make the engines produce and where to direct that force so your plane moves to your destination?

An incredible number of intricate calculations need to be done to enable a successful flight, and the basis for all of them is simple mechanics.

This chapter explains the multiple movements of objects. It looks at how movement can be described and recorded, and then moves on to explaining why movement happens. It covers velocity and acceleration, including how to calculate these in different situations.

We only consider objects moving at speeds that could be encountered in everyday life. At these speeds (much less than the speed of light) Sir Isaac Newton succinctly described three laws of motion. With knowledge of basic geometry, we can identify aspects of movement in each dimension.

Newton's laws of motion have been constantly under test by scientists ever since he published them in 1687. Within the constraints established by Einstein in the early twentieth century, Newton's laws have always correctly described the relationships between data collected. You may have a chance to confirm Newton's laws in experiments of your own. With modern ICT recording of data, the reliability of such experiments is now much improved over traditional methods.

MATHS SKILLS FOR THIS CHAPTER

- **Units of measurement** (*e.g. the newton, N*)
- **Using Pythagoras' theorem, and the angle sum of a triangle** (*e.g. finding a resultant vector*)
- **Using sin, cos and tan in physical problems** (*e.g. resolving vectors*)
- **Using angles in regular 2D structures** (*e.g. interpreting force diagrams to solve problems*)
- **Changing the subject of an equation** (*e.g. re-arranging the kinematics equations*)
- **Substituting numerical values into algebraic equations** (*e.g. calculating the acceleration*)
- **Plotting two variables from experimental or other data, understanding that $y = mx + c$ represents a linear relationship and determining the slope of a linear graph** (*e.g. verifying Newton's second law experimentally*)
- **Estimating, by graphical methods as appropriate, the area between a curve and the x-axis and realising the physical significance of the area that has been determined** (*e.g. using a speed–time graph*)

What prior knowledge do I need?

- Using a stopwatch to measure times
- Measuring and calculating the speed of objects
- Gravity making things fall down
- Measuring forces, calculating resultant forces
- The motion of objects as a result of forces acting on them

What will I study in this chapter?

- The definitions of and equations for: speed, distance, displacement, time, velocity, acceleration
- Graphs of motion over time
- The classification of scalars and vectors
- Adding and resolving vectors
- Newton's laws of motion
- Kinematics equations
- Moments (turning forces)

What will I study later?

Topic 1B

- Kinetic energy and gravitational potential energy
- Transferring between gravitational potential energy and kinetic energy
- Work and power

Topic 1C

- Momentum and the principle of conservation of momentum

Topic 2A

- Fluid movements and terminal velocity

Topic 3A

- Wave movements

Topic 5 (Book 2: IAL)

- The meaning and calculation of impulse (A level)

1A 1 VELOCITY AND ACCELERATION

SPECIFICATION REFERENCE 1.3.1 1.3.4

LEARNING OBJECTIVES

- Explain the distinction between scalar and vector quantities.
- Distinguish between speed and velocity and define acceleration.
- Calculate values using equations for velocity and acceleration.

▲ **fig A** These runners are accelerating to a high speed.

Movement is fundamental to the functioning of our universe. Whether you are running to catch a bus or want to calculate the speed required for a rocket to travel to Mars or the kinetic energy of an electron in an X-ray machine, you need to be able to work out how fast things are moving.

RATE OF MOVEMENT

One of the simplest things we can measure is how fast an object is moving. You can calculate an object's **speed** if you know the amount of time taken to move a certain distance:

$$\text{speed (m s}^{-1}\text{)} = \frac{\text{distance (m)}}{\text{time (s)}}$$

$$v = \frac{d}{t}$$

However, the calculation for speed will only tell you how fast an object is moving. Often it is also vitally important to know in what direction this movement is taking the object. When you include the direction in the information about the rate of movement of an object, this is then known as the **velocity**. So, the velocity is the rate of change of **displacement**, where the distance in a particular direction is called the 'displacement'.

$$\text{velocity (m s}^{-1}\text{)} = \frac{\text{displacement (m)}}{\text{time (s)}}$$

$$\mathbf{v} = \frac{\mathbf{s}}{t}$$

OR

$$\mathbf{v} = \frac{\Delta \mathbf{s}}{\Delta t}$$

LEARNING TIP

The upper case Greek letter delta, Δ, is used in mathematics to indicate a change in a quantity. For example, $\Delta \mathbf{s}$ means the change in the displacement of an object, to be used here to calculate the velocity of the object.

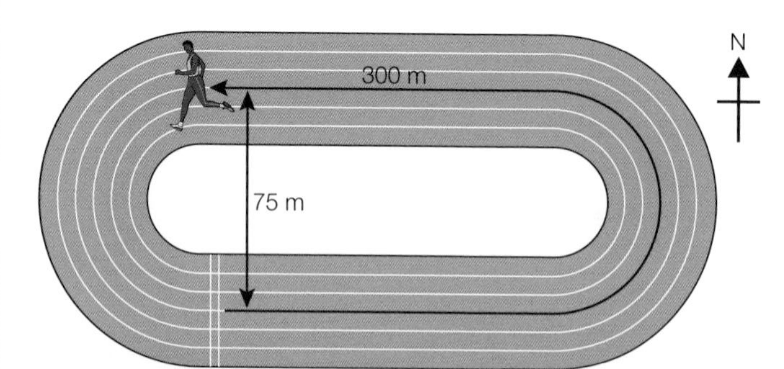

▲ **fig B** The displacement due north is only 75 m, whilst the actual distance this athlete has run is 300 m. So the velocity due north is much less than the actual speed.

DID YOU KNOW?

The froghopper, a 6 mm long insect, can accelerate at 4 000 m s^{-1}.

A quantity for which the direction must be stated is known as a **vector**. If direction is not important, the measurement is referred to as a **scalar** quantity. Therefore, velocity is a vector and speed is a scalar; distance is a scalar and displacement is a vector.

Scalar and vector quantities are not limited to measurements related to movement. Every measured quantity can be classified to include the direction (vector, e.g. force) or as being sufficiently stated by its magnitude only (scalar, e.g. mass).

LEARNING TIP

Vector notation means that vectors are written in **bold** type to distinguish them from scalar variables.

AVERAGE AND INSTANTANEOUS SPEED

In most journeys, it is unlikely that speed will remain constant throughout. As part of his training programme, an athlete in **fig A** wants to keep a record of his speed for all races. From rest, before the starting gun starts the race, he accelerates to a top speed. However, the race timing will be made from start to finish, and so it is most useful to calculate an average speed over the whole race. **Average speed** is calculated by dividing the total distance for a journey by the total time for the journey. Thus it averages out the slower and faster parts of the journey, and even includes stops.

Instantaneous speed can be an important quantity, and we will look at how to measure it in the next topic.

▲ **fig C** Most speed checks look at instantaneous speed, but CCTV allows police to monitor average speed over a long distance.

ACCELERATION

Acceleration is defined as the rate of change of velocity. Therefore, it must include the direction in which the speed is changing, and so acceleration is a vector quantity. The equation defining acceleration is:

$$\text{acceleration (m s}^{-2}) = \frac{\text{change in velocity (m s}^{-1})}{\text{time taken to change the velocity (s)}}$$

$$a = \frac{v - u}{t}$$

OR

$$a = \frac{\Delta v}{\Delta t}$$

where u is the initial velocity and v is the final velocity.

The vector nature of acceleration is very important. One of the consequences is that if an object changes only the direction of its velocity, it is accelerating, *while remaining at a constant speed.* Similarly, deceleration represents a negative change in velocity, and so could be stated as a negative acceleration.

EXAM HINT

Whilst accelerations can (very briefly) be extraordinarily high, like that for the electron in question 3(b), no speed or velocity can ever be greater than the speed of light, which is 3×10^8 m s^{-1}. If you calculate a speed that is higher than this, check your calculation again as it must be wrong.

CHECKPOINT

SKILLS PROBLEM SOLVING

1. The athlete in **fig B** has taken 36 seconds from the start to reach the 300 m mark as shown. Calculate:

(a) his average speed during this 36 seconds

(b) his average velocity due north during this 36 seconds

(c) his average velocity due east during this 36 seconds.

2. A driver in a car travelling at about 40.2 km h^{-1} sees a cat run onto the road ahead.

(a) Convert 40.2 km h^{-1} into a speed in m s^{-1}.

(b) The car travels 16.5 m whilst the driver is reacting to the danger. What is his reaction time?

(c) The car comes to a stop in 2.5 s. What is its deceleration?

3. An electron in an X-ray machine is accelerated from rest to half the speed of light in 1.7×10^{-15} s. Calculate:

(a) the speed the electron reaches in m s^{-1}

(b) the acceleration the electron experiences.

SUBJECT VOCABULARY

speed the rate of change of distance:

$$\text{speed (m s}^{-1}) = \frac{\text{distance (m)}}{\text{time (s)}}$$

$$v = \frac{d}{t}$$

velocity the rate of change of displacement:

$$\text{velocity (m s}^{-1}) = \frac{\text{displacement (m)}}{\text{time (s)}}$$

$$v = \frac{s}{t} \text{ OR } v = \frac{\Delta s}{\Delta t}$$

displacement the vector measurement of distance in a certain direction

vector a quantity that must have both magnitude and direction

scalar a quantity that has magnitude only

average speed speed for a whole journey, calculated by dividing the total distance for a journey by the total time for the journey:

$$\text{average speed (m s}^{-1}) = \frac{\text{total distance (m)}}{\text{total time (s)}}$$

instantaneous speed the speed at any particular instant in time on a journey, which can be found from the gradient of the tangent to a distance–time graph at that time

acceleration the vector defined as the rate of change of velocity:

$$\text{acceleration (m s}^{-2}) = \frac{\text{change in velocity (m s}^{-1})}{\text{time taken to change the velocity (s)}}$$

$$a = \frac{v - u}{t} \text{ OR } a = \frac{\Delta v}{\Delta t}$$

LEARNING OBJECTIVES

- Interpret displacement–time graphs, velocity–time graphs and acceleration–time graphs.
- Make calculations from these graphs.
- Understand the graphical representations of accelerated motion.

One of the best ways to understand the movements of an object whilst on a journey is to plot a graph of the position of the object over time. Such a graph is known as a displacement–time graph. A **velocity–time graph** will also provide detail about the movements involved. A velocity–time graph can be produced from direct measurements of the velocity or generated from calculations made using the displacement–time graph.

DISPLACEMENT–TIME GRAPHS

If we imagine a boat trip on a river, we could monitor the location of the boat over the hour that it travels for and plot the displacement–time graph for these movements. Depending on what information we want the graph to provide, it is often simpler to draw a distance–time graph in which the direction of movement is ignored.

The graphs shown in **fig A** are examples of plotting position against time, and show how a distance–time graph cannot decrease with time. A displacement–time graph could have parts of it in the negative portions of the *y*-axis, if the movement went in the opposite direction at some points in time.

The simplest thing we could find from these graphs is how far an object has moved in a certain time. For example, in **fig A**, both the graphs show that in the first 15 minutes the boat moved 150 m. Looking at the time from 40 to 48 minutes, both show that the boat travelled 120 m, but the displacement–time graph is in the negative region of the *y*-axis, showing the boat was moving down river from the starting point – the opposite direction to the places it had been in the first 40 minutes.

During the period from 20 to 25 minutes, both graphs have a flat line at a constant value, that shows no change in the distance or displacement. This means the boat was not moving – a flat line on a distance–time (d–t) graph means the object is stationary. From 20 to 25 minutes on the velocity–time ($\boldsymbol{v}$–t) graph of this journey (see **fig B**) the line would be at a velocity of $0\,\mathrm{m\,s^{-1}}$.

SPEED AND VELOCITY FROM *d*–*t* GRAPHS

The **gradient** of the d–t graphs in **fig A** will tell us how fast the boat was moving. Gradient is found from the ratio of changes in the *y*-axis divided by the corresponding change on the *x*-axis, so:

for a distance–time graph:

$$\text{gradient} = \frac{\text{distance (m)}}{\text{time (s)}} = \text{speed (m s}^{-1}\text{)}$$

$$v = \frac{d}{t}$$

for a displacement–time graph:

$$\text{gradient} = \frac{\text{displacement (m)}}{\text{time (s)}} = \text{velocity (m s}^{-1}\text{)}$$

$$\boldsymbol{v} = \frac{\Delta s}{\Delta t}$$

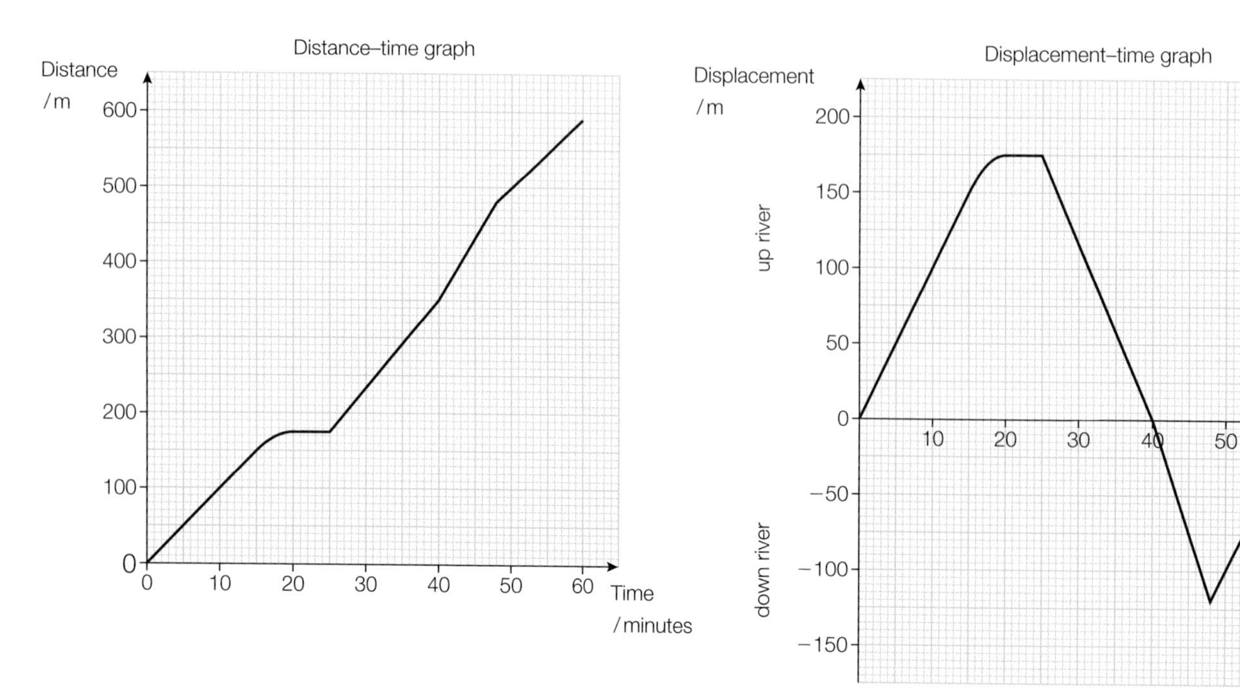

▲ **fig A** A comparison of the displacement–time graph of the boat trip up and down a river with its corresponding distance–time graph.

For example, the first 15 minutes of the boat trip in **fig A** represents a time of 900 seconds. In this time, the boat travelled 150 m. Its velocity is:

$$v = \frac{\Delta s}{\Delta t} = \frac{150}{900} = 0.167 \text{ m s}^{-1} \textit{ up river}$$

VELOCITY–TIME GRAPHS

A velocity–time graph will show the velocity of an object over time. We calculated that the velocity of the boat on the river was 0.167 m s^{-1} up river for the first 15 minutes of the journey. Looking at the graph in **fig B**, you can see that the line is constant at $+0.167 \text{ m s}^{-1}$ for the first 15 minutes.

Also notice that the velocity axis includes negative values, so that the difference between travelling up river (positive *y*-axis values) and down river (negative *y*-axis values) can be represented.

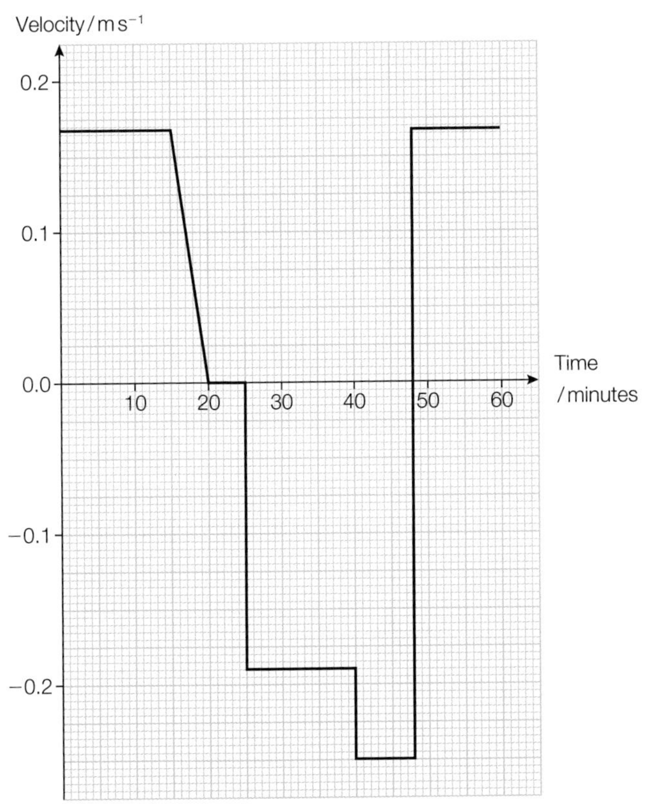

▲ **fig B** Velocity–time graph of the boat trip.

ACCELERATION FROM *v*–*t* GRAPHS

Acceleration is defined as the rate of change in velocity.

In order to calculate the gradient of the line on a ***v***–*t* graph, we must divide a change in velocity by the corresponding time difference. This exactly matches with the equation for acceleration:

$$\text{gradient} = \frac{\Delta v}{\Delta t} = \frac{v - u}{t} = \text{acceleration}$$

For example, between 15 and 20 minutes on the graphs, the boat slows evenly to a stop. The acceleration here can be calculated as the gradient:

$$\text{gradient} = \frac{\Delta v}{\Delta t} = \frac{v - u}{t} = \frac{0 - 0.167}{5 \times 60} = \frac{-0.167}{300} = -0.0006 \text{ m s}^{-2}$$

So the acceleration is: $\boldsymbol{a} = -0.6 \times 10^{-3} \text{ m s}^{-2}$.

DISTANCE TRAVELLED FROM *v*–*t* GRAPHS

Speed is defined as the rate of change in distance:

$$v = \frac{d}{t}$$

$$\therefore \quad d = v \times t$$

As the axes on the ***v***–*t* graph represent velocity and time, an area on the graph represents the multiplication of velocity × time, which gives distance. So to find the distance travelled from a ***v***–*t* graph, find the area between the line and the *x*-axis.

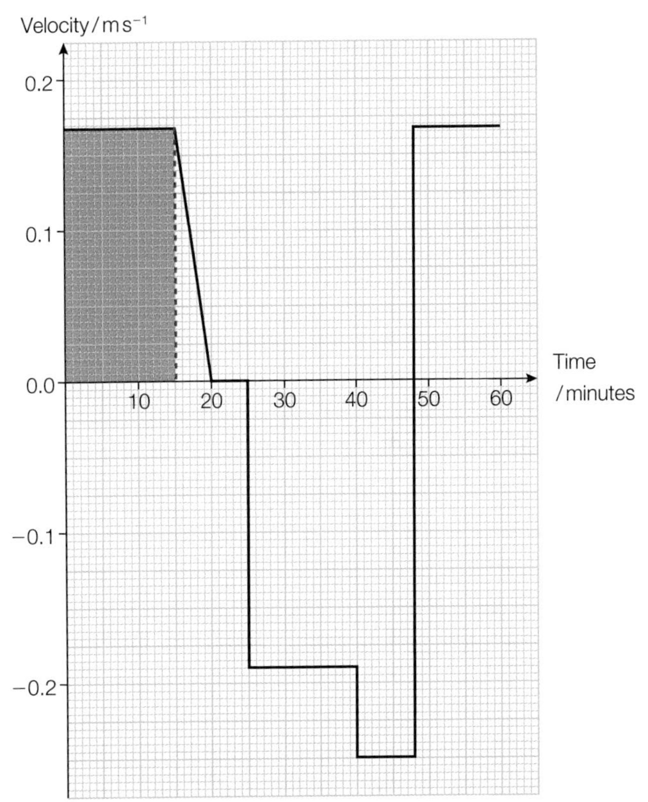

▲ **fig C** In the first 15 minutes (900 seconds) the distance travelled by the boat moving at 0.167 m s^{-1} is given by the area between the line and the *x*-axis: $d = \boldsymbol{v} \times t = 0.167 \times 900 = 150 \text{ m}$.

If we are only interested in finding the distance moved, this also works for a negative velocity. You find the area from the line up to the time axis. This idea will still work for a changing velocity. Find the area under the line and you have found the distance travelled. For example, from 0 to 20 minutes, the area under the line, all the way down to the *x*-axis, is a trapezium, so we need to find that area. To calculate the whole distance travelled in the journey for the first 40 minutes, we would have to find the areas under the four separate stages (0–15 minutes; 15–20 minutes; 20–25 minutes; and 25–40 minutes) and then add these four answers together.

PRACTICAL SKILLS CP1

Finding the acceleration due to gravity by multiflash photography

Using a multiflash photography technique, or a video recording that can be played back frame by frame, we can observe the falling motion of a small object such as a marble (see **fig D**). We need to know the time between frames.

From each image of the falling object, measure the distance it has fallen from the scale in the picture. A carefully drawn distance–time graph will show a curve as the object accelerates. From this curve, take regular measurements of the gradient by drawing tangents to the curve. These gradients show the instantaneous speed at each point on the curve.

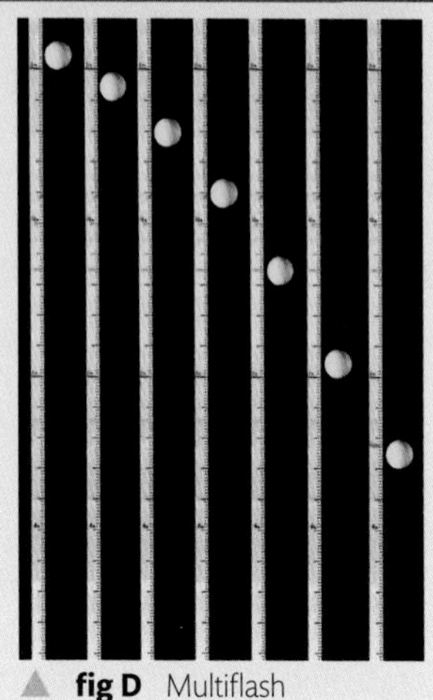

▲ **fig D** Multiflash photography allows us to capture the accelerating movement of an object falling under gravity.

Plotting these speeds on a velocity–time graph should show a straight line, as the acceleration due to gravity is a constant value. The gradient of the line on this ***v***–*t* graph will be the acceleration due to gravity, ***g***.

Safety Note: Persons with medical conditions such as epilepsy or migraine may be adversely affected by multiflash photography.

ACCELERATION–TIME GRAPHS

Acceleration–time graphs show how the acceleration of an object changes over time. In many instances the acceleration is zero or a constant value, in which case an acceleration–time (***a***–*t*) graph is likely to be of relatively little interest. For example, the object falling in our investigation above will be accelerated by gravity throughout. Assuming it is relatively small, air resistance will be minimal, and the ***a***–*t* graph of its motion would be a horizontal line at $\boldsymbol{a} = -9.81\,\mathrm{m\,s^{-2}}$. Compare this with your results to see how realistic it is to ignore air resistance.

For a larger object falling for a long period, such as a skydiver, then the acceleration will change over time as the air resistance increases with speed.

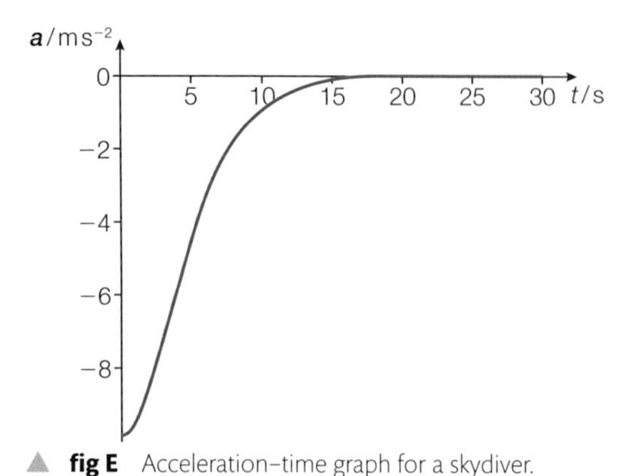

▲ **fig E** Acceleration–time graph for a skydiver.

The weight of a skydiver is constant, so the resultant force will be decreasing throughout, which means that the acceleration will also reduce (see **Section 1A.5**). The curve would look like that in **fig E**.

See **Section 2A.4** for more details on falling objects and terminal velocity.

EXAM HINT

Remember that the gradient of a distance–time graph represents speed or velocity. So if the line is curved, the changing gradient indicates a changing speed, which you can describe as the same as the changes in gradient.

CHECKPOINT

SKILLS INTERPRETATION (OF GRAPHICAL DATA)

1. Describe in as much detail as you can, including calculated values, what happens in the bicycle journey shown on the *d*–*t* graph in **fig F**.

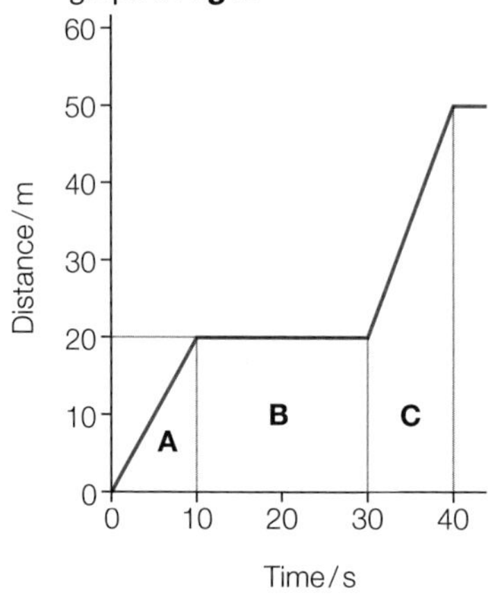

▲ **fig F** Distance–time graph of a bike journey.

2. Describe in as much detail as you can, including calculated values, what happens in the car journey shown on the ***v***–*t* graph in **fig G**.

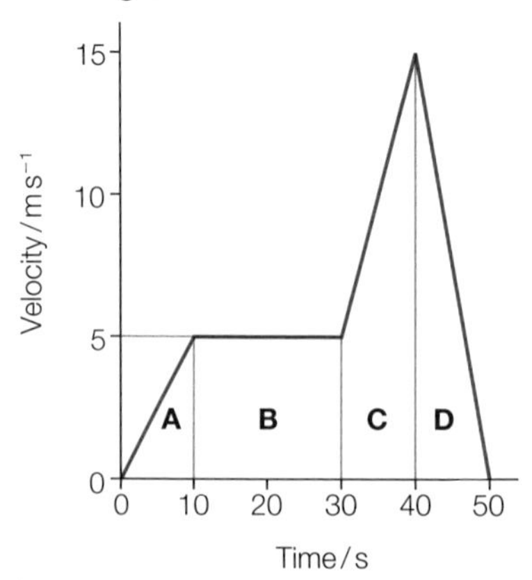

▲ **fig G** Velocity–time graph of a car journey.

3. From **fig B**, calculate the distance travelled by the boat from 40 to 60 minutes.

SUBJECT VOCABULARY

displacement–time graph a graph showing the positions visited on a journey, with displacement on the *y*-axis and time on the *x*-axis.
velocity–time graph a graph showing the velocities on a journey, with velocity on the *y*-axis and time on the *x*-axis.
gradient the slope of a line or surface

1A 3 ADDING FORCES

SPECIFICATION REFERENCE 1.3.6 1.3.8

LEARNING OBJECTIVES

- Add two or more vectors by drawing.
- Add two perpendicular vectors by calculation.

Forces are vectors. This means that measuring their magnitude is important, but equally important is knowing the direction in which they act. In order to calculate the overall effect of multiple forces acting on the same object, we can use vector addition to work out the **resultant force**. This resultant force can be considered as a single force that has the same effect as all the individual forces combined.

ADDING FORCES IN THE SAME LINE

If two or more forces are acting along the same line, then combining them is simply a case of adding or subtracting their magnitudes depending on their directions.

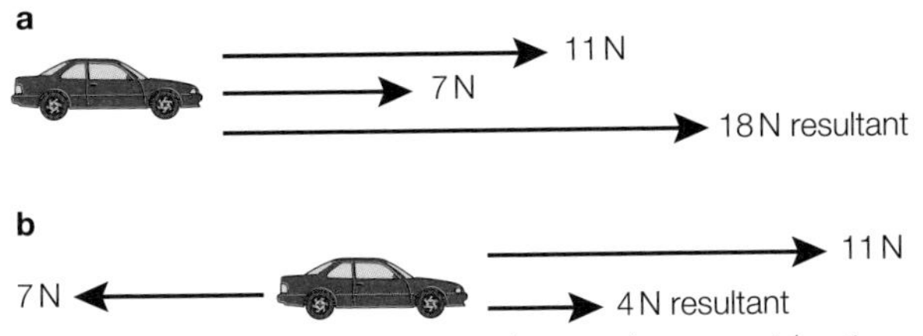

▲ **fig A** Adding forces in the same line requires a consideration of their comparative directions.

ADDING PERPENDICULAR FORCES

The effect on an object of two forces that are acting at right angles (perpendicular) to each other will be the vector sum of their individual effects. We need to add the sizes with consideration for the directions in order to find the resultant.

▲ **fig B** These two rugby players are each putting a force on their opponent. The forces are at right angles, so the overall effect would be to move him in a third direction, which we could calculate.

MAGNITUDE OF THE RESULTANT FORCE

To calculate the resultant magnitude of two perpendicular forces, we can draw them, one after the other, as the two sides of a right-angled triangle and use Pythagoras' theorem to calculate the size of the hypotenuse.

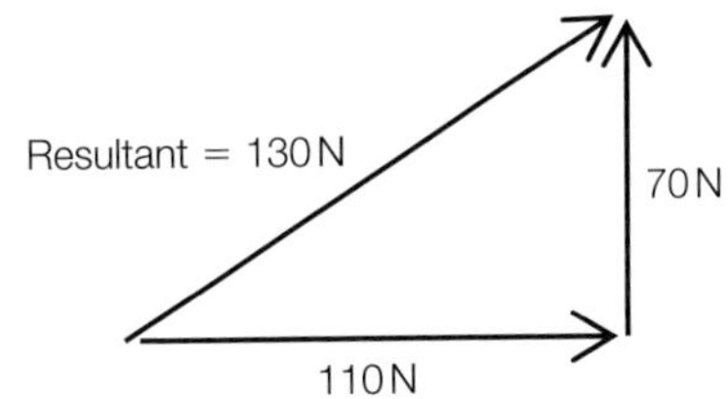

▲ **fig C** The resultant force here is calculated using Pythagoras' theorem: $F = \sqrt{(70^2 + 110^2)} = 130\text{ N}$

DIRECTION OF THE RESULTANT FORCE

As forces are vectors, when we find a resultant force it must have both magnitude and direction. For perpendicular forces (vectors), trigonometry will determine the direction.

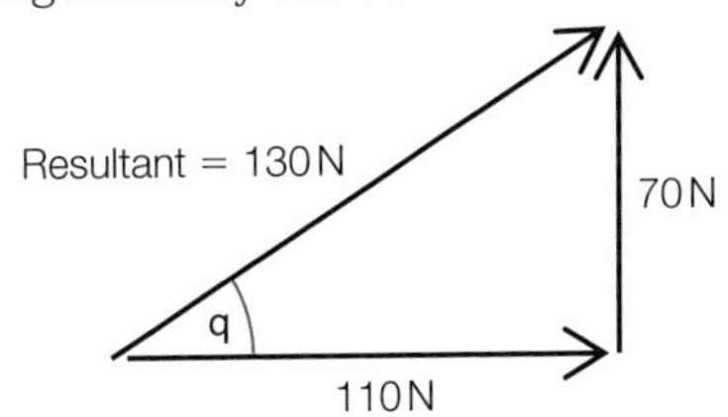

▲ **fig D** The resultant force here is at an angle up from the horizontal of: $\theta = \tan^{-1}\left(\frac{70}{110}\right) = 32°$

EXAM HINT

Always take care to state where the angle for a vector's direction is measured. For example, in **fig D**, the angle should be stated as 32° *up from the horizontal*. This is most easily expressed on a diagram of the situation, where you draw in the angle.

ADDING TWO NON-PERPENDICULAR FORCES

The geometry of perpendicular vectors makes the calculation of the resultant simple. We can find the resultant of any two vectors by drawing one after the other, and then the resultant will be the third side of the triangle from the start of the first one to the end of the second one. A scale drawing of the vector triangle will allow measurement of the size and direction of the resultant.

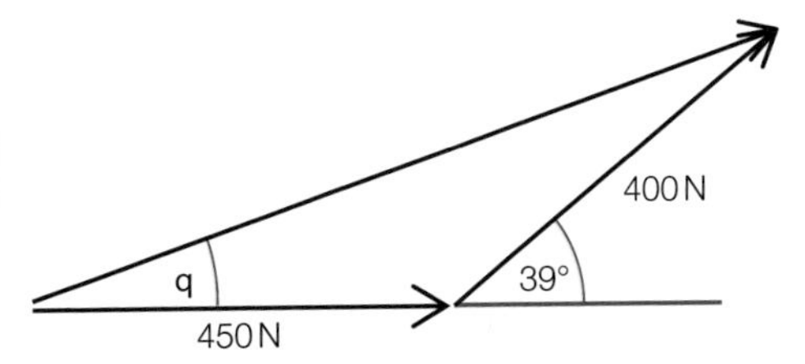

▲ **fig E** The resultant force here can be found by scale drawing the two forces, and then measurement of the resultant on the drawing using a ruler and a protractor.

THE PARALLELOGRAM RULE

There is another method for finding the resultant of two non-perpendicular forces (or vectors) by scale drawing, which can be easier to use. This is called the parallelogram rule. Draw the two vectors to scale – at the correct angle and scaled so their length represents the magnitude – starting from the same point. Then draw the same two vectors again parallel to the original ones, so that they form a parallelogram, as shown in **fig F**. The resultant force (or vector) will be the diagonal across the parallelogram from the starting point.

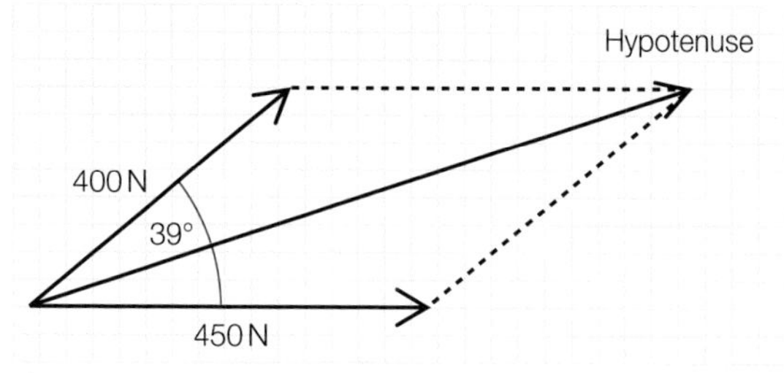

▲ **fig F** Finding the resultant vector using the parallelogram rule.

EXAM HINT

The vector addition rules shown on these pages work for all vectors, not just forces. They are useful only for co-planar vectors, which means vectors that are in the same plane. If we have more than two vectors that are in more than one plane, add two vectors together first, in their plane, and then add the resultant to the next vector using these rules again. Keep doing this until all the vectors have been added in.

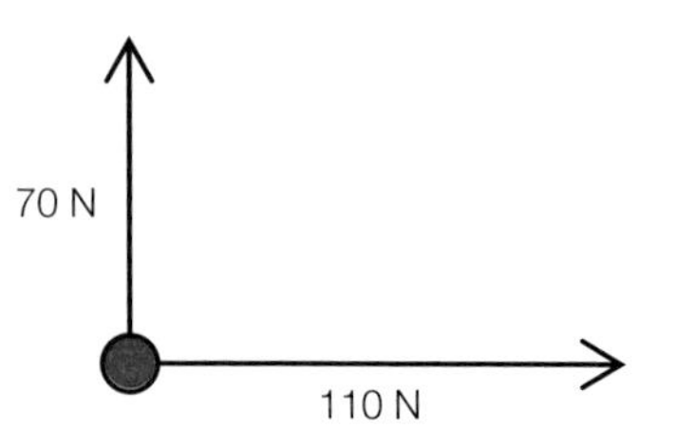

▲ **fig G** Free-body force diagram of a rugby player (red circle). The forces from the tacklers are marked on as force arrows.

FREE-BODY FORCE DIAGRAMS

If we clarify what forces are acting on an object, it can be simpler to calculate how it will move. To do this, we usually draw a **free-body force diagram**, which has the object isolated, and all the forces that act on it drawn in at the points where they act. Forces acting on other objects, and those other objects, are not drawn. For example, **fig G** could be said to be a free-body force diagram of the rugby player being tackled in **fig B**, and this would lead us to draw **fig C** and **fig D** to make our resultant calculations.

SKILLS ANALYSIS

CHECKPOINT

1. Work out the resultant force on a toy car if it has the following forces acting on it:
   - rubber band motor driving forwards 8.4 N
   - air resistance 0.5 N
   - friction 5.8 N
   - child's hand pushing forward 10 N.
2. As a small plane accelerates to take off, the lift force on it is 6000 N vertically upwards, whilst the thrust is 2800 N horizontally forwards. What is the resultant of these forces on the plane?
3. Draw a free-body force diagram of yourself sitting on your chair.
4. (a) Draw the scale diagram of **fig E**, and work out what the resultant force would be.
   (b) Use the parallelogram rule, as in **fig F**, to check your answer to part (a).
5. In order to try and recover a car stuck in a muddy field, two tractors pull on it. The first acts at an angle of 20° left of the forwards direction with a force of 2250 N. The second acts 15° to the right of the forwards direction with a force of 2000 N. Draw a scale diagram of the situation and find the resultant force on the stuck car.

SUBJECT VOCABULARY

resultant force the total force (vector sum) acting on a body when all the forces are added together accounting for their directions

free-body force diagram diagram showing an object isolated, and all the forces that act on it are drawn in at the points where they act, using arrows to represent the forces

1A 4 MOMENTS

SPECIFICATION REFERENCE 1.3.8 1.3.15 1.3.16

LEARNING OBJECTIVES

- Calculate the moment of a force.
- Apply the principle of moments.
- Find the centre of gravity of an object.

Forces on an object could act so that the object does not start to move along, but instead rotates about a fixed pivot. If the object is fixed so that it cannot rotate, it will bend.

THE MOMENT OF A FORCE

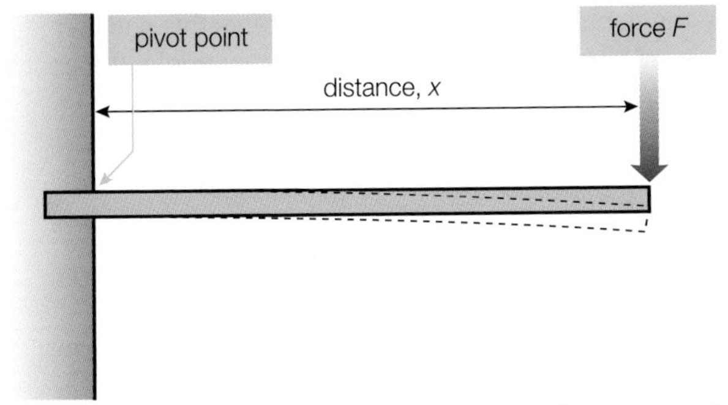

▲ **fig A** A force acts on a beam fixed at a point. The moment of a force causes rotation or, in this case, bending.

The tendency to cause rotation is called the moment of a force. It is calculated from:

moment (Nm) = force (N) × perpendicular distance from the pivot to the line of action of the force (m)

$$\text{moment} = Fx$$

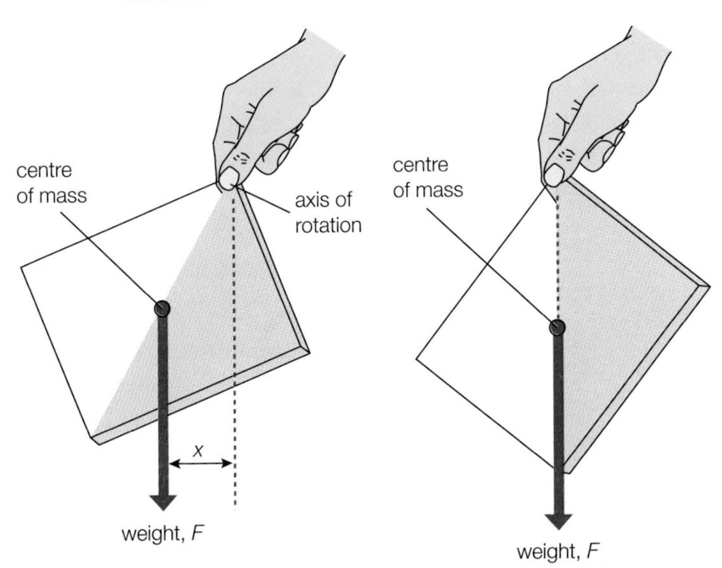

▲ **fig B** The calculation of moment only considers the perpendicular distance between the line of action of the force and the axis of rotation, through the pivot point. When free to rotate, a body will turn in the direction of any net moment.

PRINCIPLE OF MOMENTS

▲ **fig C** Balanced moments create an equilibrium situation.

If we add up all the forces acting on an object and the resultant force, accounting for their directions, is zero, then the object will be in **equilibrium**. Therefore it will remain stationary or, if it is already moving, it will carry on moving at the same velocity. The object could keep a constant velocity, but if the moments on it are not also balanced, it could be made to start rotating. The **principle of moments** tells us that if the total of all the moments trying to turn an object clockwise is equal to the total of all moments trying to turn an object anticlockwise, then it will be in rotational equilibrium. This means it will either remain stationary, or if it is already rotating it will continue at the same speed in the same direction.

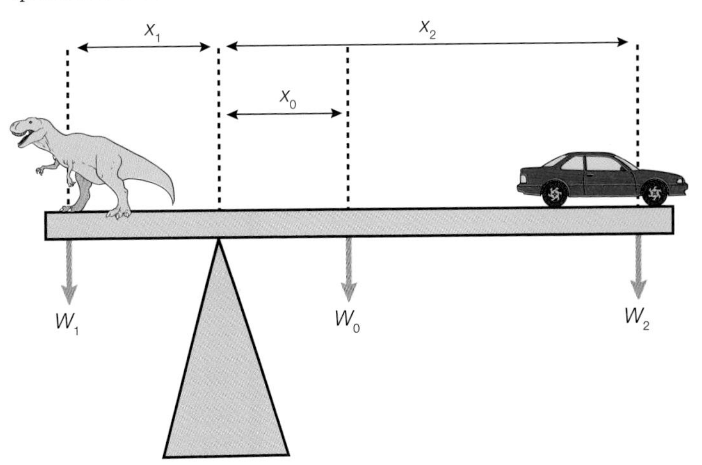

▲ **fig D** As the metre-long beam is balanced, the sum of all the clockwise moments must equal the sum of all the anticlockwise moments.

LEARNING TIP

The clockwise moments and the anticlockwise moments must all be taken about the same pivot point.

EXAM HINT

Note that the steps and layout of the solution in this worked example are suitable for moments questions in the exam.

WORKED EXAMPLE

In **fig D**, we can work out the weight of the beam if we know all the other weights and distances. The beam is uniform, so its weight will act from its centre. The length of the beam is 100 cm. So if x_1 = 20 cm, then x_0 must be 30 cm, and x_2 = 80 cm. The dinosaur (W_1) weighs 5.8 N and the toy car's weight (W_2) is 0.95 N.

In equilibrium, principle of moments: sum of anticlockwise moments = sum of clockwise moments

$$W_1x_1 = W_0x_0 + W_2x_2$$

$$5.8 \times 0.20 = W_0 \times 0.30 + 0.95 \times 0.80$$

$$\therefore \quad W_0 = \frac{1.16 - (0.76)}{0.30}$$

$$W_0 = 1.3\text{ N}$$

LEARNING TIP

In order to calculate the sum of the moments in either direction, each individual moment must be calculated first and these individual moments then added together. The weights and/or distances *cannot* be added together and this answer used to calculate some sort of combined moment.

LEARNING TIP

You can consider the terms 'centre of gravity' and 'centre of mass' to mean the same thing. They are exactly the same for objects that are small compared to the size of the Earth.

CENTRE OF GRAVITY

The weight of an object is caused by the gravitational attraction between the Earth and each particle contained within the object. The sum of all these tiny weight forces appears to act from a single point for any object, and this point is called the **centre of gravity**. For a symmetrical object, we can calculate the position of its centre of gravity, as it must lie on every line of symmetry. The point of intersection of all lines of symmetry will be the centre of gravity. **Fig E** illustrates this with two-dimensional shapes, but the idea can be extended into three dimensions. For example, the centre of gravity of a sphere is at the sphere's centre.

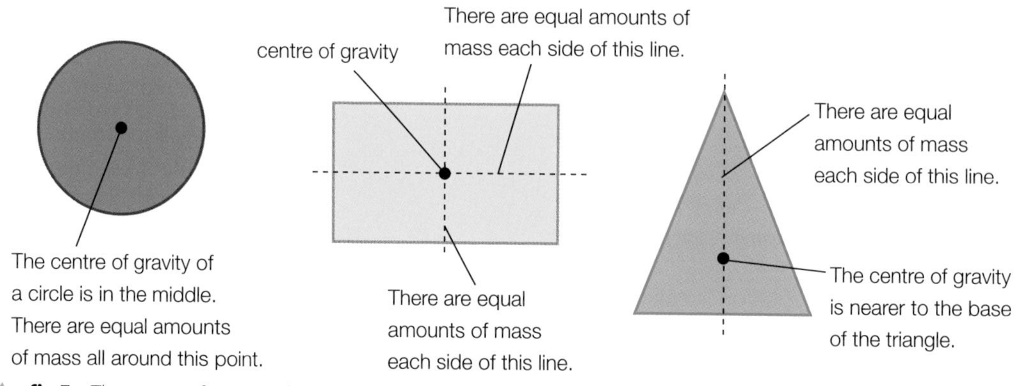

▲ **fig E** The centre of gravity of a symmetrical object lies at the intersection of all lines of symmetry.

IRREGULAR OBJECTS

The centre of gravity of an irregularly shaped object will still follow the rule that it is the point at which its weight appears to act on the object. A Bunsen burner, for example, has a heavy base. As such, the centre of gravity is low down near that concentration of mass, as there will be a greater attraction by the Earth's gravity to this large mass.

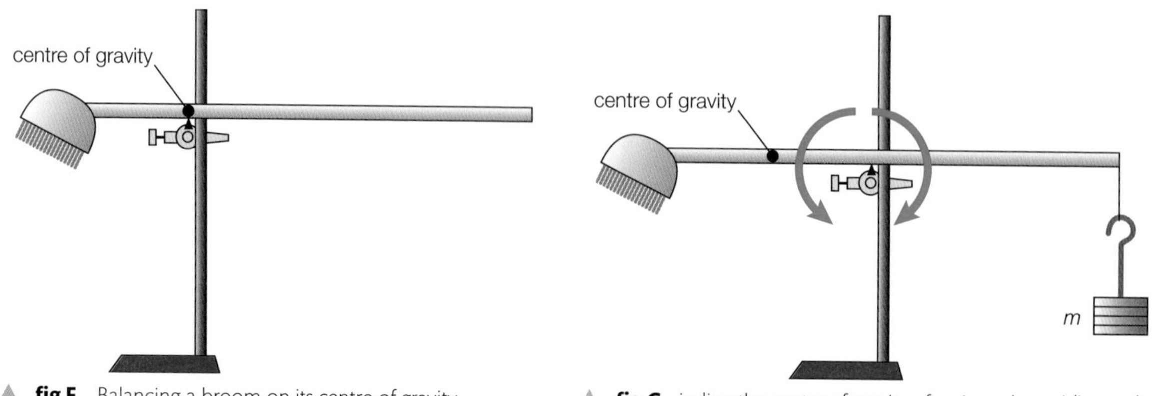

▲ **fig F** Balancing a broom on its centre of gravity.

▲ **fig G** inding the centre of gravity of an irregular rod (broom).

PRACTICAL SKILLS

Finding the centre of mass of an irregular rod

In this investigation, we use the principle of moments to find the centre of gravity of a broom. It is not easy to estimate/determine the (location of the) centre of gravity by looking at a broom because it is not a symmetrical object. With the extra mass at the brush head end, the centre of gravity will be nearer that end.

If you can balance the broom on the edge of a thick metal ruler, then the centre of gravity must lie above the ruler edge. As the perpendicular distance from the line of action to the weight is zero, the moment is zero so the broom sits in equilibrium.

You will probably find it difficult to balance the broom exactly, so you can use an alternative method. First you measure the mass of the broom (M) using a digital balance. Then you use a set of hanging masses (of mass m) to balance the broom more in the middle of the handle, as in **fig G**. When the broom is balanced, you measure the distance (d) from the hanging masses to the pivot. You calculate the distance (x) from the pivot to the centre of gravity of the broom using the principle of moments:

clockwise moment = anticlockwise moment

$$mg \times d = Mg \times x$$

$$\therefore \quad x = \frac{md}{M}$$

Note: Do not get into the habit of using only the mass in moments calculations, as the definition is *force* times distance. It just happens that in this case g cancels on each side.

Safety Note: Securely clamp the base of the stand to the bench. Do not use a very heavy broom as you would need a heavier counterweight which could fall and cause injury.

CHECKPOINT

1. What is the moment of a 252 N force acting on a solid object at a perpendicular distance of 1.74 m from an axis of rotation of the object?
2. A child and his father are playing on a seesaw, see **fig H**. They are exactly balanced when the boy (mass 46 kg) sits at the end of the seesaw, 2.75 m from the pivot. If his father weighs 824 N, how far is he from the pivot?

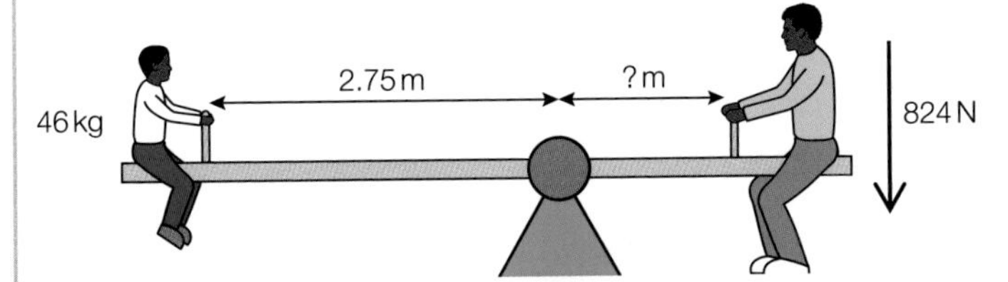

▲ **fig H**

3. The weight of the exercise book in the left-hand picture in **fig B** causes a rotation so it moves towards the second position. Explain why it does not continue rotating but comes to rest in the position of the second picture.
4. If the same set-up as shown in **fig D** was used again, but the toy car was replaced with a banana weighing 1.4 N, find out where the banana would have to be positioned for the beam to balance – calculate the new x_3.

SUBJECT VOCABULARY

equilibrium the situation for a body where there is zero resultant force and zero resultant moment. It will have zero acceleration

principle of moments a body will be in equilibrium if the sum of clockwise moments acting on it is equal to the sum of the anticlockwise moments

centre of gravity the point through which the weight of an object appears to act

5 NEWTON'S LAWS OF MOTION

SPECIFICATION REFERENCE 1.3.9 1.3.10 1.3.12

LEARNING OBJECTIVES

- Recall Newton's laws of motion and use them to explain the acceleration of objects.
- Make calculations using Newton's second law of motion.
- Identify pairs of forces involved in Newton's third law of motion.

Sir Isaac Newton was an exceptional thinker and scientist. His influence over science in the West is still enormous, despite the fact that he lived from 1642 to 1727. He was a professor at Cambridge University, a member of the British Parliament, and a president of the respected scientific organisation, the Royal Society, in London. Probably his most famous contribution to science was the development of three simple laws governing the movement of objects subject to forces.

▲ **fig A** A portrait painting of Sir Isaac Newton

NEWTON'S FIRST LAW OF MOTION

If an object is stationary there needs to be a resultant force on it to make it move. We saw how to calculate resultant forces in **Section 1A.3**. If the object is already moving then it will continue at the same speed in the same direction unless a resultant force acts on it. If there is no resultant force on an object – either because there is zero force acting or all the forces balance out – then the object's motion is not changed.

NEWTON'S SECOND LAW OF MOTION

This law tells us how much an object's motion will be changed by a resultant force. For an object with constant mass, it is usually written mathematically:

$$\Sigma F = ma$$

$$\text{resultant force (N)} = \text{mass (kg)} \times \text{acceleration (m s}^{-2}\text{)}$$

For example, this relationship allows us to calculate the acceleration due to gravity (g) if we measure the force (F) accelerating a mass (m) downwards.

$$F = ma = mg$$

$$\therefore \quad g = \frac{F}{m}$$

▲ **fig B** A stationary object will not move unless it is acted upon by a resultant force.

PRACTICAL SKILLS

Newton's second law investigation

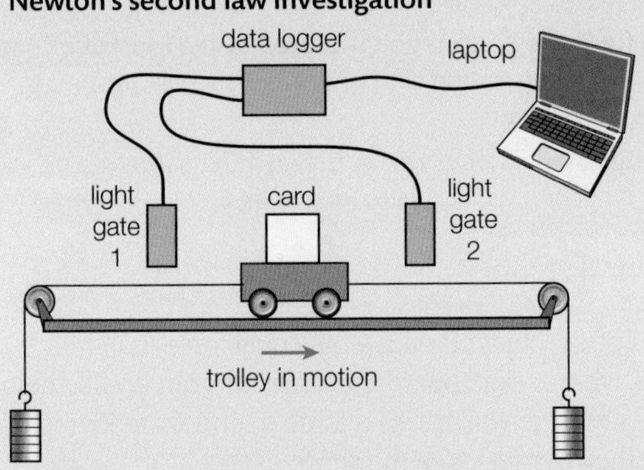

▲ **fig C** Experimental set-up for investigating the relationship between ***F***, ***m*** and ***a***.

You can use the set-up shown in **fig C**, or a similar experiment on an air track. This set-up measures the acceleration for various values of the resultant force that acts on the trolley whilst you keep its mass constant (**table A**). By plotting a graph of acceleration against resultant force, a straight line will show that acceleration is proportional to the resultant force. You could also plot a graph for varying masses of trolley whilst you keep the resultant force constant (**table B**).

Safety Note: Place heavy trolleys and runways where they cannot slide or fall off benches and cause injuries. Place air track 'blowers' on the floor. Secure the hose so that it cannot come loose and blow dirt and dust into eyes.

FORCE / N	ACCELERATION / m s^{-2}
0.1	0.20
0.2	0.40
0.3	0.60
0.4	0.80
0.5	1.00
0.6	1.20

table A Values of acceleration for different forces acting on a trolley.

MASS / kg	ACCELERATION / m s^{-2}
0.5	1.00
0.6	0.83
0.7	0.71
0.8	0.63
0.9	0.55
1.0	0.50

table B Values of acceleration resulting from an applied force of 0.5 N when the mass of the trolley is varied.

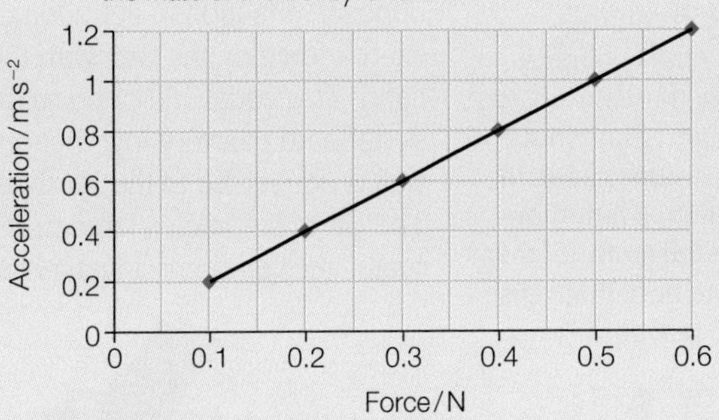

▲ **fig D** Graph of results from the first investigation into Newton's second law. Acceleration is directly proportional to force.

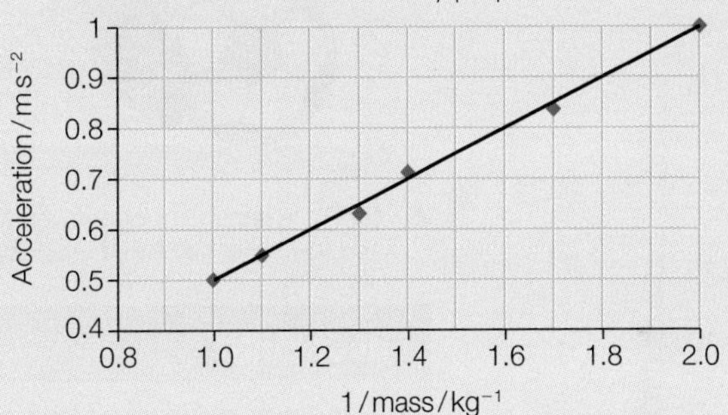

▲ **fig E** Graph of results from the second investigation into Newton's second law. The *x*-axis represents the derived data of '1/mass', so the straight best-fit line shows that acceleration is inversely proportional to mass. Note that there is no need for a graph to start at the origin: choose axes scales that will best show the pattern in the data by making the points fill the graph paper.

Experimental verification of Newton's second law is well established. The investigation shown in **fig C** demonstrates that: $\boldsymbol{a} = \frac{\boldsymbol{F}}{m}$

LEARNING TIP

Straight-line graphs

Physicists always try to arrange their experimental data into graphs that produce a straight best-fit line. This proves a linear relationship between the experimental variables, and can also give us numerical information about the quantities involved.

The equation for a straight line is:

$$y = mx + c$$

'*m*' in the equation is the gradient of the straight line, and '*c*' is the value on the *y*-axis where the line crosses it, known as the *y*-intercept.

If we plot experimental data on a graph and get a straight best-fit line, then this proves the quantities we plotted on *x* and *y* have a linear relationship. It is referred to as 'directly proportional' (or simply 'proportional') only if the line also passes through the origin, meaning that $c = 0$.

The graphs in **fig D** and **fig E** demonstrate experimental verification of Newton's second law. In each case, the third variable in the equation was kept constant as a control variable. For example, in **fig D** the straight best-fit line, showing that $y \propto x$, proves that $\boldsymbol{a} \propto \boldsymbol{F}$, and the gradient of the best-fit line would represent the reciprocal of the mass that was accelerated and was kept constant throughout, as a control variable. In this example $c = 0$, which means that the proportional relationships are simple:

$$y = mx$$

$$\boldsymbol{a} = \frac{\boldsymbol{F}}{m}$$

As both the above equations represent the graph in **fig D**, it follows that the gradient, *m*, equals the reciprocal of the mass, $1/m$. It is just coincidence that the symbol '*m*' for gradient, and '*m*' for mass are the same letter in this example.

Note that graphs in physics are causal relationships. In **fig D**, the acceleration is caused by the force. It is very rare in physics that a graph would represent a statistical correlation, and so phrases such as 'positive correlation' do not correctly describe graphs of physics experiments.

Similarly, it is very rare that a graph of a physics experiment would be correctly drawn if the points are joined 'dot-to-dot'. In most cases a best-fit line should be drawn, as has been done in **fig E**.

LEARNING TIP

Identifying Newton's third law of force pairs can confuse people. To find the two forces, remember they must always act on different objects, and must always have the same cause. In **fig F**, it is the mutual repulsion of electrons in the atoms of the football and the boot that cause the action and reaction.

NEWTON'S THIRD LAW OF MOTION

'When an object A causes a force on another object B, then object B causes an equal force in the opposite direction to act upon object A.' For example, when a skateboarder pushes off from a wall, they exert a force on the wall with their hand. At the same time, the wall exerts a force on the skateboarder's hand. This equal and opposite reaction force is what they can feel with the sense of touch, and as the skateboard has very low friction, the wall's push on them causes acceleration away from the wall. As the wall is connected to the Earth, the Earth and wall combination will accelerate in the opposite direction. The Earth has such a large mass that its acceleration can't be noticed. That's Newton's second law again: acceleration is inversely proportional to mass; huge mass means tiny acceleration.

▲ **fig F** When the boot puts a force on the football, the football causes an equal and opposite force on the boot. The footballer can feel the kick because they feel the reaction force from the ball on their toe.

SKILLS ANALYSIS

CHECKPOINT

1. In terms of Newton's laws of motion:
   (a) Explain why this book will sit stationary on a table.
   (b) Describe and explain what will happen if your hands then put an upwards force on the book that is greater than its weight.
   (c) Explain why you feel the book when your hands put that upwards force on it.
2. (a) Calculate the gradient of the best-fit line on the graph in **fig D** and thus work out the mass of the trolley that was accelerated in the first investigation.
   (b) State what quantity the gradient of the line on the graph in **fig E** represents. Calculate the value of that quantity.
3. Calculate the acceleration in each of the following cases:
   (a) A mass of 12.0 kg experiences a resultant force of 785 N.
   (b) A force of 22.2 N acts on a 3.1 kg mass.
   (c) A 2.0 kg bunch of bananas is dropped. The bunch weighs 19.6 N.
   (d) During a tackle, two footballers kick a stationary ball at the same time, with forces acting in opposite directions. One kick has a force of 210 N, the other has a force of 287 N. The mass of the football is 430 g.

SUBJECT VOCABULARY

Newton's first law of motion an object will remain at rest, or in a state of uniform motion, until acted upon by a resultant force

Newton's second law of motion if an object's mass is constant, the resultant force needed to cause an acceleration is given by the equation:

$$\sum F = ma$$

Newton's third law of motion for every action, there is an equal and opposite reaction

6 KINEMATICS EQUATIONS

SPECIFICATION REFERENCE 1.3.1 CP1

CP1 LAB BOOK PAGE 6

LEARNING OBJECTIVES

- Recall the simple kinematics equations.
- Calculate unknown variables using the kinematics equations.

Kinematics is the study of the movement of objects. We can use equations to find out details about the motion of objects accelerating in one dimension.

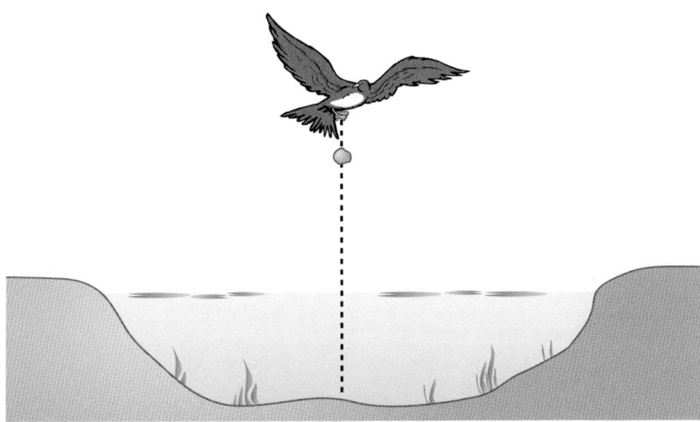

fig A Kinematics is the study of the description of the motion of objects. The equations could be used to make calculations about the stone falling through the air, and separately about its motion through the water. The acceleration in air and in water will be different as the resultant force acting in each will be different.

ZERO ACCELERATION

If an object has no resultant force acting on it then it does not accelerate. This is **uniform motion**. In this situation, calculations on its motion are very easy, as they simply involve the basic velocity equation:

$$v = \frac{s}{t}$$

The velocity is the same at the beginning and end of the motion, and if we need to find the displacement travelled, it is a simple case of multiplying velocity by time:

$$s = v \times t$$

CONSTANT ACCELERATION

There are equations that allow us to work out the motion of an object that is moving with a constant acceleration. For these kinematics equations, the first step is to define the five variables used:

$\boldsymbol{s}$ – displacement (m)
$\boldsymbol{u}$ – initial velocity ($\mathrm{m\,s^{-1}}$)
$\boldsymbol{v}$ – final velocity ($\mathrm{m\,s^{-1}}$)
$\boldsymbol{a}$ – acceleration ($\mathrm{m\,s^{-2}}$)
t – time (s)

Each equation uses four of the variables, which means that if we know the values of any three variables, we can find out the other two.

ACCELERATION REDEFINED

By re-arranging the equation that defined acceleration, we come to the usual expression of the first kinematics equation:

$$\boldsymbol{v} = \boldsymbol{u} + \boldsymbol{a}t$$

For example, if a stone is dropped off a cliff (see **fig B**) and takes three seconds to hit the ground, what is its speed when it does hit the ground?

Identify the three things we know:

- falling under gravity, so $\boldsymbol{a} = \boldsymbol{g} = 9.81\ \mathrm{m\,s^{-2}}$ (constant acceleration, so the kinematics equations can be used)
- starts at rest, so $\boldsymbol{u} = 0\ \mathrm{m\,s^{-1}}$
- time to fall $t = 3$ s

$$\boldsymbol{v} = \boldsymbol{u} + \boldsymbol{a}t = 0 + 9.81 \times 3 = 29.43$$

$$\boldsymbol{v} = 29.4\ \mathrm{m\,s^{-1}}$$

EXAM HINT

Often the acceleration will not be clearly stated, but the object is falling under gravity, so:

$$\boldsymbol{a} = \boldsymbol{g} = 9.81\ \mathrm{m\,s^{-2}}$$

EXAM HINT

Often the initial velocity will not be explicitly stated, but the object starts 'at rest'. This means it is stationary at the beginning, so:

$$\boldsymbol{u} = 0\ \mathrm{m\,s^{-1}}$$

DISTANCE FROM AVERAGE SPEED

As the kinematics equations only work with *uniform* acceleration, the average speed during any acceleration will be halfway from the initial velocity to the final velocity. Therefore the distance travelled is the average speed multiplied by the time:

$$\boldsymbol{s} = \frac{(\boldsymbol{u} + \boldsymbol{v})}{2} \times t$$

For example, for the same stone dropping off the cliff as in the previous example, we could work out how high the cliff is.

Identify the three things we know:

- final velocity came to be $\boldsymbol{v} = 29.4\ \mathrm{m\,s^{-1}}$
- starts at rest, so $\boldsymbol{u} = 0\ \mathrm{m\,s^{-1}}$
- time to fall $t = 3$ s

$$\boldsymbol{s} = \frac{(\boldsymbol{u} + \boldsymbol{v})}{2} \times t = \frac{(0 + 29.4)}{2} \times 3 = 44.1$$

$$\boldsymbol{s} = 44.1\ \mathrm{m}$$

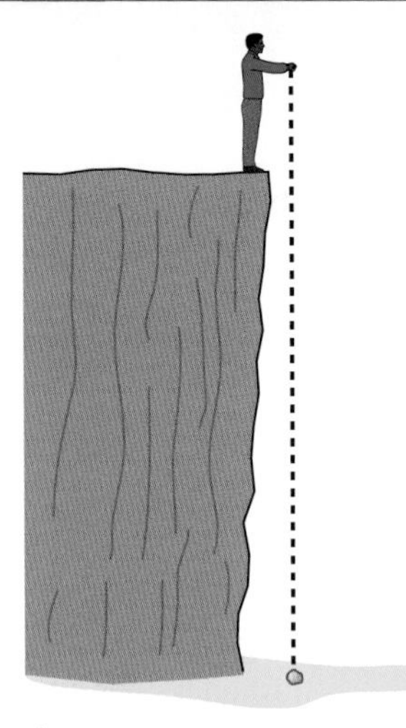

▲ **fig B** Calculations about falling objects are very common.

COMBINING EQUATIONS A

We can combine the equations

$$v = u + at$$

and

$$s = \frac{(u + v)}{2} \times t$$

By substituting the first of these equations into the second, we get the combination equation:

$$s = \frac{(u + (u + at))}{2} \times t = \frac{(2ut + at^2)}{2}$$

$$s = ut + \tfrac{1}{2}at^2$$

We can use this equation to check again how high the stone was dropped from:

- falling under gravity, so $\boldsymbol{a} = \boldsymbol{g} = 9.81\,\text{m s}^{-2}$
- starts at rest, so $\boldsymbol{u} = 0\,\text{m s}^{-1}$
- time to fall $t = 3\,\text{s}$

$$s = ut + \tfrac{1}{2}at^2 = (0 \times 3) + (\tfrac{1}{2} \times 9.81 \times 3^2) = 44.1$$

$$s = 44.1\,\text{m}$$

Notice that the answer must come out the same, as we are calculating for the same cliff. This highlights the fact that we can use whichever equation is most appropriate for the information given.

COMBINING EQUATIONS B

$$s = \frac{(u + v)}{2} \times t$$

$$\therefore \quad t = \frac{2s}{(u + v)}$$

and $v = u + at$

By substituting the first of these equations into the second, we get the combination equation:

$$v = u + a \times \frac{2s}{(u + v)}$$

$$\therefore \quad v(u + v) = u(u + v) + 2as$$

$$\therefore \quad vu + v^2 = u^2 + uv + 2as \qquad \textit{vu = uv so subtract from each side}$$

$$v^2 = u^2 + 2as$$

Check again what the stone's final velocity would be:

- falling under gravity, so $\boldsymbol{a} = \boldsymbol{g} = 9.81\,\text{m s}^{-2}$
- starts at rest, so $\boldsymbol{u} = 0\,\text{m s}^{-1}$
- height to fall $\boldsymbol{s} = 44.1\,\text{m}$

$$v^2 = u^2 + 2as = 0^2 + (2 \times 9.81 \times 44.1) = 865$$

$$\therefore \quad v = \sqrt{865} = 29.4$$

$$v = 29.4\,\text{m s}^{-1}$$

Notice that the answer must come out the same as previously calculated, and this again highlights that there are many ways to reach the answer.

KINEMATICS EQUATION	QUANTITY NOT USED
$v = u + at$	distance
$s = \frac{(u + v)}{2} \times t$	acceleration
$s = ut + \frac{1}{2}at^2$	final velocity
$v^2 = u^2 + 2as$	time

table A Each of the kinematics equations is useful, depending on the information we are given. If you know three quantities, you can always find a fourth by identifying which equation links those four quantities and re-arranging that equation to find the unknown.

EXAM HINT

The kinematics equations are only valid if there is a constant acceleration. If the acceleration is changing they cannot be used.

PRACTICAL SKILLS CP1

Finding the acceleration due to gravity by freefall

A system for timing the fall of an object under gravity can allow us to measure the acceleration due to gravity. In this experiment, we measure the time taken by a falling object to drop under gravity from a certain height, and then alter the height and measure again.

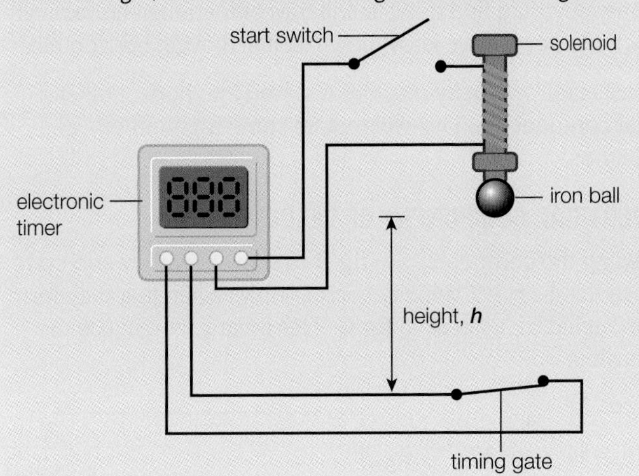

fig C The freefall time of an object from different heights allows us to find the acceleration due to gravity, ***g***.

If we vary the height from which the object falls, the time taken to land will vary. The kinematics equations tell us that:

$$s = ut + \frac{1}{2}at^2$$

As it always starts from rest, $u = 0$ throughout; the acceleration is that caused by gravity, g; and the distance involved is the height from which it is released, h. Thus:

$$h = \frac{1}{2}gt^2$$

$$\therefore \quad t^2 = \frac{2h}{g}$$

Compare this equation with the equation for a straight-line graph:

$$y = mx + c$$

We plot a graph of h on the x-axis and t^2 on the y-axis to give a straight best-fit line. The gradient of the line on this graph will be $2/g$, from which we can find g.

We could find a value for g by taking a single measurement from this experiment and using the equation to calculate it:

$$g = \frac{2h}{t^2}$$

However, a single measurement in any experiment is subject to uncertainty from both random and systematic errors. We can reduce such uncertainties significantly by taking many readings and plotting a graph, which leads to much more reliable conclusions.

Safety Note: Secure the tall stand holding the solenoid so that it cannot topple over.

CHECKPOINT

1. What is the final velocity of a bike that starts at $4\,\text{m s}^{-1}$ and has zero acceleration act on it for 10 seconds?
2. How far will the bike in question 1 travel in the 10-second time period?
3. Calculate the acceleration in each of the following cases:
   (a) $v = 22\,\text{m s}^{-1}$; $u = 8\,\text{m s}^{-1}$; $t = 2.6\,\text{s}$
   (b) a ball starts at rest and after 30 m its velocity has reached $4.8\,\text{m s}^{-1}$
   (c) in 15 s, a train moves 100 m from rest.
4. The bird in **fig A** drops the stone from a height of 88 m above the water surface. Initially, the stone has zero vertical velocity. How long will it take the stone to reach the surface of the pond? Assume air resistance is negligible.
5. If the stone in **fig A** enters the water at $41.6\,\text{m s}^{-1}$, and takes 0.6 s to travel the 3 metres to the bottom of the pond, what is its average acceleration in the pond water?

SUBJECT VOCABULARY

kinematics the study of the description of the motion of objects

uniform motion motion when there is no acceleration:

$$\text{velocity (m s}^{-1}) = \frac{\text{displacement (m)}}{\text{time (s)}}$$

$$v = \frac{s}{t}$$

LEARNING OBJECTIVES

- Explain that any vector can be split into two components at right angles to each other.
- Calculate the values of the component vectors in any such right-angled pair (resolution).

We have seen that vectors that act perpendicular to each other can be combined to produce a single vector that has the same effect as the two separate vectors. If we start with a single vector, we can find a pair of vectors at right angles to each other that would combine to give our single original vector. This reverse process is called **resolution** or **resolving vectors**. The resolved pair of vectors will both start at the same point as the original single vector.

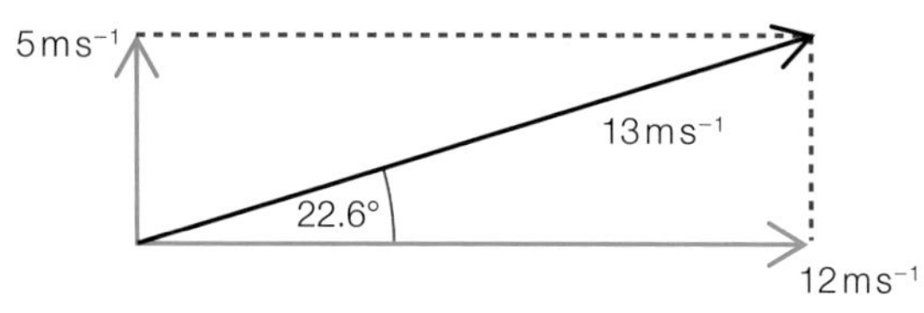

fig A The original velocity vector of 13 m s^{-1} has been resolved into a horizontal and a vertical velocity vector, which would separately be the same overall effect. An object moving right at 12 m s^{-1} and up at 5 m s^{-1} will move 13 metres each second at an angle of 22.6° up and right from the horizontal.

RESOLVING VECTORS CALCULATIONS

In order to resolve a vector into a pair at right angles, we must know all the details of the original vector. This means we must know its size and direction. The direction is most commonly given as an angle to either the vertical or the horizontal. This is useful, as we most commonly want to split the vector up into a horizontal and vertical pair.

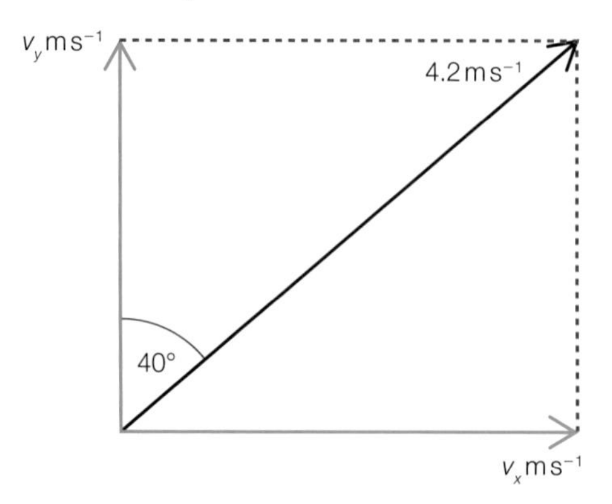

fig B Resolving a velocity vector into vertical and horizontal components.

In the example of **fig B**, a basketball is thrown into the air at an angle of 40° to the vertical. In order to find out if it will go high enough to reach the hoop, kinematics calculations could be done on its vertical motion. However, this can only be done if we can isolate the vertical component of the basketball's motion. Similarly, we could find out if it will travel far enough horizontally to reach the hoop if we know how fast it is moving horizontally.

The basketball's velocity must be resolved into horizontal and vertical components. This will require some trigonometrical calculations.

THE VERTICAL COMPONENT OF VELOCITY

Redrawing the components in **fig B** to show how they add up to produce the 4.2 m s^{-1} velocity vector, shows again that they form a right-angled triangle, as in **fig C**. This means we can use the relationship:

$$\cos 40° = \frac{\mathbf{v}_y}{4.2}$$

$$\therefore \quad \mathbf{v}_y = 4.2 \times \cos 40° = 4.2 \times 0.766$$

$$\therefore \quad \mathbf{v}_y = 3.2 \text{ m s}^{-1}$$

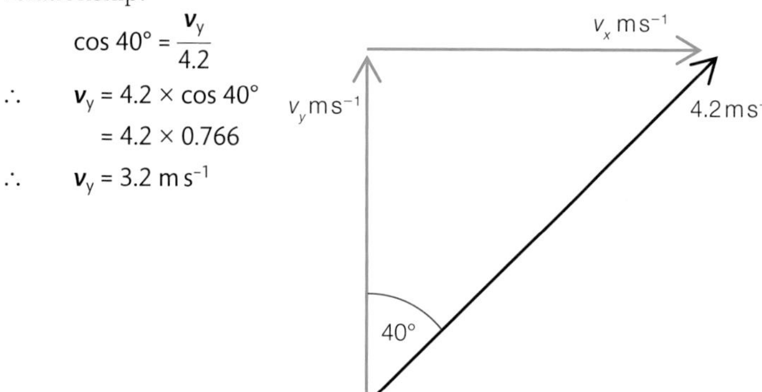

fig C Finding the components of velocity.

THE HORIZONTAL COMPONENT OF VELOCITY

Similarly, for the horizontal component, we can use the relationship:

$$\sin 40° = \frac{\mathbf{v}_x}{4.2}$$

$$\therefore \quad \mathbf{v}_x = 4.2 \times \sin 40° = 4.2 \times 0.643$$

$$\therefore \quad \mathbf{v}_x = 2.7 \text{ m s}^{-1}$$

LEARNING TIP

$\sin \theta°$ = opposite/hypotenuse

$\cos \theta°$ = adjacent/hypotenuse

$\tan \theta°$ = opposite/adjacent

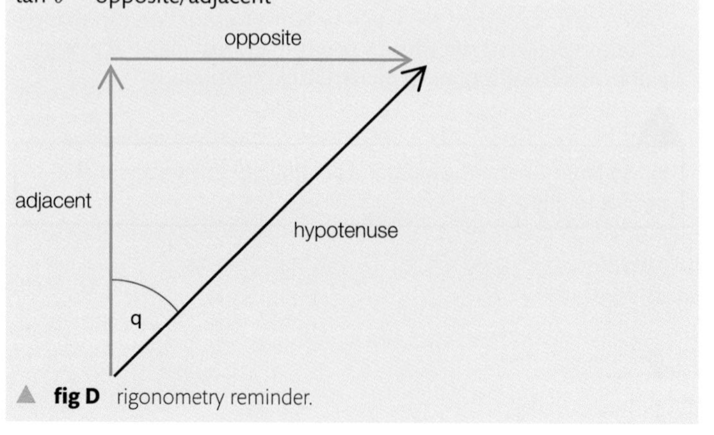

fig D rigonometry reminder.

LEARNING TIP

Resolving vectors could also be achieved using a scale diagram. Using this method in rough work may help you to get an idea as to what the answer should be in order to check that your calculations are about right.

ALTERNATIVE RESOLUTION ANGLES

If we know the velocity of an object, we have seen that we can resolve this into a pair of velocity vectors at right angles to each other. The choice of direction of the right-angle pair, though, is arbitrary, and can be chosen to suit a given situation.

Imagine a submarine descending underwater close to an angled seabed.

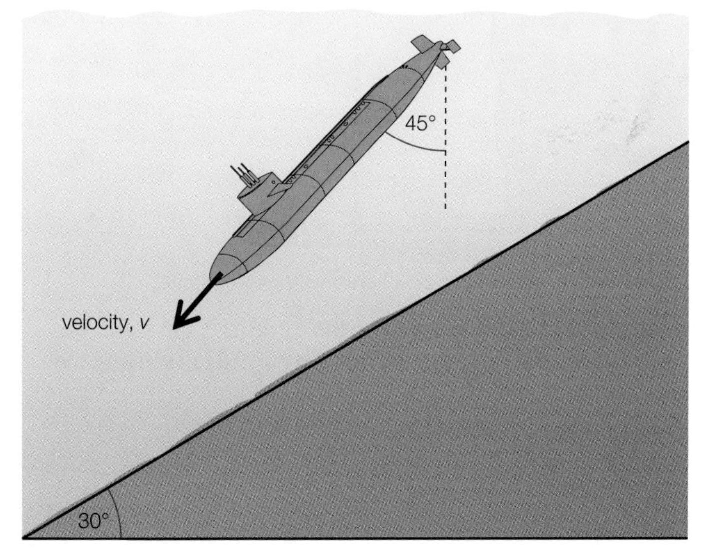

▲ **fig E** Resolving submarine velocity vectors helps to avoid collision.

The velocity could be resolved into a pair of vectors, horizontal and vertical.

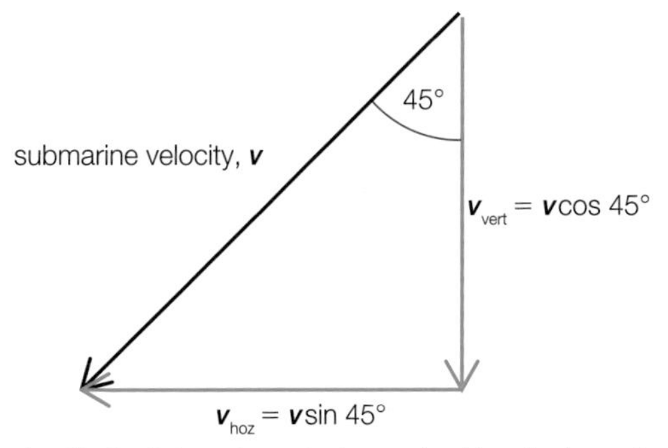

▲ **fig F** Submarine velocity resolved into horizontal and vertical components.

However, the submarine commander is likely to be most interested in knowing how quickly the submarine is approaching the seabed. This could be found by resolving the velocity of the submarine into a component parallel with the seabed, and the right-angled pair with that will be a component perpendicular to the seabed. It is this $\boldsymbol{v}_{perp}$ that will tell the submarine commander how quickly he is approaching the seabed.

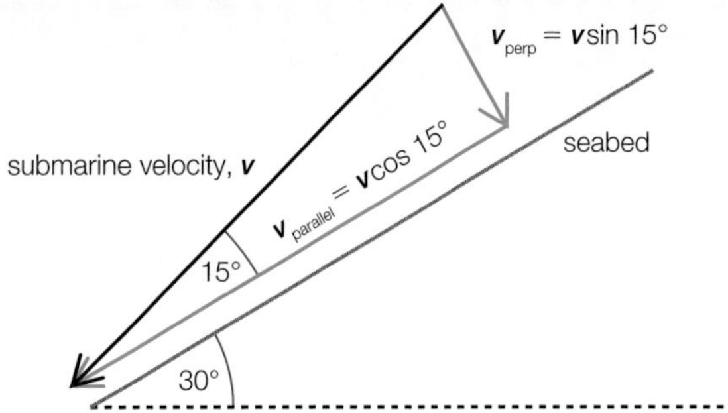

▲ **fig G** Submarine velocity resolved into components parallel and perpendicular to the seabed.

LEARNING TIP

The vectors produced by resolution can be at any angle with respect to the original vector, as long as they are perpendicular to each other and their resultant equals the original vector. In order to allow for every pair to add up to the original vector, their magnitudes will vary as the angle changes.

CHECKPOINT

SKILLS INTERPRETATION

1. (a) On graph paper, draw a velocity vector for a stone fired from a **catapult** at 45° to the horizontal. Your arrow should be 10 cm long, representing a velocity of 10 m s^{-1}. Draw onto your diagram the horizontal and vertical components that would make up the overall velocity. Use a ruler to measure the size of the horizontal and vertical components, and convert these lengths into metres per second using the same scaling.
   (b) Find the horizontal and vertical velocity components for this catapult stone by calculation, and compare with your answers from part (a).
2. A javelin is thrown at 16 m s^{-1} at an angle of 35° up from the horizontal. Calculate the horizontal and vertical components of the javelin's motion.
3. A ladder is leant against a wall, at an angle of 28° to the wall. The 440 N force from the floor acts along the length of the ladder. Calculate the horizontal and vertical components of the force from the floor that act on the bottom of the ladder.
4. A plane is flying at 240 m s^{-1}, on a bearing of 125° from due north. Calculate its velocity component due south, and its velocity component due east.

SUBJECT VOCABULARY

resolution or **resolving vectors** the determination of a pair of vectors, at right angles to each other, that sums to give the single vector they were resolved from
catapult a device that can throw objects at high speed

1A 8 PROJECTILES

SPECIFICATION REFERENCE 1.3.1 1.3.7

LEARNING OBJECTIVES

- Apply kinematics equations to moving objects.
- Apply the independence of horizontal and vertical motion to objects moving freely under gravity.
- Combine horizontal and vertical motion to calculate the movements of projectiles.

Objects thrown or fired through the air generally follow **projectile** motion. Here we are going to combine ideas from the various earlier sections in order to solve questions about projectiles. Resolving vectors showed that the actions in each of two perpendicular directions are wholly independent. This means we can use Newton's laws of motion and the kinematics equations separately for the horizontal and vertical motions *of the same object*. This will allow us to calculate its overall motion in two dimensions.

We only consider the motion after the force projecting an object has finished – for example, after a cannonball has left the cannon. Air resistance is ignored in these calculations, so the only force acting will be the object's weight. Thus, all vertical motion will follow kinematics equations, with gravity as the acceleration. There will be no horizontal force at all, which means no acceleration and therefore $\boldsymbol{v} = \boldsymbol{s}/t$.

LEARNING TIP

Vectors acting at right angles to each other act independently. Their combined effect can be calculated using vector addition, but they can also be considered to act separately and at the same time. This would cause independent effects, which themselves could then be combined to see an overall effect.

LEARNING TIP

Physics is holistic: you can apply ideas from one area of physics to other areas of physics. Remember that physics is a set of rules that attempts to explain everything in the universe. Every rule should therefore be universally applicable.

HORIZONTAL THROWS

If an object is thrown horizontally, it will start off with zero vertical velocity. However, gravity will act on it so that its motion will curve downwards in a parabola shape, like the stone in **fig A**.

In the example of **fig A**, a stone is kicked horizontally off a cliff with a velocity of 8.2 m s^{-1}. How much time is the stone in flight? How far does it get away from the cliff by the time it lands?

Horizontal and vertical motions are totally independent. Here the vertical component of velocity is initially zero, but the stone accelerates under gravity. Uniform acceleration means the kinematics equations can be used.

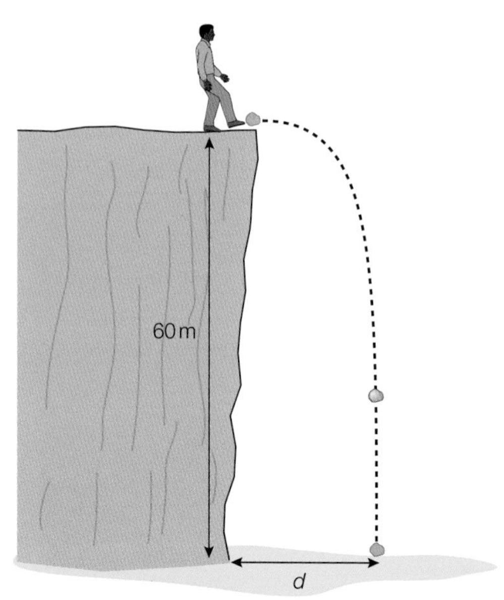

▲ **fig A** Vertical acceleration on a horizontally moving stone.

The time to hit the beach, t, will be the same as if the stone was simply dropped. We know $\boldsymbol{u} = 0\text{ m s}^{-1}$; $\boldsymbol{a} = -9.81\text{ m s}^{-2}$; and the height fallen, $\boldsymbol{s} = -60\text{ m}$.

$$s = ut + \tfrac{1}{2}at^2$$

$$u = 0 \therefore ut = 0$$

$$\therefore \quad s = \tfrac{1}{2}at^2$$

$$\therefore \quad t = \sqrt{\frac{2s}{a}} = \sqrt{\frac{(2 \times -60)}{-9.81}}$$

$$t = 3.5\text{ s}$$

Horizontally, there is no accelerating force once the stone is in flight, so it has a constant speed. Thus, to find the distance travelled horizontally, $\boldsymbol{d}$:

$$v = \frac{d}{t}$$

$$\therefore \quad d = v \times t$$

$$d = 8.2 \times 3.5$$

$$d = 28.7\text{ m}$$

RECOMBINING VELOCITY COMPONENTS

In the example of the stone kicked from the cliff to the beach, we might also want to calculate the final velocity of the stone on landing. This means adding vertical and horizontal components into their resultant.

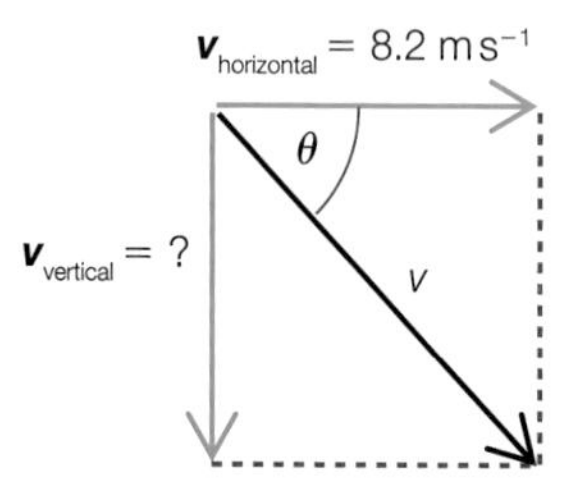

▲ **fig B** The stone's final velocity is the resultant of its vertical and horizontal components.

In the example of **fig A**, what is the velocity of the stone when it hits the beach?

The horizontal velocity was given as $8.2\,m\,s^{-1}$. To calculate the vertical velocity: $u = 0\,m\,s^{-1}$; $a = -9.81\,m\,s^{-2}$; and $s = -60\,m$.

$$v^2_{vertical} = u^2 + 2as = 0^2 + (2 \times -9.81 \times -60) = 1177.2$$

$$\therefore \quad v_{vertical} = \sqrt{1177.2} = -34.3$$

$$v_{vertical} = -34.3\,m\,s^{-1}$$

Pythagoras' theorem gives the magnitude of the final velocity:

$$v = \sqrt{(8.2^2 + 34.3^2)}$$

$$\therefore \quad v = 35.3\,m\,s^{-1}$$

Trigonometry will give the angle at which the stone is flying on impact with the beach:

$$\tan\theta = \frac{34.3}{8.2}$$

$$\therefore \quad \theta = 77°$$

The stone's velocity when it lands on the beach is 35.3 metres per second at an angle of 77° down from the horizontal.

EXAM HINT

We use the usual approach that up is positive throughout this section. You can choose any direction to be negative, but you must refer to that direction as negative at every step of the calculation. If you are not consistent, your calculation will be incorrect.

VERTICAL THROWS

Imagine throwing a ball to a friend. The ball goes up as well as forwards. One common idea in these calculations is that an object thrown with a vertical upwards component of motion will have a symmetrical trajectory. At the highest point, the vertical velocity is momentarily zero. Getting to this point will take half of the time of the whole flight.

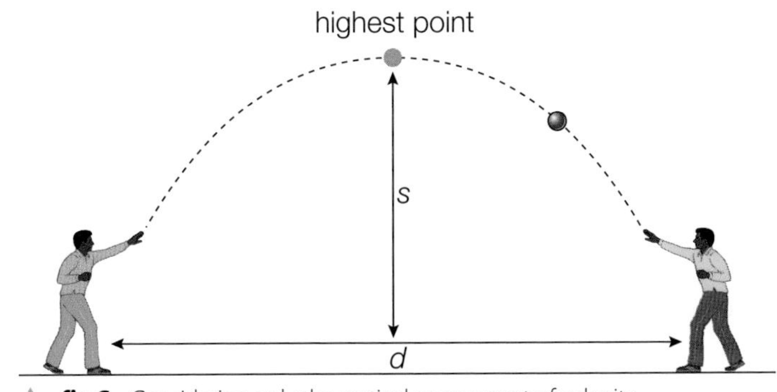

▲ **fig C** Considering only the vertical component of velocity.

In the example of **fig C**, a ball is thrown with a vertical velocity component of $5.5\,m\,s^{-1}$. How much time is it in flight? How high does it get?

These questions would have the same answer if a person threw the ball vertically up and caught it again themselves. This again highlights the *independence* of horizontal and vertical motions. It may be that an initial velocity at an angle is quoted, so that we need to resolve the velocity vector into its horizontal and vertical components in order to know that here $v_{vertical} = 5.5\,m\,s^{-1}$.

Consider the second question first: uniform acceleration under gravity means $a = -9.81\,m\,s^{-2}$ and the kinematics equations can be used. We know $u = 5.5\,m\,s^{-1}$; at the top of the path, $v = 0\,m\,s^{-1}$; and we want to find the height, s.

$$v^2 = u^2 + 2as$$

$$\therefore \quad s = \frac{v^2 - u^2}{2a} = \frac{0^2 - (5.5)^2}{2 \times -9.81}$$

$$s = 1.54\,m$$

Note that 1.54 metres is actually the height the ball reaches above the point of release at which it left the hand – the point where its initial speed was $5.5\,m\,s^{-1}$ – but this is often ignored in projectiles calculations.

The time of flight for the ball will be just the time taken to rise and fall vertically. We find the time to reach the highest point, and then double that value. We know $u = 5.5\,m\,s^{-1}$; at the top of the path, $v = 0\,m\,s^{-1}$; and we want to find the time, t.

$$v = u + at$$

$$\therefore \quad t = \frac{v - u}{a} = \frac{0 - (5.5)}{-9.81}$$

$$t = 0.56\,s$$

So the overall time of flight will be 0.56 seconds doubled: total time = 1.12 s.

EXAM HINT

For virtually all projectile motion calculations, we assume that there is no air resistance, so the only force acting is gravity, vertically downwards.

CHECKPOINT

SKILLS PROBLEM SOLVING

1. A boy throws a ball vertically at a velocity of $4.8\,m\,s^{-1}$.
   (a) How long is it before he catches it again?
   (b) What will be the ball's greatest height above the point of release?
2. The boy in question 1 now throws his ball horizontally out of a high window with a velocity of $3.1\,m\,s^{-1}$.
   (a) How long will it take to reach the ground 18 m below?
   (b) How far away, horizontally, should his friend stand in order to catch the ball?
3. ▶ A basketball is thrown with a velocity of $6.0\,m\,s^{-1}$ at an angle of 40° to the vertical, towards the hoop.
   (a) If the hoop is 0.90 m above the point of release, will the ball rise high enough to go in the hoop?
   (b) If the centre of the hoop is 3.00 m away, horizontally, from the point of release, explain whether or not you believe this throw will score in the hoop. Support your explanation with calculations.

SUBJECT VOCABULARY

projectile a moving object on which the only force of significance acting is gravity. The trajectory is thus pre-determined by its initial velocity

THE BATTLE OF AGRA

SKILLS CRITICAL THINKING, PROBLEM SOLVING, ANALYSIS, INTERPRETATION, ADAPTIVE LEARNING, PRODUCTIVITY, ETHICS, COMMUNICATION, ASSERTIVE COMMUNICATION

Agra Fort was built in the 11th Century, although the present structure was built in 1573. In this activity, you need to imagine attacking the fort using a cannon that fires a cannonball as a projectile.

STUDENT ESSAY

▲ **fig A** Agra Fort is now an UNESCO World Heritage Site.

In this section, I will use some basic mechanics to answer a question: could the Mughal Empire artillery really have attacked Agra Fort in the way described previously? The nineteenth-century source material suggests that the fortress was under siege by the Mughals for three months and 'battered by artillery'. However, the current walls bear little in the way of obvious battle scars.

Looking at **fig B**, the question that needs to be answered here is:

'How high up the front wall of the fortress will the cannonball hit?' This height is marked on **fig B** as '*H*'.

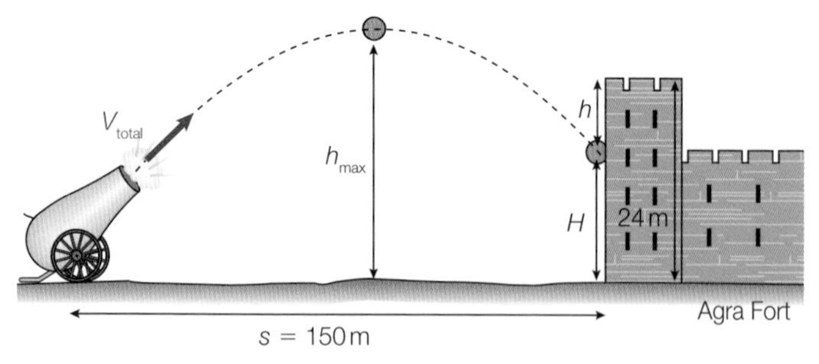

▲ **fig B** The trajectory of a cannonball fired towards Agra Fort. We assume the cannonball leaves the cannon at ground level.

In addition to the layout shown in **fig B**, we need information about the initial velocity of the cannonball. The cannon explosion could act for 0.05 s to accelerate the cannonball (mass = 12 kg) with a force of 9300 N. It causes the cannonball to leave the cannon at an angle of 45° to the horizontal.

Steps to the answer

We can work out what calculations are required to solve this problem, by working back from the answer we want to find. The fundamental idea is that the parabola trajectory would be symmetrical if the flight was not interrupted by crashing into the fortress wall.

1. To find the height up the wall from the ground, we will need to work out how far down from the cannonball's maximum height it falls:

   $$H = h_{max} - h$$

2. To find h, we need to know the time of flight, t_{total} so we can divide this into a time to reach h_{max}, and time left to fall height h. We will use vertical gravitational acceleration to calculate the vertical drop in that remaining time:

   $$t_{total} = \frac{s}{v_{horizontal}}$$

   From **fig B**, we can see that $s = 150$ m.

3. $v_{horizontal}$ can be found by resolving the velocity to give the horizontal component:

   $$v_{horizontal} = v_{total} \times \cos 45°$$

4. The overall velocity will come from the cannon's acceleration of the cannonball:

   $$v = u + at$$

   where $u = 0\ \text{m s}^{-1}$, and the question tells us that the explosion acts for 0.05 seconds.

5. Newton's second law of motion gives us the acceleration caused by the sling:

   $$a = \frac{F}{m}$$

Calculate the answer by reversing these steps:

The acceleration caused by the explosion:

$$a = \frac{F}{m}$$

$a =$

SCIENCE COMMUNICATION

1 The extract opposite is a draft for a university essay about the Mughal siege of 1857. Consider the extract and comment on the type of writing being used. For example, think about whether this is a scientist reporting the results of their experiments, a scientific review of data, a newspaper or a magazine-style article for a specific audience. Try and answer the following questions:

(a) How can you tell that the author is doubtful about the historical source material?

(b) What is the purpose of this mathematical analysis, for its inclusion in this essay?

PHYSICS IN DETAIL

Now we will look at the physics in detail. You may need to combine concepts from different areas of physics to work out the answers.

2 Complete the calculation steps, in reverse as suggested, in order to find out the answer, H:

(a) the acceleration caused by the explosion
(b) overall velocity that the cannonball is projected from the cannon
(c) horizontal and vertical components of the velocity
(d) time of flight found from the horizontal travel
(e) time to reach maximum height using vertical motion
(f) remaining flight time from maximum height
(g) height fallen from the maximum in the remaining flight time
(h) final answer, H.

3 State **two** assumptions that have been made in these calculations.

4 Calculate what difference there would be in the answer if the cannon was loaded with different cannonballs of masses 10 kg and 14 kg. Note from **fig B** that the fortress walls are 24 m high. Comment on these answers.

5 If the available supply of cannonballs offered very variable masses, how might the Mughals be able to overcome the problems shown in question 4.

ACTIVITY

Imagine the writer of this essay is a friend of yours, and he has come to you for help with the calculations as he is not an experienced scientist. His section 'Steps to the answer' was taken from a research source about a different fortress under siege. Write an email to Claus to explain the calculations required in each step.

INTERPRETATION NOTE

Once you have answered the calculation questions below, decide whether you think the Mughal siege happened as the author suggests.

THINKING BIGGER TIP

Inside a cannon, an explosion exerts a force on the cannonball to fire it out of the cannon.

INTERPRETATION NOTE

You can assume that the writer understands mathematics, and is generally intelligent – a student who could have done A Level physics but preferred arts subjects.

1A EXAM PRACTICE

1 Quantities can be scalar or vector. Select the row of the table that correctly states a scalar quantity and a vector quantity.

	Scalar quantity	Vector quantity
A	acceleration	mass
B	mass	weight
C	speed	distance
D	weight	speed

[1]

(Total for Question 1 = 1 mark)

2 How is the kinetic energy, E_k, of a car related to its speed, v?

A $E_k \; v$

B $E_k \; v^2$

C $E_k \; \sqrt{v}$

D $E_k \; \frac{1}{v}$

[1]

(Total for Question 2 = 1 mark)

3 The unit of force is the newton. One newton is equivalent to:

A 0.1 kg

B $1\,\text{kg}\,\text{m}\,\text{s}^{-1}$

C $1\,\text{kg}\,\text{m}\,\text{s}^{-2}$

D $1\,\text{m}\,\text{s}^{-2}$

[1]

(Total for Question 3 = 1 mark)

4 A ball is thrown vertically upwards at a speed of $11.0\,\text{m}\,\text{s}^{-1}$. What is the maximum height it reaches?

A 0.561 m

B 1.12 m

C 6.17 m

D 12.3 m

[1]

(Total for Question 4 = 1 mark)

5 Calculate the moment exerted on the nut by the spanner shown in the diagram.

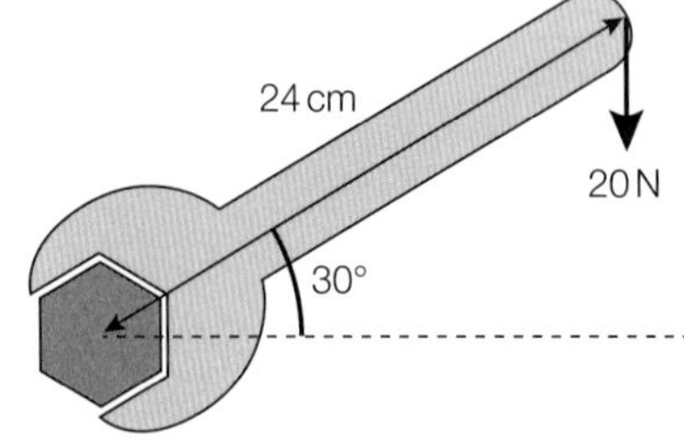

A 2.4 N m

B 4.2 N m

C 4.8 N m

D 420 N m

[1]

(Total for Question 5 = 1 mark)

6 (a) What is meant by a vector quantity? [1]

(b) A car is driven around a bend at a constant speed. Explain what happens to its velocity. [2]

(Total for Question 6 = 3 marks)

7 You are asked to determine the acceleration of free fall at the surface of the Earth, ***g***, using a free fall method in the laboratory.

(a) Describe the apparatus you would use, the measurements you would take and explain how you would use them to determine ***g***. [6]

(b) Give **one** precaution you would take to ensure the accuracy of your measurements. [1]

(Total for Question 7 = 7 marks)

8 The graph shows how displacement varies with time for an object that starts from rest with constant acceleration.

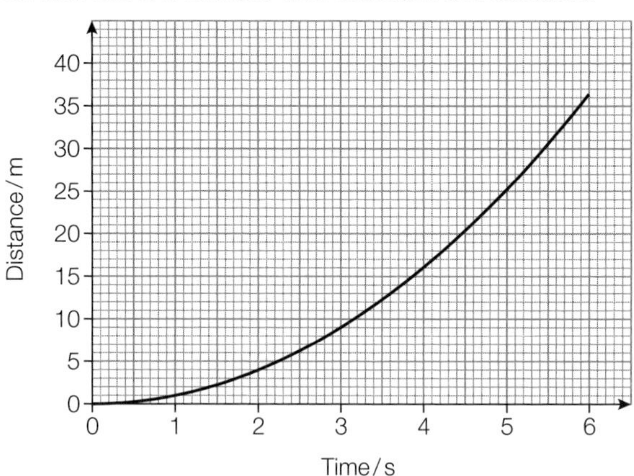

(a) Use the distance–time graph to determine the speed of the object at a time of 4.0 s. [3]

(b) Calculate the acceleration. [2]

(Total for Question 8 = 5 marks)

9 The photograph shows a sequence of images of a bouncing tennis ball.

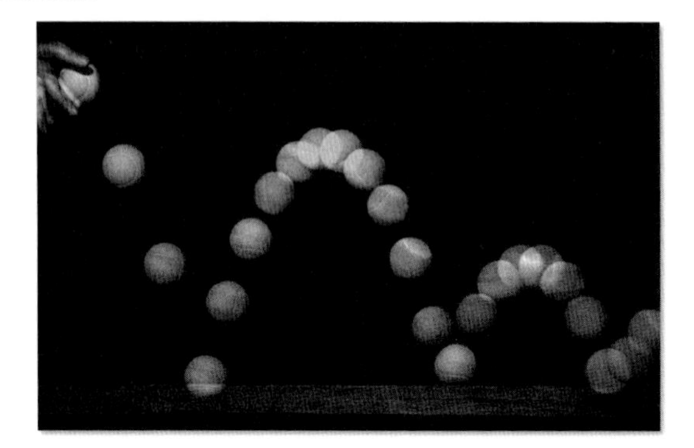

A student plots the following graph and claims that it shows the vertical motion of the ball in the photograph.

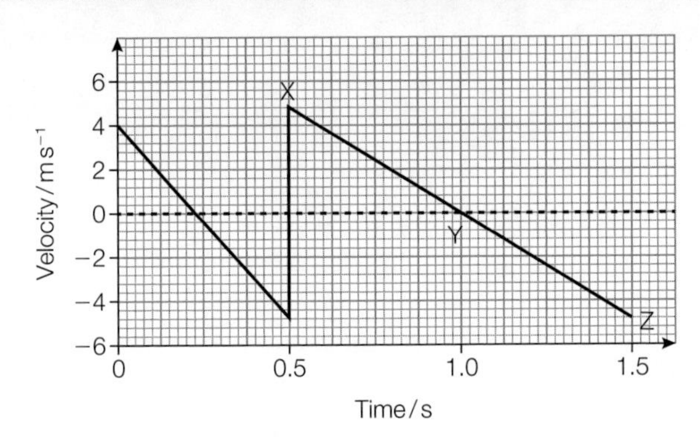

(a) Without carrying out any calculations, describe how the following can be found from the graph
   (i) the vertical distance travelled by the ball between 0.5 s and 1.0 s
   (ii) the acceleration at Y. [2]

(b) The graph contains several errors in its representation of the motion of the ball
   Explain two of these errors. [4]

(Total for Question 9 = 6 marks)

10 There has been a proposal to build a train tunnel underneath the Atlantic Ocean from England to America. The suggestion is that in the future the trip of 5000 km could take as little as one hour.

Assume that half the time is spent accelerating uniformly and the other half is spent decelerating uniformly with the same magnitude as the acceleration.

(a) Show that the acceleration would be about $2\,m\,s^{-2}$. [2]

(b) Calculate the maximum speed. [2]

(c) Calculate the resultant force required to decelerate the train.
   mass of train = 4.5×10^5 kg [2]

(Total for Question 10 = 6 marks)

11 During a lesson on Newton's laws of motion, a student says, 'We don't really need to bother with Newton's first law because it is included in his second law'.
State Newton's first two laws of motion and explain how Newton's second law includes the first law. [5]

(Total for Question 11 = 5 marks)

12 The diagram shows an arrangement used to launch a light foam rocket at a school science competition.

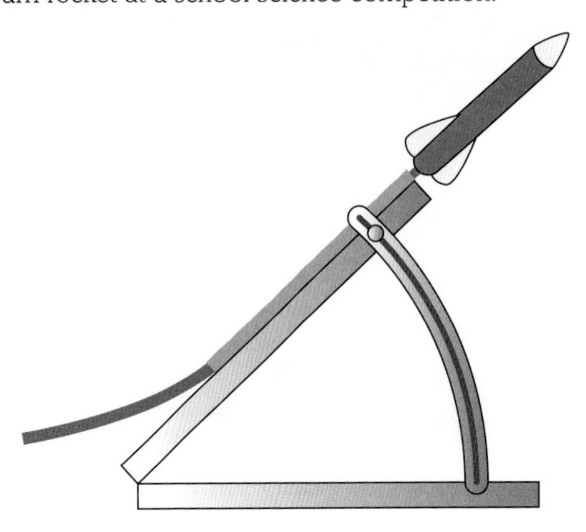

The rocket is launched at the level of one end of a long table and lands at the other end at the same level. The students measure the horizontal distance travelled by the rocket and the time of flight.

(a) The rocket travels 1.88 m in a time of 0.88 s.
   (i) Show that the horizontal component of the initial velocity of the rocket is about $2\,m\,s^{-1}$. [2]
   (ii) Show that the vertical component of the initial velocity of the rocket is about $4\,m\,s^{-1}$. [2]
   (iii) Calculate the initial velocity of the rocket. [4]

(b) The students obtained their data by filming the flight. When they checked the maximum height reached by the rocket they found it was less than the height predicted using this velocity.
   (i) Suggest why the maximum height reached was less than predicted. [1]
   (ii) Give **two** advantages of filming the flight to obtain the data. [2]

(Total for Question 12 = 11 marks)

TOPIC 1 MECHANICS

CHAPTER 1B ENERGY

Chapter 1A finished with a discussion of the motion of a projectile cannonball's flight. An alternative way of considering changes in movements of objects affected by gravity is to follow what happens to their energy. An object held up has an amount of gravitational potential energy as a result of its position. This energy can be transferred to kinetic, or movement, energy if the object falls. Humans have evolved over millions of years to avoid situations in which they might fall a great height, as such falls are generally dangerous. The kinetic energy that humans have gained from falling could be transferred to other stores of energy in their bodies, through large forces, and cause injuries.

The effect of gravity on the movement of an object should be considered in relation to the energy a body may possess or transfer. There are equations for calculating kinetic energy and gravitational potential energy, and the transfer of energy when a force is used to cause the transfer. These formulae and Newton's laws can be used together to work out everything we might want to know about the movement of any everyday object in any everyday situation.

Whilst it is difficult for scientists to describe or identify the exact nature of energy, the equations that describe energy relationships have always worked correctly.

MATHS SKILLS FOR THIS CHAPTER

- **Units of measurement** (*e.g. the joule, J*)
- **Changing the subject of an equation** (*e.g. finding the velocity of a falling object*)
- **Substituting numerical values into algebraic equations** (*e.g. calculating the power used*)
- **Plotting two variables from experimental or other data, understanding that** $y = mx + c$ **represents a linear relationship and determining the slope of a linear graph** (*e.g. finding the acceleration due to gravity experimentally*)
- **Using angles in regular 2D and 3D structures** (*e.g. finding the angle with which to calculate work done*)
- **Using sin, cos and tan in physical problems** (*e.g. calculating the work done by a force acting at an angle*)

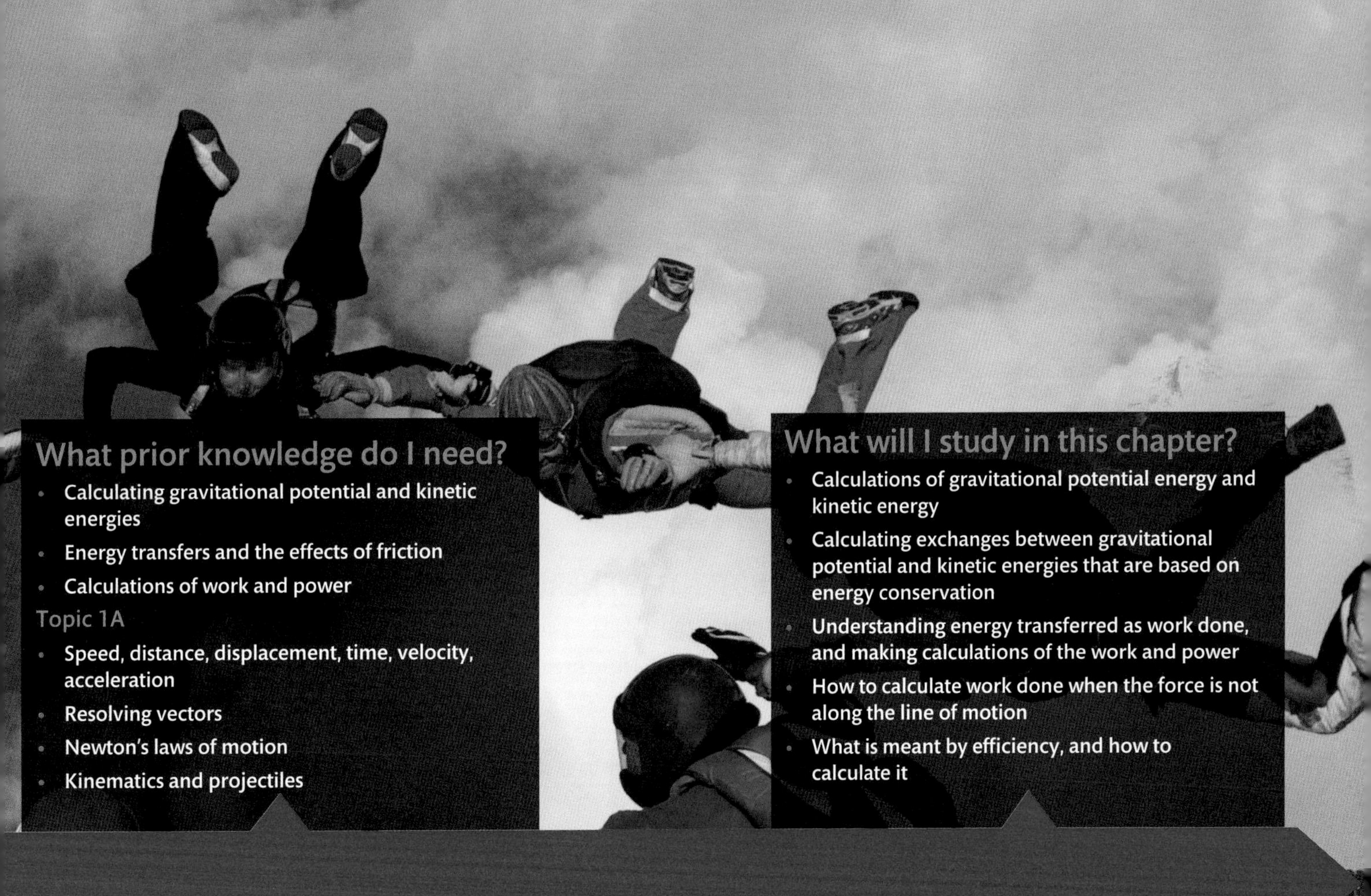

What prior knowledge do I need?

- Calculating gravitational potential and kinetic energies
- Energy transfers and the effects of friction
- Calculations of work and power

Topic 1A

- Speed, distance, displacement, time, velocity, acceleration
- Resolving vectors
- Newton's laws of motion
- Kinematics and projectiles

What will I study in this chapter?

- Calculations of gravitational potential energy and kinetic energy
- Calculating exchanges between gravitational potential and kinetic energies that are based on energy conservation
- Understanding energy transferred as work done, and making calculations of the work and power
- How to calculate work done when the force is not along the line of motion
- What is meant by efficiency, and how to calculate it

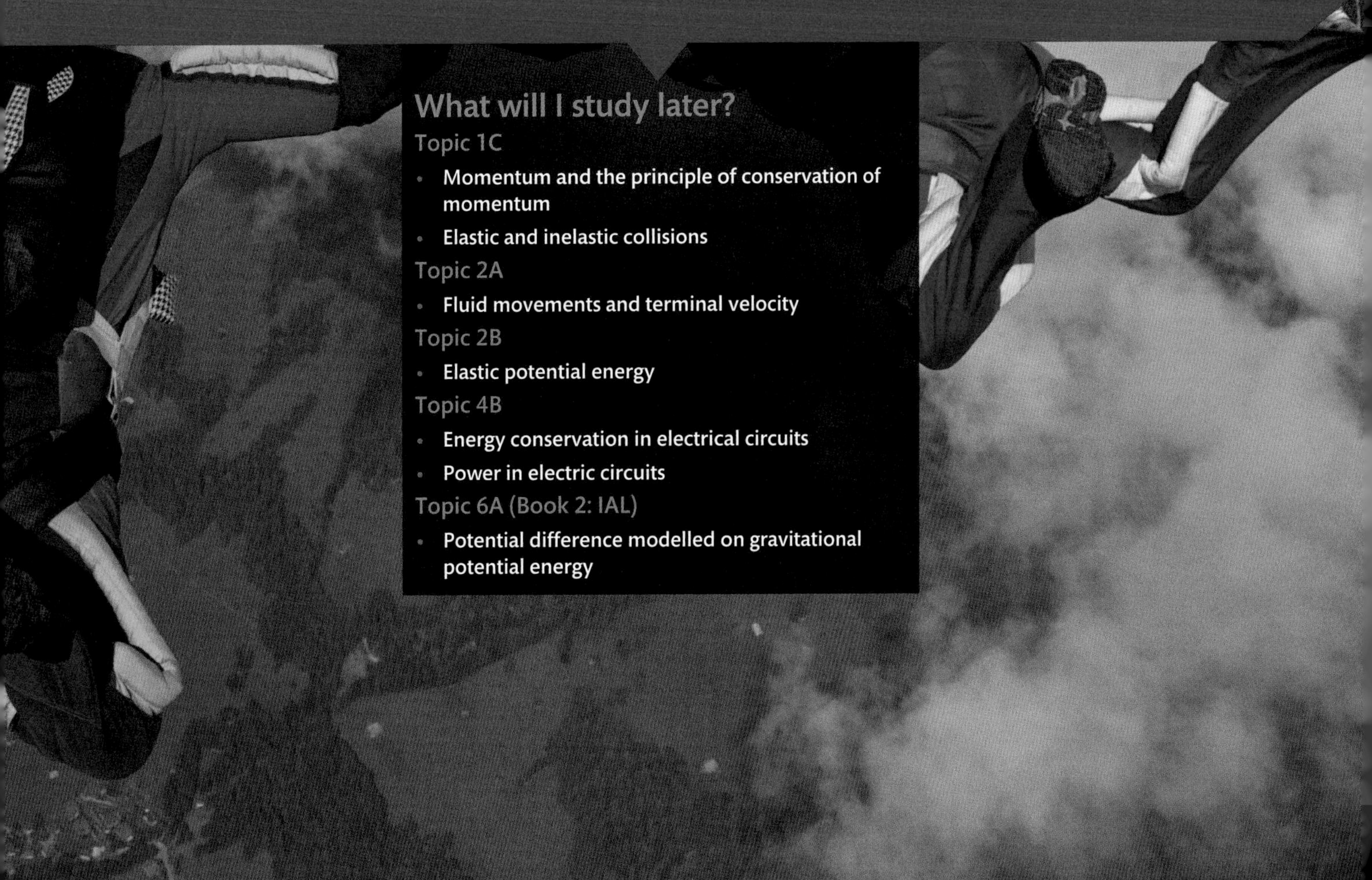

What will I study later?

Topic 1C

- Momentum and the principle of conservation of momentum
- Elastic and inelastic collisions

Topic 2A

- Fluid movements and terminal velocity

Topic 2B

- Elastic potential energy

Topic 4B

- Energy conservation in electrical circuits
- Power in electric circuits

Topic 6A (Book 2: IAL)

- Potential difference modelled on gravitational potential energy

1B 1 GRAVITATIONAL POTENTIAL AND KINETIC ENERGIES

SPECIFICATION REFERENCE

1.3.18 1.3.19 1.3.20 CP1

CP1 LAB BOOK PAGE 6

LEARNING OBJECTIVES

- Calculate transfers of gravitational potential energy near the Earth's surface.
- Calculate the kinetic energy of a body.
- Calculate exchanges between gravitational potential and kinetic energies, based on energy conservation.

▲ **fig A** The gravitational potential energy transferred to kinetic energy for a falling coconut can be a significant hazard in tropical countries.

Gravitational potential energy (E_{grav}) is the energy an object has by virtue of its position in a gravitational field. **Kinetic energy (E_k)** is the energy an object has by virtue of its movement. As objects rise or fall, gravitational potential energy can be transferred to kinetic energy and kinetic energy can be transferred to gravitational potential energy.

GRAVITATIONAL POTENTIAL ENERGY

Gravitational potential energy (gpe) can be calculated using the equation:

$$\text{gpe (J)} = \text{mass (kg)} \times \text{gravitational field strength (N kg}^{-1}\text{)} \times \text{height (m)}$$

$$E_{grav} = m\boldsymbol{g}h$$

Usually, the gpe is considered as a change caused by a change in height, for example, the change in gpe when you lift an object onto a shelf. This alters the equation slightly to consider transfers to or from gpe:

$$\Delta E_{grav} = m\boldsymbol{g}\Delta h$$

A brick of mass 2.2 kg is lifted vertically through a height of 1.24 m. The gpe gained is calculated as follows:

$$\Delta E_{grav} = m\boldsymbol{g}\Delta h$$

$$\Delta E_{grav} = 2.2 \times 9.81 \times 1.24$$

$$\Delta E_{grav} = 26.8\,\text{J}$$

Writing the formula this way suggests that the gravitational field strength is a fixed value. The gravitational field strength is a measure of the pull of gravity by a planet at a distance from its centre. This is not actually constant, as the strength of the gravitational field experienced by a mass is inversely proportional to the square of the distance from the planet's centre. However, close to the Earth's surface, over small scales, such as the heights that humans deal with in everyday life, it is an acceptably close approximation to say $\boldsymbol{g}$ is fixed at 9.81 N kg^{-1}.

DID YOU KNOW?

The strength of the Earth's gravitational field at the top of the Eiffel Tower is less than a hundredth of a per cent smaller than at the foot of the Eiffel Tower.

KINETIC ENERGY

Kinetic energy can be calculated using the equation:

$$\text{kinetic energy (J)} = \tfrac{1}{2} \times \text{mass (kg)} \times (\text{speed})^2\ (\text{m}^2\,\text{s}^{-2})$$

$$E_k = \tfrac{1}{2} \times m \times v^2$$

For example, a large jumbo jet plane might have a cruising speed of 900 km h^{-1} and a flight mass of 400 tonnes. What would its kinetic energy be?

First convert into SI units:

$$v = 900\,\text{km h}^{-1} = 900\,000\,\text{m h}^{-1} = \frac{9 \times 10^5}{60 \times 60} = 250\,\text{m s}^{-1}$$

$$m = 400 \times 1000 = 4 \times 10^5\,\text{kg}$$

$$E_k = \tfrac{1}{2} \times m \times v^2$$

$$\therefore \quad E_k = \tfrac{1}{2} \times 4 \times 10^5 \times 250^2$$

$$E_k = 1.25 \times 10^{10}\,\text{J} = 12.5\,\text{GJ}$$

▲ **fig B** A jumbo jet plane has a lot of kinetic energy.

TRANSFER BETWEEN E_{grav} AND E_K

The principle of conservation of energy tells us that we can never lose any energy or gain energy out of nowhere. In any energy transfer, we must have the same total energy before and after the transfer. Gravitational potential energy can be transferred to kinetic energy if an object falls to a lower height. Alternatively, an object thrown upwards will slow down as its kinetic energy is transferred to gpe. In either case:

$$\Delta E_{grav} = mg\Delta h = \frac{1}{2}mv^2 = E_k$$

Depending on the situation, it can often be useful to divide out the mass that appears on both sides of this equation. This allows a convenient calculation of how fast an object will be travelling after falling a certain distance from rest:

$$v = \sqrt{2g\Delta h}$$

or how high an object could rise if projected upwards at a certain speed:

$$\Delta h = \frac{v^2}{2g}$$

LEARNING TIP

The fact that mass divides out to give the relationship $v = \sqrt{2gh}$ confirms Galileo's idea that objects will all fall to the ground at the same rate regardless of their mass.

However, remember that all of the relationships shown in this section assume that no energy is lost through friction or air resistance, and that this can be an important factor when some objects fall.

▲ **fig C** The Burj Khalifa tower in Dubai.

For example, how fast would a coin hit the ground if it were dropped from the top of Burj Khalifa tower in Dubai, which is 830 m tall?

$$v = \sqrt{2g\Delta h}$$

$$v = \sqrt{2(9.81 \times 830)} = \sqrt{16285}$$

$$v = 128\text{ m s}^{-1}$$

Another example: how high would water from a fountain rise if it were ejected vertically upwards from a spout at 13.5 m s^{-1}?

$$\Delta h = \frac{v^2}{2g}$$

$$\Delta h = \frac{13.5^2}{2(9.81)} = \frac{182.25}{19.62}$$

$$h = 9.29\text{ m}$$

▲ **fig D** Fountain designers need to be able to calculate gpe and kinetic energy.

PRACTICAL SKILLS

CP1

Finding *g* from energy conservation

There are a number of different experimental methods for finding the gravitational field strength. One example of these relies on the transfer of gravitational potential energy to kinetic energy. In this experiment, we measure the velocity that has been reached by a falling object after it has fallen under gravity from a certain height, and then alter the height and measure the velocity again.

If we vary the height from which the object falls, the gravitational potential energy is different at each height. This gpe will all be transferred to kinetic energy.

$$mg\Delta h = \frac{1}{2}mv^2$$

As we saw previously, this equation can be simplified to give:

$$v^2 = 2g\Delta h$$

Compare this equation with the equation for a straight-line graph: $y = mx + c$

If we plot a graph of Δh on the x-axis and v^2 on the y-axis, we will get a straight best-fit line. The gradient of the line on this graph will be twice the gravitational field strength, $2g$, from which we can find g.
We could find a value for g by taking a single measurement from this experiment and using the equation to calculate it:

$$g = \frac{v^2}{2\Delta h}$$

However, as mentioned in **Section 1A.6**, a single measurement in any experiment is prone to uncertainty from both random and systematic errors. The reliability of the conclusions is significantly improved with multiple readings and graphical analysis.

fig E The freefall velocity of an object from different heights allows us to find the gravitational field strength, g.

!

Safety Note: Secure the tall stand so that it cannot topple over. The object must be positioned so that it cannot cause injury as it falls.

SKILLS CREATIVITY

CHECKPOINT

1. Estimate the speed at which a coconut from the tree in **fig A** would hit the sand.
2. How fast would a fountain need to squirt its water upwards to reach a height of 15 m?
3. How fast would a snowboarder be moving if he slid down a slope dropping a vertical height of 45 m?
4. How high will a 48 kg trampolinist rise if he leaves the trampoline at a speed of 6.1 m s^{-1}?
5. What assumption must you make in order to answer all of the above questions?

SUBJECT VOCABULARY

gravitational potential energy (E_{grav}) the energy an object stores by virtue of its position in a gravitational field:

gpe (J) = mass (kg) × gravitational field strength (N kg^{-1}) × height (m)

$E_{grav} = mgh$ OR $\Delta E_{grav} = mg\Delta h$

kinetic energy (E_k) the energy an object stores by virtue of its movement:

kinetic energy (J) = $\frac{1}{2}$ × mass (kg) × (speed)^2 (m^2 s^{-2})

$$E_k = \frac{1}{2} \times m \times v^2$$

1B 2 WORK AND POWER

SPECIFICATION REFERENCE 1.3.17 1.3.21 1.3.22

LEARNING OBJECTIVES

- Calculate energy transferred as work done, including when the force is not along the line of motion.
- Calculate the power of an energy transfer.
- Explain efficiency and be able to calculate.

We often make assumptions that allow simplification of calculations in physics. In general, these assumptions make little difference as they are chosen to ignore effects which have a very small impact on the actual real world answers. An example of this is with the transfer of gravitational potential energy to kinetic energy, where the effects of air resistance are ignored.

It is important to remember that the principle of **conservation of energy** insists that no energy can be lost in any scenario. Even if we were to consider the loss of kinetic energy to heating of the air through air resistance, the total amount of energy would be constant; it would simply have transferred to different stores.

WORK

In physics, the phrase 'doing work' has a specific meaning to do with energy use. The amount of **work done** means the amount of **energy** transferred, so work is measured in joules.

In general terms, we can express any energy transfer as work done. For example, a 15 W light bulb working for 10 seconds transfers 150 J of electrical energy as heat and light – it does 150 J of work. In any situation where we know how to calculate the energy before and after, we can calculate the energy transferred and thus the work done.

▲ **fig A** Work is done by transferring gravitational potential energy to a heavy object.

FORCING WORK

If energy is transferred mechanically by means of a force, then the amount of work done can be calculated simply:

work done (J) = force (N) × distance moved in the direction of the force (m)

$$\Delta W = F\Delta s$$

▲ **fig B** 'Work' on a building site.

In the example of **fig B**, a brick of mass 2.2 kg is lifted vertically against its weight through a height of 1.24 m. The work done is:

$$\Delta W = F\Delta s$$

$$\Delta W = mg \times \Delta s$$

$$\therefore \quad W = 2.2 \times 9.81 \times 1.24$$

$$W = 26.8\,\text{J}$$

Note that this is the same amount of energy as we calculated for the gravitational potential energy of this same brick undergoing the same lift in **Section 1B.1**.

EXAM HINT

Beware of confusing ΔW for work with W for weight, and W as the abbreviation for watts.

LEARNING TIP

An object that gains gravitational potential energy is having work done on it against its weight force.

weight = mass × gravitational field strength

$W = mg$

work = force × distance = weight × height = $mg\Delta h$

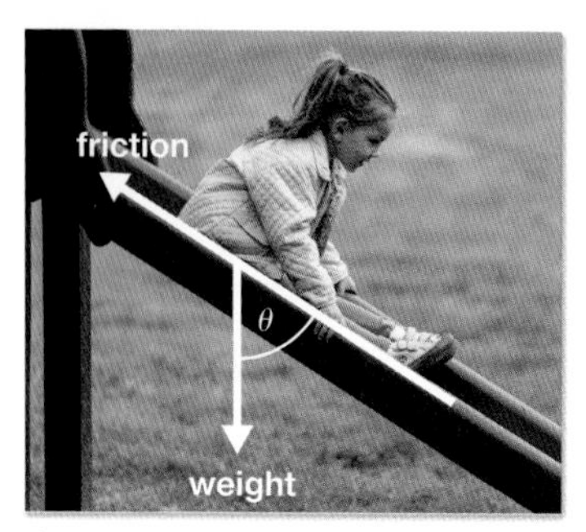

▲ **fig C** Gravity working against friction.

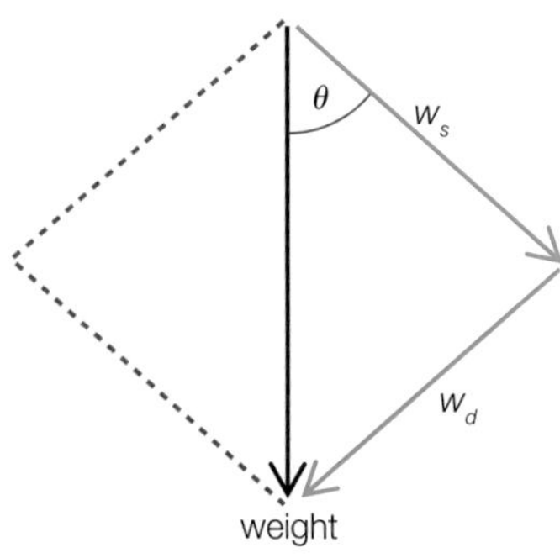

▲ **fig D** The weight force is the resultant of its components along the slide and perpendicular to it.

WORK DONE BY FORCES AT AN ANGLE

In the bricklaying example on the previous page, the direction of movement was exactly in line with the force lifting the brick. This is an unusual situation, and usually the force doing work will be at an angle to the direction of movement. In **fig C**, gravity pulls the child down the slide but it must work against friction. The weight acts vertically downwards, but the friction acts up the slide. Friction always acts in the exact opposite direction to movement.

Assuming that the child slides at constant velocity, Newton's first law of motion tells us that this means the friction is exactly balanced by the component of gravity pulling the child down the slope. So to find the component of the weight force that is acting down the slope, we will need to resolve it in the directions of down the slope and perpendicular to it.

The weight component working down the slope equals the friction, $\boldsymbol{F}$:

$$\boldsymbol{F} = m\boldsymbol{g}\cos\theta$$

The work done is force multiplied by the distance travelled along the line of the force, so here:

$$work = \Delta \boldsymbol{s} \times m\boldsymbol{g}\cos\theta$$

This example shows us the general formula for calculating the work done when there is an angle between the force and the distance along which we are measuring:

$$\Delta W = \boldsymbol{F}\Delta\boldsymbol{s}\cos\theta$$

POWER

Power is defined as the rate of energy transfer. This may be done with reference to work done.

$$\text{power (W)} = \frac{\text{energy transferred (J)}}{\text{time for the energy transfer (s)}} \qquad P = \frac{E}{t}$$

$$\text{power (W)} = \frac{\text{work done (J)}}{\text{time for the work to be done (s)}} \qquad P = \frac{\Delta W}{t}$$

Remember:

work (J) = force (N) × distance moved in the direction of the force (m)

$$\Delta W = \boldsymbol{F}\Delta\boldsymbol{s}$$

So:

$$\text{power (W)} = \frac{\text{force (N)} \times \text{distance moved (m)}}{\text{time for the force to move (s)}}$$

$$P = \frac{\boldsymbol{F}\Delta\boldsymbol{s}}{t}$$

For example, the power of a forklift truck lifting a 120 kg crate vertically up 5.00 m in 4.0 seconds would be calculated as:

$$P = \frac{\boldsymbol{F}\Delta\boldsymbol{s}}{t} = \frac{m\boldsymbol{g}\Delta s}{t}$$

$$P = \frac{120 \times 9.81 \times 5}{4.0}$$

$$P = 1470\text{ W} = 1.47\text{ kW}$$

EXAM HINT

1 kilowatt (kW) = 1000 watts (W) = 1×10^3 W

1 megawatt (MW) = 1000 kW = 1×10^6 W

1 gigawatt (GW) = 1000 MW = 1×10^6 kW = 1×10^9 W

1 terawatt (TW) = 1000 GW = 1×10^6 MW = 1×10^9 kW = 1×10^{12} W

EFFICIENCY

From the equations above, we can calculate the work and power generated in different situations. If most of the energy is not actually transferred to a store that is useful to us, then the activity may be a waste of energy. The ability of a machine to transfer energy usefully is called **efficiency**.

Efficiency is defined mathematically as:

$$\text{efficiency} = \frac{\text{useful work done}}{\text{total energy input}}$$

$$\text{efficiency} = \frac{\text{useful energy output}}{\text{total energy input}}$$

If we remember that power is energy divided by time, then measuring the energy flows in a machine for a fixed amount of time means that we can write a power version of the efficiency equation:

$$\text{efficiency} = \frac{\text{useful energy output/time}}{\text{total energy input/time}}$$

$$\text{efficiency} = \frac{\text{useful power output}}{\text{total power input}}$$

The answer will be a decimal between zero and one. It is common to convert this to a percentage value (multiply the decimal by 100).

WORKED EXAMPLE

If the forklift truck referred to above lifted the crate when supplied with electrical energy from its battery at a rate of 3000 joules per second, what is its efficiency?

$$\text{efficiency} = \frac{\text{useful power output}}{\text{total power input}}$$

$$\text{efficiency} = \frac{1470}{3000}$$

$$\text{efficiency} = 0.49 = 49\,\%$$

EXAM HINT

Note that the steps and layout of the solution in this worked example are suitable for efficiency questions in the exam.

CHECKPOINT

SKILLS PROBLEM SOLVING

1. In these two situations, who does more work, and by how much?
   (a) A lioness carries a 2.8 kg cub 4.60 m up a tree.
   (b) An eagle lifts a 1.4 kg rabbit 8.25 m up to her nest.
2. Calculate the work done by the tension in the kite string in **fig E** over the distance shown.

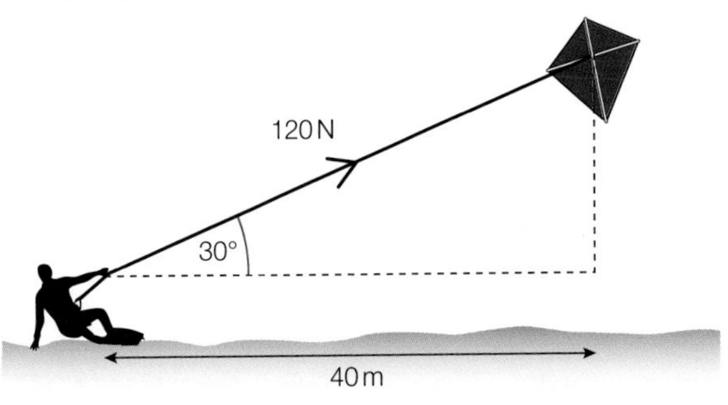

fig E Moving at an angle to the force doing work.

3. (a) What is the power of an electric motor, in watts, if it lifts 500 g, through 80 cm in 20 seconds?
   (b) What is the motor's efficiency if the electricity supplied it with a total of 12 J to make the lift?
4. In loading a delivery van, the driver pushed a 15 kg crate for 5 metres up a ramp. He had to push with a force of 132 N and the vertical height gained was 1.3 m. What was the efficiency of pushing this crate up the ramp onto the back of the van?

SUBJECT VOCABULARY

conservation of energy the rule that requires that energy can never be created or destroyed

work done in a mechanical system. This is the product of a force and the distance moved in the direction of the force:

work done (J) = force (N) × distance moved in the direction of the force (m)

$$\Delta W = F\Delta s$$

energy the property of an object that gives it the capacity to do work. A change in the amount of energy of an object could be equated to work being done, even if this is not mechanical – a change in the heat energy of a sample of gas, for example

power the rate of energy transfer:

$$P = \frac{E}{t} = \frac{\Delta W}{t}$$

efficiency the ability of a machine to transfer energy usefully:

$$\text{efficiency} = \frac{\text{useful energy output}}{\text{total energy input}}$$

$$\text{efficiency} = \frac{\text{useful power output}}{\text{total power input}}$$

1B THINKING BIGGER

THE MECHANICS OF SOCCER

SKILLS CRITICAL THINKING, PROBLEM SOLVING, ANALYSIS, INTERPRETATION, ADAPTIVE LEARNING, PRODUCTIVITY, COMMUNICATION

Soccer is the most popular sport in the world in terms of numbers of people playing. It is a fast moving, skilful sport in which the ball may fly at over 200 kilometres per hour.

In this activity, we will consider how mechanics can apply to events in soccer.

COACHING BOOKLET

GOALSCORING

In this section, we will look at some of the science behind shots on goal.

fig A Antoine Griezmann kicks a shot at goal.

If we want to calculate how fast a soccer ball is moving after it has been kicked from stationary (for example, after a penalty shot), we need to think about its acceleration by using Newton's second law. A standard soccer ball has a mass of 0.40 kg. If the foot applies a force of 350 N for a twentieth of a second, we can work out the answer:

$$F = ma \text{ so } a = \frac{F}{m}$$

$$a = \frac{350}{0.40} \quad \therefore a = 875\ \text{m s}^{-2}$$

$$v = u + at$$

The ball is stationary before it is kicked

$$\text{so } u = 0\ \text{m s}^{-1}, a = 875\ \text{m s}^{-2}, t = 0.05\ \text{s}$$

$$v = 0 + 875\ (0.05)$$

$$v = 44\ \text{m s}^{-1}$$

fig B Tiago Volpi tries to save a penalty kick.

A penalty kick is taken from a spot 11 metres from the goal. If we assume zero drag forces, we can calculate the longest time the goalkeeper has to react to this shot after it leaves the foot.

We know the start velocity and the distance, so this is a straightforward question.

$$v = \frac{s}{t} \qquad \text{so } zt = \frac{s}{v}$$

$$\therefore\ t = \frac{11}{44} \qquad t = 0.25\ \text{s}$$

If the goalkeeper reacts quickly enough to catch the ball, how far will his hands be pushed backwards in order to stop the ball? The maximum decelerating force his arms can provide to slow the ball is 2000 newtons. Here we should consider the removal of all the ball's kinetic energy as work being done.

$$E_k = \tfrac{1}{2}mv^2 = \tfrac{1}{2}(0.40) \times (44)^2 = 387\ \text{J}$$

$$\Delta W = F\Delta s$$

$$\Delta s = \frac{\Delta W}{F} = \frac{387}{2000} = 0.194\ \text{m}$$

So the goalkeeper's hands are pushed back just under 20 cm in order to stop the ball. According to Newton's third law, when the hands produce a force of 2000 N on the ball, the ball exerts a force of 2000 N on the hands.

From *Soccer is Mad Easy*, a booklet aimed at new coaches, particularly teachers who may also need to teach International A Level Physical Education

SCIENCE COMMUNICATION

1 The extract opposite is from a booklet to teach soccer coaches the theory behind the sport, especially for reference in International A Level Physical Education lessons, which have a significant amount of theory in them. Consider the extract and comment on the type of writing that is used. Try and answer the following questions:

(a) How has the author maintained the relevance of the calculations for the reader?

(b) Discuss the level of difficulty of the calculations, with reference to the target audience and the purpose of the text.

INTERPRETATION NOTE

Think about the complexity and level of difficulty of these calculations compared with those that you have to do in this physics course. Also consider that the numbers used in the examples are for top players – what level of player are these coaches probably dealing with?

PHYSICS IN DETAIL

Now we will look at the physics in detail. Some of these questions link to topics earlier in this book, so you may need to combine concepts from different areas of physics to work out the answers.

2 Identify, and comment on the validity of, the assumptions that have been made in these calculations.

3 A free kick is awarded 32 m from the goal line. The striker kicks it so that the ball leaves the ground at a 5° angle and a speed of 40 m s^{-1}. How long does the goalkeeper have to react before the ball reaches the goal line?

4 The study of mechanics in sport is a popular and often profitable new area of scientific study.

(a) Describe how a sports scientist could use electronic equipment to collect data to study the movement of players and equipment over time.

(b) Explain why technological developments have made the data collected more valid and reliable than other traditional methods of studying mechanics.

ACTIVITY

Junior soccer rules for young children use a reduced pitch, 27.5 m by 18 m, and a ball with a mass of 320 g. Write a similar section for a new version of this coaching manual which is aimed at junior soccer coaches.

THINKING BIGGER TIP

You will need to consider the strength of small children in order to estimate the forces they can give with a kick.

1B EXAM PRACTICE

1 The definition of the watt comes from which equation?

A $P = \frac{E}{t}$

B $P = E \times t$

C $\Delta W = F\Delta s$

D $\Delta W = F\Delta \cos\theta$

[1]

(Total for Question 1 = 1 mark)

2 A 305 g ball is thrown vertically upwards to reach a height of 5.55 m. How much potential energy has it gained at that height?

A 16.6 J

B 16.6 kJ

C 1690 J

D 16 600 J

[1]

(Total for Question 2 = 1 mark)

3 Efficiency can be calculated using values of energy, or values of power.

Select the row of the table that correctly gives the expressions for calculating efficiency in terms of either energy or power:

		Energy expression	Power expression
A	efficiency =	$\frac{\text{total energy input}}{\text{useful energy output}}$	$\frac{\text{useful power input}}{\text{total power output}}$
B	efficiency =	$\frac{\text{useful energy output}}{\text{total energy input}}$	$\frac{\text{total power input}}{\text{useful power output}}$
C	efficiency =	$\frac{\text{total energy input}}{\text{useful energy output}}$	$\frac{\text{useful power input}}{\text{total power output}}$
D	efficiency =	$\frac{\text{useful energy output}}{\text{total energy input}}$	$\frac{\text{useful power output}}{\text{total power input}}$

[1]

(Total for Question 3 = 1 mark)

4 A horse pulls a carriage of weight 5600 N with a force of 80 N for a distance of 1.2 km around New York's Central Park. How much work is done by the horse?

A 6.7 kJ

B 96 kJ

C 538 kJ

D 6720 kJ

[1]

(Total for Question 4 = 1 mark)

5 The photograph shows a wind turbine. Kinetic energy of the wind is transferred to electrical energy as the turbine blades rotate.

(a) Explain why we can say that the wind is doing work on the blades. [2]

(b) The area swept out by one blade, as it turns through 360°, is 6000 m^2. Wind at a speed of 9 m s^{-1} passes the turbine.

(i) Show that the volume of air passing through this area in 5 seconds is about 300 000 m^3. [2]

(ii) Calculate the mass of this air.
Density of air = 1.2 kg m^{-3} [2]

(iii) Calculate the kinetic energy of this mass of air. [2]

(iv) Betz's law states that a turbine cannot usefully transfer more than 59% of the kinetic energy of the wind.

Use this law to find the maximum output of the wind turbine. [2]

(c) Suggest a reason why it is not possible to usefully transfer 100% of the kinetic energy of the wind. [1]

(d) Suggest the limitations of using wind turbines to provide power. [2]

(Total for Question 5 = 13 marks)

6 One account of the origin of the term *horsepower* is as follows. In the eighteenth century, James Watt manufactured steam engines. He needed a way to demonstrate the benefits of these compared to the horses they replaced. He did some calculations based on horses walking in circles to turn a mill wheel.

Watt observed that a horse could turn the wheel 144 times in one hour. The horse travelled in a circle of radius 3.7 m and exerted a force of 800 N.

(a) Show that the work done by the horse in turning the wheel through one revolution was about 20 000 J. [3]

(b) Calculate the average power of the horse in SI units. [3]

(Total for Question 6 = 6 marks)

7 In a demonstration of energy transfer, a large pendulum is made by suspending a 7.0 kg bowling ball on a long piece of wire.

A student is invited to pull back the ball until it just touches her nose and then to release it and stand perfectly still while waiting for the ball to return.

The following instructions are given:

Do not push the ball – just release it. Do not move your face before the ball returns.

Explain this demonstration and the need for these instructions. [6]

(Total for Question 7 = 6 marks)

8 The photograph shows a lawnmower being used to cut grass.

(a) (i) In order to push the lawnmower, a minimum force of 650 N must be applied to the handle of the lawnmower at an angle of 42° to the horizontal.
Show that the horizontal component of the force is about 500 N. [2]

(ii) The lawnmower is used to cut 15 strips of grass, each 7 m long.

Calculate the work done by the person pushing the lawnmower. [2]

(b) This photograph shows a lawnmower with the top section of the handle horizontal.

Explain how this changes the minimum force required to push the lawnmower. [2]

(Total for Question 8 = 6 marks)

9 Metrology is the science of measurement and World Metrology Day is May 20th. In 2010, the day was used to celebrate the 50th anniversary of the SI system.

A metrologist from the National Physical Laboratory said on a radio programme that the SI system uses units that everyone can understand. He stated the following example.

'If you hold an apple in your hand it's about a *newton*, if you raise it through one metre that's about a *joule* and if you do it in one second that's about a *watt*.'

Assuming that the apple has a mass of 100 g, explain and justify the statements made about the three words in italics. [6]

(Total for Question 9 = 6 marks)

TOPIC 1 MECHANICS

CHAPTER 1C MOMENTUM

Collisions can be devastating. Vehicle safety is a very important area of technological research, and much of the science is based on the concept of momentum and momentum transfer. Momentum is a property of a moving object. It is larger if the mass is greater and if the object moves faster. To change the momentum of an object requires a force, and this is how car crashes are so damaging. Larger forces cause more damage, whether to the vehicle or its passengers.

To reduce the forces involved in transferring momentum we need to know how momentum transfers and how to calculate the forces depending on the momentum transfer needed.

Not all momentum changes are dangerous. Rocket science is generally based on maximising the forces caused by a transfer of momentum, by maximising that momentum transfer. This will create maximum acceleration in order to move the rocket fast enough to gain enough gravitational potential energy to leave the Earth. This chapter will show you how the properties of a moving object can tell us its momentum, and how to calculate the transfer of momentum, and the forces involved in changing the movement of an object.

MATHS SKILLS FOR THIS CHAPTER

- **Units of measurement** (*e.g. the unit for momentum,* $kg\,m\,s^{-1}$)
- **Changing the subject of an equation** (*e.g. finding the velocity of an object after collision*)
- **Substituting numerical values into algebraic equations** (*e.g. calculating the momentum*)
- **Plotting two variables from experimental data** (*e.g. observing changes in momentum over time*)
- **Using sin, cos and tan in physical problems, and making calculations using them** (*e.g. resolving a momentum vector*)

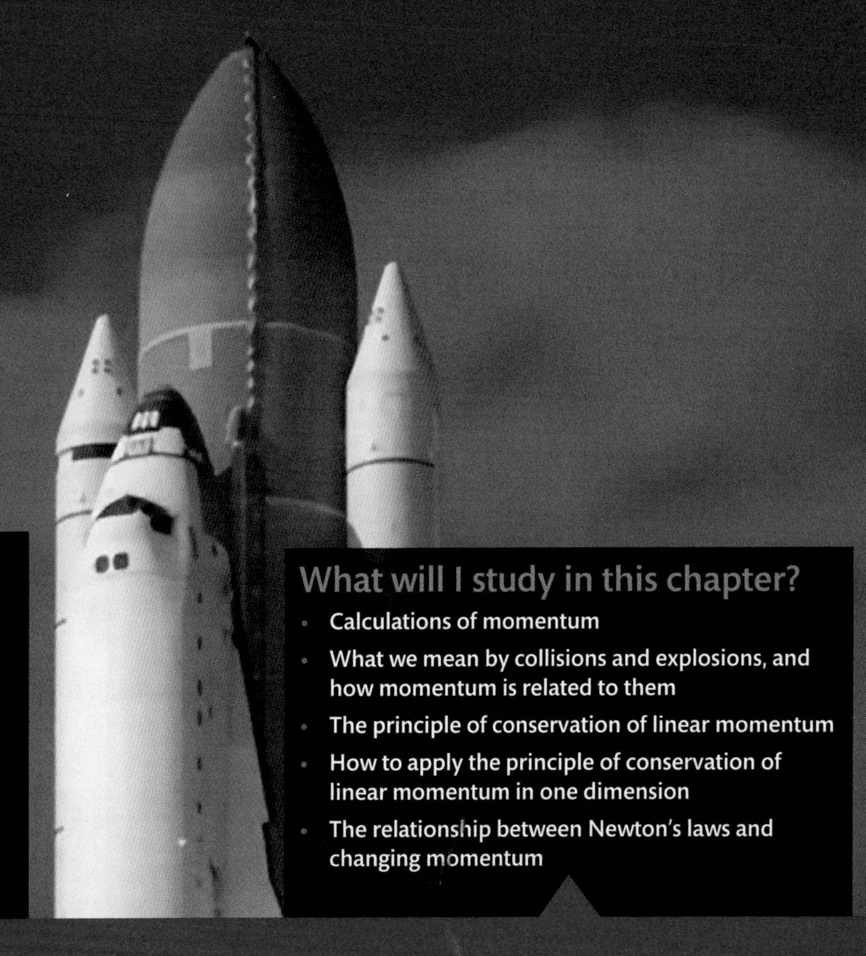

What prior knowledge do I need?

Topic 1A

- Ideas about stopping distances of cars, and the safety features in vehicles
- Speed, distance, displacement, time, velocity, acceleration
- Resolving vectors
- Newton's laws of motion and the kinematics equations

What will I study in this chapter?

- Calculations of momentum
- What we mean by collisions and explosions, and how momentum is related to them
- The principle of conservation of linear momentum
- How to apply the principle of conservation of linear momentum in one dimension
- The relationship between Newton's laws and changing momentum

What will I study later?

Topic 2A

- Fluid movements and terminal velocity

Topic 2B

- Stress and strain and the deformation of solids

Topic 5A (Book 2: IAL)

- The conservation of momentum in two dimensions
- Elastic and inelastic collisions

Topic 7B (Book 2: IAL)

- The relationship between momentum and the kinetic energy of a particle
- The importance of particle momentum in the design of accelerators

1C 1 MOMENTUM

SPECIFICATION REFERENCE 1.3.9 1.3.13

LEARNING OBJECTIVES

- Calculate the momentum of an object.
- Explain how momentum is gained or lost.

MOMENTUM

Momentum is a measure of an object's motion. It is quite difficult to define momentum in words, but it gives an idea of what will be required to stop the object moving. The best definition is mathematical:

$$\text{momentum (kg m s}^{-1}) = \text{mass (kg)} \times \text{velocity (m s}^{-1})$$

$$p = m \times v$$

As momentum is the product of mass (a scalar) and velocity (a vector), momentum is a vector. This means its direction is very important and must be remembered. The direction will be the same as that of its velocity.

▲ **fig A** Which object moves with the greatest momentum?

EXAM HINT

Note that the steps and layout of the solution in this worked example are suitable for momentum questions in the exam.

WORKED EXAMPLE

An athletics hammer (see **fig A**) has a mass of 7.26 kg (men's competition standard) and can be released at speeds in excess of 25 m s^{-1}. Its momentum at 25.0 m s^{-1} would be:

$p = m \times v$

$p = 7.26 \times 25.0$

$p = 182 \text{ kg m s}^{-1}$

A baseball has a mass of 145 grams. A fast pitcher can throw it at 40 m s^{-1}. If this baseball is released at 40 m s^{-1}, its momentum is:

$p = m \times v$

$p = 0.145 \times 40$

$p = 5.8 \text{ kg m s}^{-1}$

These example show that it is much more difficult to stop a well-thrown athletics hammer than to stop a baseball. Think about what 'more difficult' means in this case.

NEWTON'S SECOND LAW OF MOTION

If we want to bring an object to rest or to accelerate it up to a certain velocity, the requirements will be different for different situations. A golf ball is accelerated in a very different way to a ferry. Think of the forces needed and the time for which they act. This brings us to another way of thinking about momentum. It is a measure of the accelerating force, and the time it is applied for, that is required to bring an object up to the speed it is moving at. Alternatively, it is the force required, and for how long, to bring a moving object to rest.

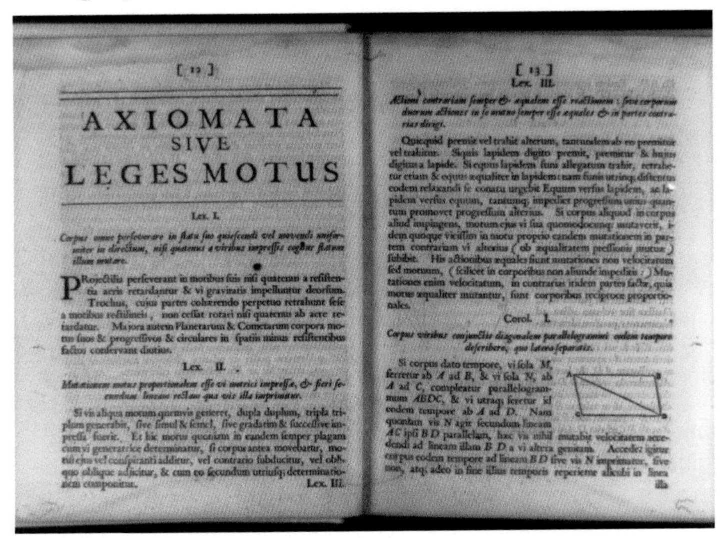

AXIOMATA
SIVE
LEGES MOTUS

Lex. I.

Lex. II.

Lex. III.

Corol. I.

▲ **fig B** Newton's laws of motion as he originally wrote them.

Newton's second law can be written mathematically as $F = ma$. In fact, that formula is only true if the mass remains constant. When Newton originally wrote his second law in the 1687 book, *Philosophiae naturalis principia mathematica*, he actually wrote it as:

The rate of change of momentum of a body is directly proportional to the resultant force applied to the body, and is in the same direction as the force.

This can be written mathematically as:

$$F = \frac{dp}{dt} = \frac{d(mv)}{dt}$$

Here, F is the applied force, and $\frac{dp}{dt}$ is the rate of change of momentum in the direction of the force.

The $\frac{d(x)}{dt}$ term is a mathematical expression meaning the rate of change of x, or how quickly x changes. However, if the quantities are not being measured over a very short timescale, we can express this using average changes:

$$F = \frac{\Delta p}{\Delta t}$$

PRACTICAL SKILLS

Investigating momentum change

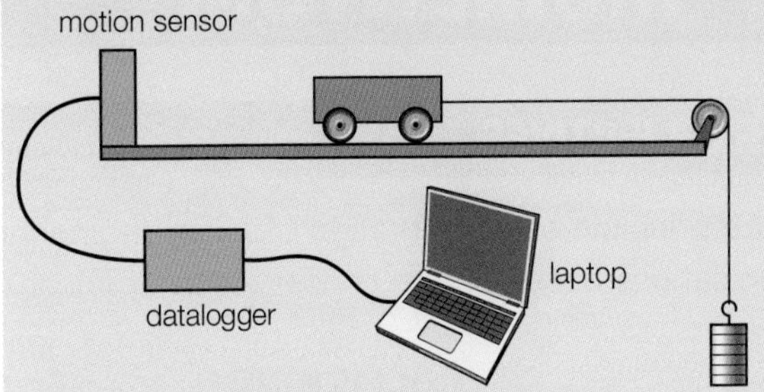

▲ **fig C** Measuring how a force changes the momentum of a trolley.

You can investigate the rate of change of momentum in the school laboratory. A trolley starts from rest and as a force acts upon it its velocity increases. If you record the trolley's movement over time, you can find the velocity each second. If you then calculate the momentum each second, you will be able to plot a graph of momentum against time. It should be a straight line. As $p = Ft$, the gradient of this line will be equal to the accelerating force.

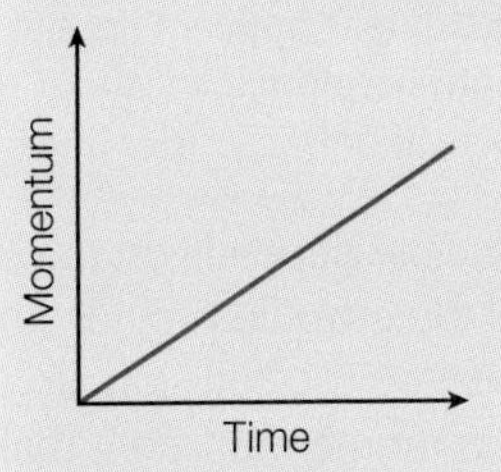

▲ **fig D** Accelerating from rest, momentum will be proportional to time.

Safety Note: Place trolleys and runways so they cannot fall and cause injuries. For large masses, place a 'catch box' filled with crumpled paper or bubbled plastic in the 'drop zone' to avoid injury to feet.

CHECKPOINT

SKILLS PROBLEM SOLVING

1. Calculate the momentum in each of these examples:
   (a) an ice skater with a mass of 64 kg glides at 3.75 m s^{-1}
   (b) a rugby player of mass 120 kg runs at a speed of 4.9 m s^{-1}
   (c) an ant of mass 5 milligrams moves at a speed of 5 centimetres per second.
2. Estimate the momentum of the motorcyclist and the skateboarder shown in **fig A**.
3. Using the ideas of Newton's second law, explain why hitting an airbag will cause less injury than if a passenger hits the dashboard.
4. Estimate the force applied by a person throwing a Frisbee.

SUBJECT VOCABULARY

momentum (kg m s^{-1}) = mass (kg) × velocity (m s^{-1})

$$p = m \times v$$

1C 2 CONSERVATION OF LINEAR MOMENTUM

SPECIFICATION REFERENCE 1.3.12 1.3.14

LEARNING OBJECTIVES

- Explain the principle of conservation of linear momentum.
- Make calculations based on the conservation of linear momentum.

COLLISIONS

When objects collide, we can use the laws of physics to calculate where the objects will go after the collision. We can use the principle of **conservation of linear momentum** to predict the motion of objects after a collision. This principle tells us that if we calculate the momentum of each object before they collide, the sum total of these momenta (accounting for their direction) will be the same as the sum total afterwards.

▲ **fig A** Newton's cradle: each time the balls collide, momentum is transferred from one to another, but the total momentum remains constant.

LEARNING TIP

The word 'linear' appears here to remind us that this is all about objects moving in straight lines. There are similar physics principles about rotating objects, but they use different equations for the calculations. In this book we will only consider linear momentum.

LEARNING TIP

Total momentum is only conserved when no external forces (such as friction) act on the system.

This principle depends on the condition that no external force acts on the objects in question. An external force would provide an additional acceleration, and the motion of the objects would not be dependent on the collision alone. As we saw in the previous section, a resultant force will cause a change in momentum, so it makes sense that momentum is only conserved if no external force acts. Imagine if a juggler's ball moving upward collided with one coming down. Momentum conservation would suggest that the one falling down would bounce back with an upward velocity after the collision. Common sense tells us that all balls will still end up back on the ground. The external force of gravity means that the principle of conservation of momentum alone cannot be used to predict the motions after the collision.

▲ **fig B** David Porter of the TCU Horned Frogs feels the full force of the conservation of momentum.

WORKED EXAMPLE

In an American football match, the stationary quarterback is tackled by a defender who dives through the air at 4 m s^{-1} and, in mid-air, grabs the quarterback and the two move quickly backwards together. Ignoring any friction effects, calculate how fast the two will move back if the tackler has a mass of 140 kg and the stationary player has a mass of 95 kg. Consider the entire situation to be happening horizontally.

Before:
Quarterback stationary so zero momentum

$p_{tackler} = mv = 140 \times 4 = 560$

momentum before = 560 kg m s^{-1}

After:

momentum after = momentum before = 560 kg m s^{-1}

$p_{both} = m_{both} \times v_{both}$

$$v_{both} = \frac{p_{both}}{m_{both}} = \frac{560}{(140 + 95)} = \frac{560}{235}$$

$v_{both} = 2.4$ m s^{-1}

EXAM HINT

For any exam question about momentum in collisions and explosions, make sure you state: 'total momentum before = total momentum after'.

LEARNING TIP

In a collision in which two objects join together to become one and move off together, they are often said to 'coalesce'.

EXPLOSIONS

fig C In this illustration, these trapeze artists are stationary. If they let go of each other, they will 'explode' – they will fly apart with equal and opposite momenta.

If a stationary object explodes, then the total momentum of all the shrapnel parts added up (taking account of the direction of their movements) must be zero. The object had zero momentum at the start, so the law of conservation of linear momentum tells us this must be the same total after the **explosion**. In physics, any such event is termed an explosion, although it may not be very dramatic. For example, if the two trapeze artists in **fig C** simply let go their hands and swing apart, they have zero total momentum before and will have equal and opposite momenta afterwards, which when added together will total zero again.

WORKED EXAMPLE

If the boy has a mass of 55 kg and steps forward at a speed of 1.5 m s^{-1}, what will happen to the boat which has a mass of 36 kg? (Ignore friction effects.)

This situation is an explosion, so:

total momentum before = total momentum after = zero

$\therefore \quad p_{boat} + p_{boy} = 0$

$\therefore \quad p_{boat} = -p_{boy}$

So when the two are added up, the total momentum is still zero.

$\therefore \quad p_{boat} = -(55 \times 1.5) = -82.5 \text{ kg m s}^{-1}$

$m_{boat} \times v_{boat} = -82.5 \text{ kg m s}^{-1}$

$v_{boat} = \frac{-82.5}{m_{boat}} = \frac{-82.5}{36}$

$v_{boat} = -2.3 \text{ m s}^{-1}$

So the boat moves at 2.3 m s^{-1} in the opposite direction to the boy.

▲ **fig D** Caution: explosions may make you wet!

EXAM HINT

Note that the steps and layout of the solution in this worked example are suitable for conservation of linear momentum questions in the exam.

PRACTICAL SKILLS

Investigating transfer of momentum

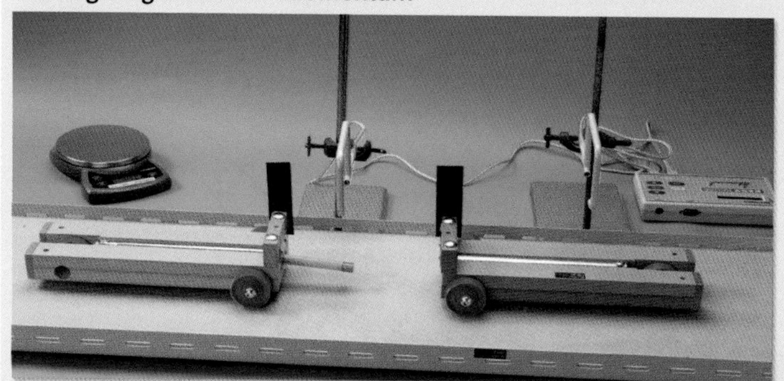

▲ **fig E** Verifying the principle of conservation of linear momentum.

Safety Note: Place trolleys and runways so they cannot fall and cause injuries.

You can investigate the transfer of momentum in collisions in the school laboratory using trolleys, or sliders on an airtrack. By recording the movement of one trolley crashing into another, you can find the momentum of each one before and after the collision. The calculation of adding up the total momenta before and after collision will allow you to prove the principle of conservation of linear momentum. Try different types of collision and trolleys with different masses. You could also try an explosion in which the trolleys come apart from a stationary position.

In experiments using trolleys, we often find that momentum is actually not conserved in the measurements we make. With airtrack collisions, the measurements match very closely or exactly with the conservation of momentum theory. What might be the reasons for this difference between the two types of experiment?

▲ **fig F** A trolley 'explosion'.

NEWTON'S THIRD LAW

Conservation of momentum is directly responsible for Newton's third law. Remember, this told us that for every force, there is an equal and opposite force. If we think of a force as a way to change momentum ($F = \frac{dp}{dt}$) then a force changing momentum in one direction must be countered by an equal and opposite one to ensure that overall momentum is conserved. For example, if the gravitational pull of the Earth causes an apple to fall from a tree, the apple gains momentum towards the Earth. For conservation of momentum, the Earth must gain an equal and opposite momentum. This is then caused by an equal and opposite gravitational force on the Earth from the apple. The huge mass of the Earth means that its acceleration cannot be noticed by us.

▲ **fig G** Conservation of momentum causes equal and opposite forces, as Newton explained in his third law of motion.

EXAM HINT

Make sure you answer the question that has been asked. In Q3, no marks will be awarded for answers that do not refer to Newton's third law.

CHECKPOINT

SKILLS ANALYSIS

1. A movie stuntman with a mass of 90 kg stands on a stationary 1 kg skateboard. An actor throws a 3.4 kg brick at the stuntman who catches it. The brick is travelling at 4.1 m s^{-1} when caught.

2. A boy in a stationary boat on a still pond has lost his oars in the water. In order to get the boat moving again, he throws his rucksack horizontally out of the boat with a speed of 4 m s^{-1}.

   Mass of boat = 60 kg; mass of boy = 40 kg; mass of rucksack = 5 kg

   (a) How fast will this action make the boat move?

   (b) If he throws the rucksack by exerting a force on it for 0.2 s, how much force does he exert?

3. How can Newton's third law explain the problem suffered by the boy stepping out of the boat in **fig D**?

4. In a stunt for an action movie, the 100 kg actor jumps from a train that is crossing a river bridge.
   On the river below, the heroine tied to a small boat is drifting towards a waterfall at 3 m s^{-1}. The small boat and heroine have a total mass of 200 kg.

   (a) If the hero times his jumps perfectly so as to land on the small boat, and his velocity is 12 m s^{-1} at an angle of 80° to the river current, what will be the velocity of the small boat immediately after his landing? Draw a vector diagram to show the momentum addition. Ignore any vertical motion.

   (b) If the waterfall is 100 m downstream, and the hero landed when the small boat was 16 m from the bank, would they drop over the fall? Assume the velocity remains constant after the hero has landed. The small boat and the waterfall are on the same side of the bridge as he jumps.

SUBJECT VOCABULARY

conservation of linear momentum the vector sum of the momenta of all objects in a system is the same before and after any interaction (collision) between the objects

explosion a situation in which a stationary object (or system of joined objects) separates into component parts, which move off at different velocities. Momentum must be conserved in explosions

SAVING HOCKEY GOALKEEPERS

SKILLS CRITICAL THINKING, PROBLEM SOLVING, ANALYSIS, REASONING/ARGUMENTATION, INTERPRETATION, ADAPTIVE LEARNING, PRODUCTIVITY, SELF-EVALUATION, COMMUNICATION, TEAMWORK

OBO is a New Zealand based company that manufactures hockey goalkeeping equipment. The following extracts from their website explain some of their testing laboratory's abilities, and report on a potential new material used in leg guards.

COMPANY WEBSITE

A LOOK INSIDE THE O LAB

In order to design and build the world's most protective and best performing goalkeeper equipment we need the facts. The O Lab is packed full of the world's most advanced impact test equipment… and a few very clever people to test and help evaluate the results.

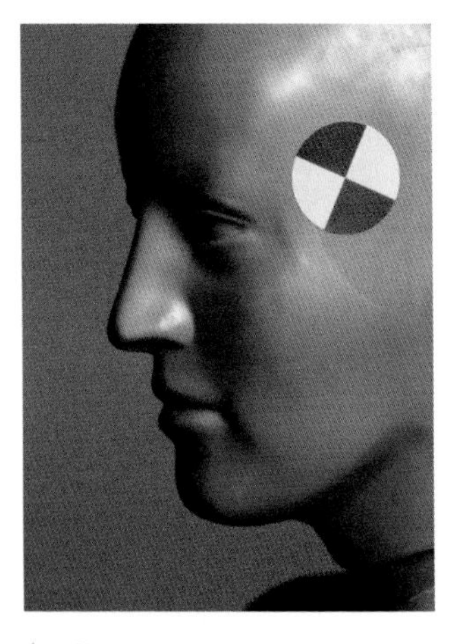

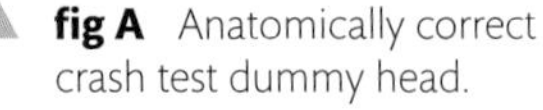

▲ **fig A** Anatomically correct crash test dummy head.

▲ **fig B** Data capture software simultaneously showing video and concussion data.

Every detail sorted by a small group of smart committed people. Video capture at speeds up to 2000 frames per second, skin contact analysis, and accurate concussion measurements.

▲ **fig C** Ball cannon capable of speeds in excess of 200 km h.

From the OBO website, www.obo.co.nz/the-o-lab

WHY OBO DON'T USE D30

A while ago, an important new protection polymer called D30 was offered to OBO. Because we are always trying to improve our products we were excited by the potential of D30 so our designer made a special trip from New Zealand to England to meet with the D30 creators. He returned home with lots of information and some samples which we tested in the purpose-built OBO impact lab... The O lab.

Our impact lab testing showed that while D30 weighed more than two and a half times the OBO polyethylene and EVA foams, it provided significantly less protection when dealing with the high speed and highly localised impact encountered with a hockey ball.

Have a look at the results on the impact graph below (the horizontal axis is speed, the vertical axis is transferred force – the higher the transferred force, the less the protection).

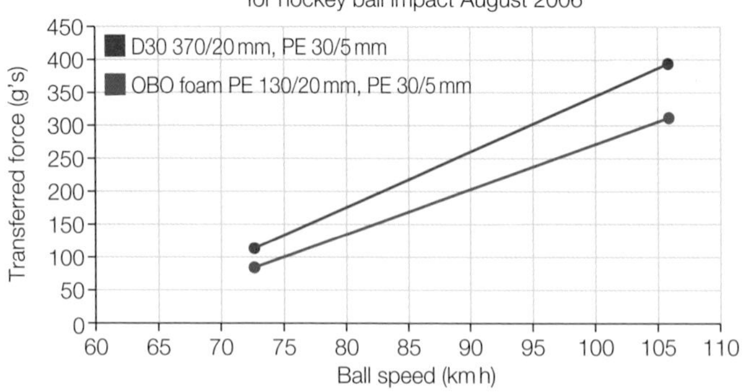

▲ **fig D** Protection comparison of D30 vs OBO foam for hockey ball impact.

SCIENCE COMMUNICATION

1 The website opposite was written to support the business activities of a sports supply company. Consider the article and comment on the type of writing being used. Think about, for example, how the physics involved is explained, and the degree of detail included. Try and answer the following questions:

(a) What range of people do you expect might read the OBO website? What would be the likely scientific background of their customers?

(b) On the actual OBO webpage. there are many more pictures, including brief videos, that we could not fit in this book. How has the ratio of text to images been chosen for the intended audience, and the website as a medium?

(c) Where units have been included, comment on the actual units chosen to measure the quantities involved.

INTERPRETATION NOTE

As you read the articles, consider where they have come from, who wrote them and whom they were written for. Established scientific publications are good sources of reliable information, whereas other resources might be less reliable for a number of reasons. Think about what makes a source reliable and why.

PHYSICS IN DETAIL

Now we will look at the physics in, or connected to, these website extracts. Some of these questions link to topics in much earlier sections of this book, so you may need to combine concepts from different areas of physics to work out the answers.

2 For the testing of the polymer D3O, calculate the range of momenta for the test balls fired from the cannon. (A standard hockey ball has a mass of 160 grams.)

3 Explain, with reference to Newton's laws:

(a) why the 'transferred force', i.e. the force felt through the foam by a goalkeeper wearing it, would be the same as the force needed to decelerate the ball

(b) why the lines on the graph show a linear relationship with a positive gradient.

4 In light of the principle of conservation of linear momentum, how can a goalkeeper remain stationary, whilst the ball's momentum is completely removed in collision with the leg pads?

ACTIVITY

Imagine you work for OBO as an international sales representative and you have to prepare a presentation to delegates at a trade show. Your presentation will need to explain in much greater scientific detail the testing that the equipment has been through in The O Lab. Prepare a questionnaire, for OBO head office in New Zealand, of questions that will give you the details you need to prepare your presentation.

THINKING BIGGER TIP

You do not need to prepare the presentation for the trade show, just your questionnaire designed to get the information you need from OBO head office to be able to prepare such a presentation.

DID YOU KNOW?

OBO use one unusual testing procedure they call the DTH test. This involves a real goalkeeper wearing the item under test. A ball is fired at the test subject and the lab researchers ask the question, 'Did That Hurt?'

1C EXAM PRACTICE

1 Which is the correct expression for calculating momentum?

A Mass × speed

B Mass ÷ speed

C Mass × velocity

D Mass ÷ speed [1]

(Total for Question 1 = 1 mark)

2 Which of the following is the correct unit for momentum?

A $kg\,s^{-1}$

B $kg\,m\,s^{-1}$

C $kg\,m\,s^{-2}$

D $kg\,m^{-1}\,s^{-1}$ [1]

(Total for Question 2 = 1 mark)

3 A hockey ball of mass 158 g is hit with a force of 2000 N so that it travels at $28.1\,ms^{-1}$. What is the ball's momentum?

A $4.4\,kg\,m\,s^{-1}$

B $62.4\,kg\,m\,s^{-2}$

C $4440\,kg\,m^{-1}\,s^{-1}$

D $8880\,kg\,m\,s^{-1}$ [1]

(Total for Question 3 = 1 mark)

4 An ice skater and his coach begin stationary and push apart from each other with a force of 75 N. The skater has a mass of 64 kg, whilst the coach weighs 804 N.

If the skater moves off to the left with a speed of $3.6\,m\,s^{-1}$, what is the velocity of the coach?

A $0.022\,m\,s^{-1}$ to the right

B $2.8\,m\,s^{-1}$ to the right

C $3.1\,m\,s^{-1}$ to the left

D $3.1\,m\,s^{-1}$ to the right [1]

(Total for Question 4 = 1 mark)

5 In an explosion, a stationary object of mass M splits into two objects which move in opposite directions, left and right, at the same speed.

Which row in the table correctly gives the mass of each of the two objects after the explosion?

	Mass of left moving object	Mass of right moving object
A	$\frac{M}{2}$	$\frac{M}{2}$
B	$\frac{M}{2}$	$2M$
C	$2M$	$2M$
D	M	M

[1]

(Total for Question 5 = 1 mark)

6 How tiny bacteria move is of interest in nanotechnology. Mycobacteria move by ejecting slime from nozzles in their bodies.

Explain the physics principles behind this form of propulsion. [4]

(Total for Question 6 = 4 marks)

7 A student is using a 'Newton's Cradle'. This consists of a set of identical solid metal balls hanging by threads from a frame so that they are in contact with each other.
He initially pulls one ball to the side as shown.

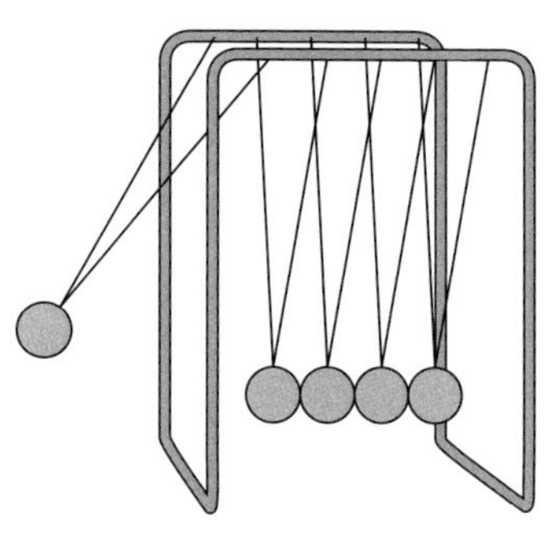

He releases the ball, it collides with the nearest stationary ball and stops. The ball furthest to the right immediately moves away. The middle three balls remain stationary.

(a) Explain what measurements the student would take and describe how he would use them to investigate whether momentum had been conserved in this event. [4]

(b) The student makes the following observations:

- the ball on the right returns and collides with a similar result; this repeats itself a number of times
- after a while, the middle balls are also moving
- shortly afterwards, the balls all come to rest.

Discuss these observations in terms of energy. [3]

(Total for Question 7 = 7 marks)

8 A student uses a motion sensor and a datalogging computer to investigate the momentum changes when a trolley is accelerated by a falling weight connected to it. Assume the trolley suffers no friction on the desk, and there is also no friction in the pulley wheel.

motion sensor

laptop

datalogger

(a) Explain how Newton's second law of motion predicts that the momentum of this trolley will change when the weight is allowed to fall freely. [2]

(b) When the weight is released, the trolley experiences a resultant accelerating force of 2.85 N and has a mass of 350 g. Calculate the rate of change of velocity of the trolley. [3]

(c) The trolley reaches a velocity of $11.1\ \mathrm{m\,s^{-1}}$, calculate its momentum at this velocity, including the correct unit. [2]

(d) Explain what is meant by the principle of conservation of momentum. [2]

The student changes the experiment so that he can collide a moving trolley with an identical stationary one.

(e) Explain the momentum and speed changes for each trolley if the moving trolley stops on collision, and the stationary one moves away. [4]

(f) Explain the momentum and speed changes involved if the trolleys join together on collision, and both move away together. [4]

(g) Explain how this experiment could be changed to investigate the momentum and speed changes in an explosion. [3]

(Total for Question 8 = 20 marks)

9 Explain how the principle of conservation of momentum in collisions is a consequence of Newton's third law of motion. [6]

(Total for Question 9 = 6 marks)

10 A hockey ball is travelling horizontally with a momentum of $0.8\ \mathrm{kg\,m\,s^{-1}}$ just before it hits a goalkeeper's leg pad. It rebounds horizontally from the leg pad with a momentum of $-1.2\ \mathrm{kg\,m\,s^{-1}}$. The graph shows the variation in the momentum of the ball during this process.

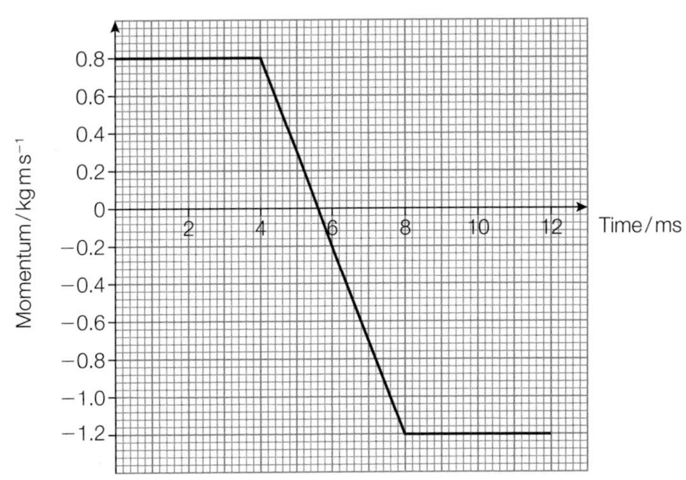

(a) Describe how the ball's momentum changes over time from 0 to 10 ms. [3]

(b) Explain in terms of Newton's laws why the momentum changes from positive to negative during the ball's collision with the leg pad. [2]

(c) What is the resultant force on the ball during the following time periods?

(i) 0–4 ms

(ii) 4–8 ms

(iii) 8–12 ms [4]

(d) Draw a new version of the graph for a collision in which the ball is initially travelling at half the speed, and for which the impact time is also halved, but the force provided by the leg pads is the same. [3]

(Total for Question 10 = 12 marks)

TOPIC 2 MATERIALS

CHAPTER 2A FLUIDS

In Topic 1A, we calculated the motion of objects. We followed Newton's laws, and assumed zero air resistance. How safe was that assumption? Urban myths describe coins dropped from high buildings that cut deep into the concrete below – are these myths actually true? A very approximate calculation, using estimates of average coin size and mass, and the strength of the concrete, suggests that from a height of 100 metres, a falling coin could penetrate tens of centimetres ... if it fell in a vacuum.

The falling coin might tumble in flight, depending upon the exact effects of air resistance. This resistance will slow the coin significantly, and it will hit the ground at a much lower speed than the theoretical estimate in the calculation mentioned above. Most likely, it will land with an impact similar to that of a hailstone.

In this chapter, you will see how gases and liquids behave when flowing, or when causing friction with moving solid objects. You will also learn about some of the factors that can affect these movements.

MATHS SKILLS FOR THIS CHAPTER

- **Units of measurement** (*e.g. the unit for density,* $kg\,m^{-3}$)
- **Visualising and representing 3D forms and finding volumes of rectangular blocks, cylinders and spheres** (*e.g. finding the volume of an object as a step towards finding its density*)
- **Changing the subject of an equation** (*e.g. re-arranging the Stokes' law equation*)
- **Solving algebraic equations** (*e.g. finding the depth at which a barge floats in a river*)
- **Substituting numerical values into algebraic equations** (*e.g. an upthrust calculation*)
- **Determining the slope of a linear graph** (*e.g. finding the viscosity from terminal velocity data*)

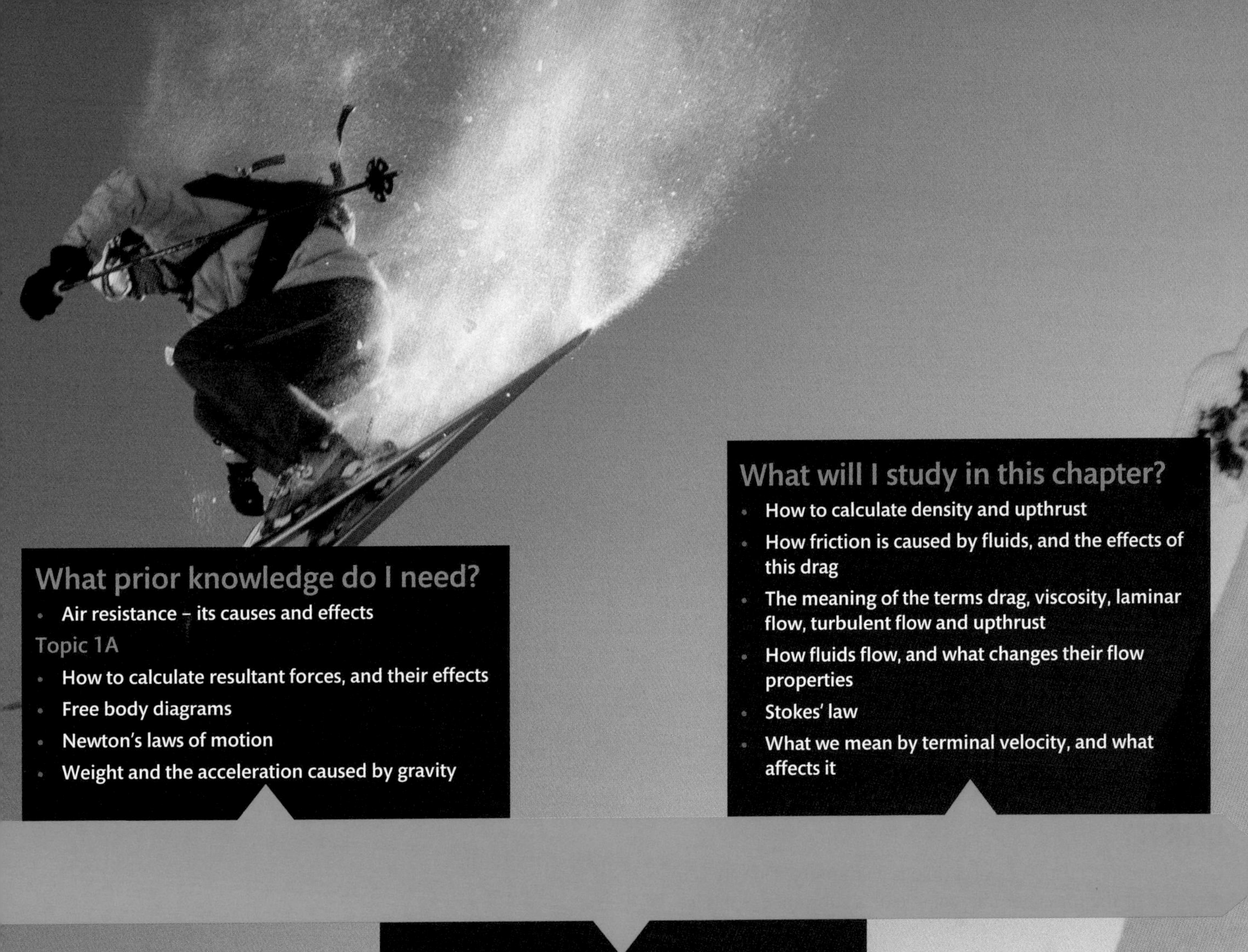

What prior knowledge do I need?

- Air resistance – its causes and effects

Topic 1A

- How to calculate resultant forces, and their effects
- Free body diagrams
- Newton's laws of motion
- Weight and the acceleration caused by gravity

What will I study in this chapter?

- How to calculate density and upthrust
- How friction is caused by fluids, and the effects of this drag
- The meaning of the terms drag, viscosity, laminar flow, turbulent flow and upthrust
- How fluids flow, and what changes their flow properties
- Stokes' law
- What we mean by terminal velocity, and what affects it

What will I study later?

Topic 8A (Book 2: IAL)

- Other thermal properties of fluids, especially gases

Topic 10A (Book 2: IAL)

- Damping of oscillations by fluid drag
- Forced vibrations, driven by turbulence in flowing fluids

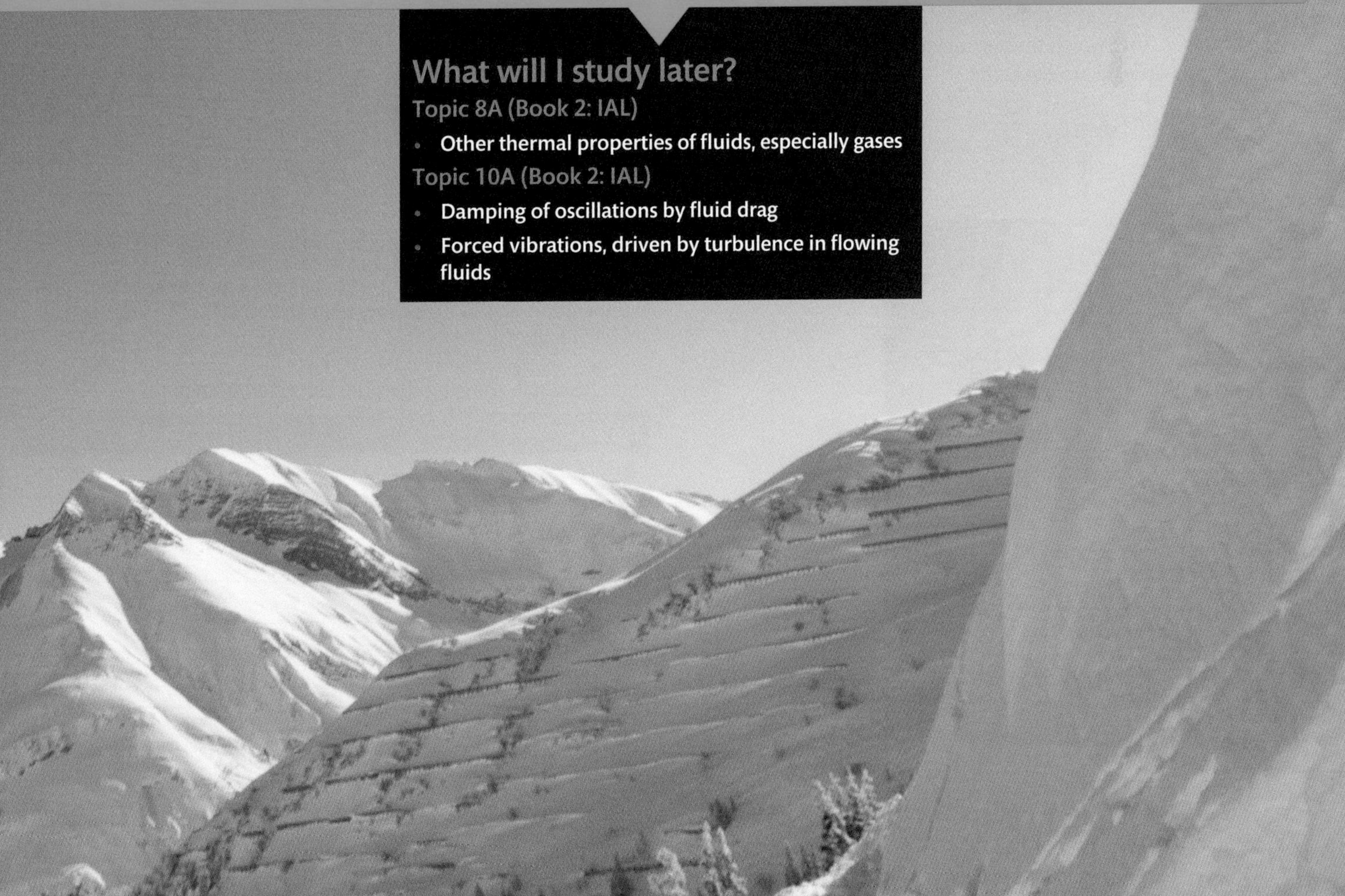

2A 1 FLUIDS, DENSITY AND UPTHRUST

SPECIFICATION REFERENCE 1.4.23 1.4.24

LEARNING OBJECTIVES

- Use the equation defining density.
- Understand how the upthrust force in a fluid can be calculated.

FLUIDS

Have you ever wondered why it is sometimes so difficult to get thick sauce out of a bottle? The answer is that the manufacturers make it thick on purpose. Market research shows that consumers enjoy a certain consistency of ketchup or mayonnaise on their fries, and producing it that thick makes the sauce flow very slowly.

This chapter will explain various aspects of the movements of fluids, including some of the ways in which fluid properties are measured. A **fluid** is defined as any substance that can flow. Normally this means any gas or liquid, but solids made up of tiny particles can sometimes behave as fluids; an example is the flow of sand through an hourglass.

DENSITY

One of the key properties of a fluid is its **density**. Density is a measure of the mass per unit volume of a substance – this is technically called 'volumic mass'. Its value depends on the mass of the particles from which the substance is made, and how closely those particles are packed:

$$\text{density (kg m}^{-3}) = \frac{\text{mass (kg)}}{\text{volume (m}^3)}$$

$$\rho = \frac{m}{V}$$

▲ **fig A** Density is very important in determining the weight of an object.

The equation for calculating density works for mixtures and pure substances, and for all states of matter. Thus, fluid density is also mass per unit volume.

WORKED EXAMPLE

A house brick is 23 cm long, 10 cm wide and 7 cm high. Its mass is 3.38 kg.

What is the brick's density?

$$\rho = \frac{m}{V}$$

$$\text{volume } V = 0.23 \times 0.10 \times 0.07 = 1.61 \times 10^{-3}\ \text{m}^3$$

$$\text{mass } m = 3.38\ \text{kg}$$

$$\text{density } \rho = \frac{3.38}{1.61 \times 10^{-3}}$$

$$\rho = 2100\ \text{kg m}^{-3}$$

At 20 °C, a child's balloon filled with helium is a sphere with a radius of 20 cm. The mass of helium in the balloon is 6 grams.
What is the density of helium at this temperature?

$$\rho = \frac{m}{V}$$

$$r = 0.20\ \text{m}$$

$$V = \tfrac{4}{3}\pi r^3 = \tfrac{4}{3}\pi \times (0.20)^3 = 0.0335\ \text{m}^3$$

$$m = 0.006\ \text{kg}$$

$$\rho = \frac{0.006}{0.0335}$$

$$\rho = 0.179 \approx 0.18\ \text{kg m}^{-3}$$

EXAM HINT

Note that the steps and layout of the solution in this worked example are suitable for density questions in the exam.

LEARNING TIP

As objects expand when they get hotter, the volume depends on the temperature, and so the density must also be affected by changes in temperature.

Here is a table showing densities for different materials.

MATERIAL	STATE	DENSITY / kg m^{-3}
air	gas (sea level, 20 °C)	1.2
pure water	liquid (4 °C)	1000
sulfuric acid (95% conc)	liquid (20 °C)	1839
cork	solid	240
ice	solid	919
window glass	solid	2579
iron	solid	7850
gold	solid	19 320

table A Examples of density values for solids, liquids and gases.

UPTHRUST

When an object is submerged in a fluid, it feels an upwards force caused by the fluid pressure – the **upthrust**. It turns out that the size of this force is equal to the weight of the fluid that has been displaced by the object. This is known as **Archimedes' principle**. If the object is completely submerged, the mass of fluid displaced is equal to the volume of the object multiplied by the density of the fluid:

$$m = V\rho$$

The weight of fluid displaced (i.e. upthrust) is then found using the relationship:

$$W = mg$$

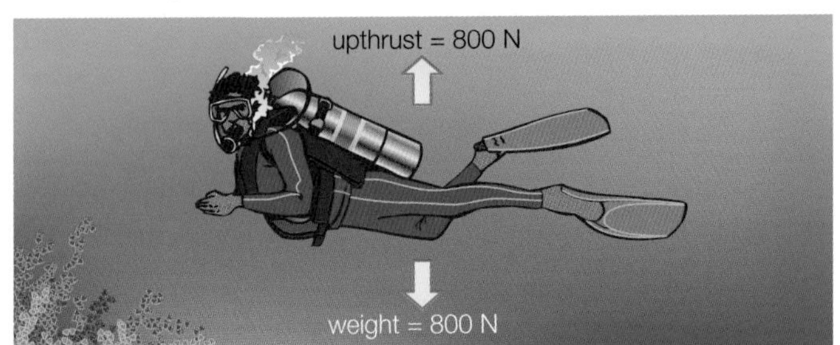

fig B Scuba diving equipment includes a buoyancy control device that can change volume to displace more or less water. This varies the upthrust and so helps the diver move up or down.

WHY DOES A BRICK SINK?

If the house brick from the example calculation of density above were dropped in a pond, it would experience an upthrust equal to the weight of water it displaced. This is simply the weight of an equal volume of water. As the density of water is 1000 kg m^{-3}, the mass of water displaced by the brick would be:

$$m = 1000\text{ kg m}^{-3} \times 1.61 \times 10^{-3}\text{ m}^3 = 1.61\text{ kg}$$

The water has a weight of:

$$W = 1.61 \times 9.81 = 15.8\text{ N}$$

so there is an upward force on the brick of 15.8 N.

If we compare the weight of the brick with the upthrust when it is submerged, the resultant force will be downwards:

weight = 3.38 × 9.81 = 33.2 N downwards

upthrust = 15.8 N upwards

resultant force = 33.2 − 15.8 = 17.6 N downwards

So, the brick will accelerate downwards within the water until it reaches the bottom of the pond, which then exerts an extra upwards force to balance the weight so the brick rests stationary on the bottom with zero resultant force.

(a)

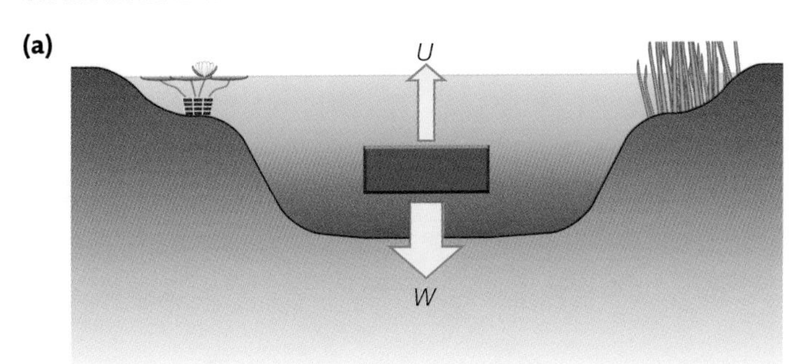

(b)

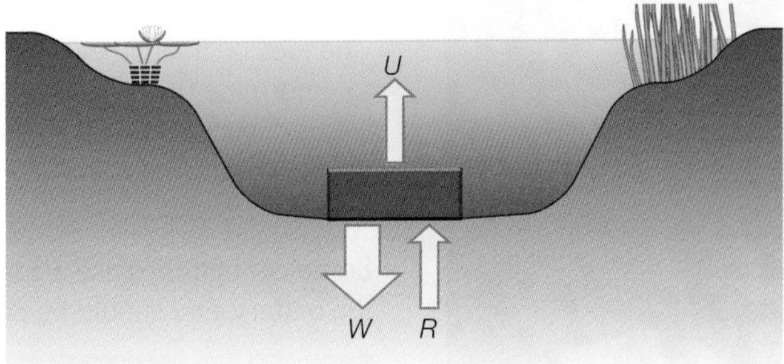

fig C **(a)** If the upthrust on an object is less than its weight, then the object will sink through a fluid; **(b)** an object will remain at rest when balanced forces act on it.

FLOATING

Imagine an object lowering into a fluid. The instant the object touches the surface of the fluid there is no upthrust, because no fluid has been displaced. As the object sinks deeper into the fluid, it displaces an increasing volume of the fluid, so increasing the upthrust acting upon it. If a point is reached when the upthrust and weight are balanced exactly, the object will stop sinking further – it will float there. So, for an object to float, it will have to sink until it has displaced its own weight of fluid.

WORKED EXAMPLE

A giant garbage barge on New York's Hudson River is 60 m long and 10 m wide. What depth of the hull will be under water if it and its cargo have a combined mass of 1500 tonnes? (Assume that the density of water in the Hudson River = 1000 kg m^{-3}.)

To float:

upthrust = weight

weight = mg = 1500 × 1000 × 9.81 = 1.47 × 10^7 N

∴ upthrust = 1.47 × 10^7 N

The upthrust is equal to the weight of the volume of water displaced by the hull:

upthrust = $\rho \times V \times g$

where:

volume V = length of hull, l × width of hull, w × depth of hull under water, d

So:

$$\text{upthrust} = 1000 \times 60 \times 10 \times d \times 9.81 = 5.89 \times 10^6 \times d$$

$$d = \frac{1.47 \times 10^7}{5.89 \times 10^6}$$

$$d = 2.5\text{ m}$$

The hull will be 2.5 m underwater.

EXAM HINT

Note that the steps and layout of the solution in this worked example are suitable for upthrust questions in the exam.

▲ **fig D** A hydrometer can check the density of battery fluids, which tells us if the fluids need changing.

THE HYDROMETER

The idea of floating at different depths is the principle behind the **hydrometer**, an instrument used to determine the density of a fluid. The device has a constant weight, so it will sink lower in fluids of lesser density. This is because a greater volume of a less-dense fluid must be displaced to balance the weight of the hydrometer. Scale markings on the narrow stem of the hydrometer indicate the density of liquid.

Some car batteries use a sulfuric acid solution. The density of this solution tells us the charge level of the battery, and should be checked by a mechanic when the car is serviced.

HYDROMETER READING (DENSITY COMPARED TO WATER)	STATE OF CHARGE
1.255–1.275	100%
1.215–1.235	75%
1.180–1.200	50%
1.155–1.165	25%
1.110–1.130	0%

table B For a particular car battery, the hydrometer readings can be compared to a table to tell us how charged the battery is.

SKILLS PROBLEM SOLVING

CHECKPOINT

1. A car battery contains 1 litre of sulfuric acid solution. The mass of the liquid in the battery is 1.265 kg. What is the density of this battery fluid? (1000 litres = 1 m^3)
2. The radius of a bowling ball is 0.11 m and its mass is 7.26 kg. What is its density
   (a) in $kg\,m^{-3}$?
   (b) in $g\,cm^{-3}$?
3. Estimate the mass of air in this room.
4. A golf ball has a diameter of 4.27 cm.
   (a) If a golf player hits the ball into a stream, what upthrust does it experience when it is completely submerged? (Assume density of water = 1000 $kg\,m^{-3}$.)
   (b) If the mass of the ball is 45 g, what is the resultant force on it when underwater?
   (c) Referring to Newton's laws of motion, explain what will happen to the submerged golf ball.
5. A ball bearing of mass 180 g is hung on a thread in oil of density 800 $kg\,m^{-3}$. Calculate the tension in the string if the density of the ball bearing is 8000 $kg\,m^{-3}$.
6. Estimate your own density.

SUBJECT VOCABULARY

fluid any substance that can flow
density a measure of the mass per unit volume of a substance
upthrust an upwards force on an object caused by the object displacing fluid
Archimedes' principle the upthrust on an object is equal to the weight of fluid displaced
hydrometer an instrument used to determine the density of a fluid

LEARNING OBJECTIVES

- Understand the terms laminar flow and turbulent flow.

If you ski down a hill, you can go faster by tucking your body into a crouching position. By presenting a smaller area to air resistance, you reduce the force slowing you down. However, speed skiers chasing world record speeds go further in their efforts to increase their speeds.

▲ **fig A** Why does this skier have such an oddly shaped helmet and adaptations to the suit's lower legs?

LAMINAR FLOW

When a fluid moves, there are two ways this can happen: **laminar flow** (also called **streamline flow**) and **turbulent flow**. In general, laminar flow occurs at lower speeds, and will change to turbulent flow as the fluid velocity increases past a certain value. The velocity at which this changeover occurs will vary depending upon the fluid in question and the shape of the area through which it is flowing.

If we take a simple example such as water flowing slowly through a pipe, it will be laminar flow. Think of the water in the pipe as several concentric cylinders from the central axis outwards to the layer of water in contact with the pipe itself. Friction between the outermost layer and the pipe wall means this layer will only be able to move slowly. The next layer in will experience friction with the slow-moving outermost layer, but this will be less than the friction between the outermost layer and the pipe. Thus, this inner layer will move faster than the outermost layer. The next layer in moves faster again, with the velocity of each layer increasing nearer the centre, where a central cylinder of water is moving the fastest.

As with most areas of scientific investigation, Isaac Newton produced much work on the subject of fluid flow. In particular, he is credited with the development of equations to describe the frictional force between the layers in streamline flow. If a liquid follows his formulae, as most common liquids do, it is known as a *Newtonian* fluid.

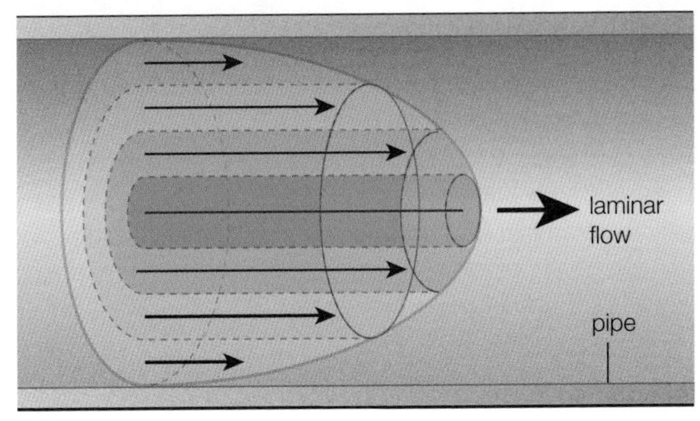

▲ **fig B** Laminar flow in a pipe shows streamlines of different but unchanging velocities.

STREAMLINING

The lines of laminar fluid flow are called **streamlines**. At any point on any one of these streamlines, the velocity of the flow will be constant over time. Remember that velocity is a vector, so this means that the water at any point in the pipe will always move in the same direction and at the same speed. The direction and/or speed may be different in different places, but at any given place direction and speed must stay constant.

In the wind tunnel in **fig C**, the smoke would flow over the car in exactly the same pattern forever if all the wind tunnel factors were kept constant. Changing the speed of the airflow in the tunnel allows designers to test how the prototype would behave at faster speeds, and at what point laminar flow changes to turbulent flow.

▲ **fig C** Smoke streamlines show laminar flow of air over a well-designed car.

In turbulent flow the fluid velocity in any given place changes over time. The flow becomes chaotic and swirling eddies form (you see eddies when water runs away through a plug hole). A poorly designed car would cause turbulent flow of air over it. In the wind tunnel the smoke trails over the car would be seen to swirl in ever-changing patterns. Turbulent flow increases the drag on a vehicle and so increases fuel consumption.

▲ **fig D** Increased speeds change streamline flow to turbulent flow.

PRACTICAL SKILLS

Investigating types of flow

Turbulent flow was first demonstrated by Osborne Reynolds in 1883 in an experiment showing coloured water flowing in a glass tube. You can set up a similar experiment to show turbulence caused by faster fluid flow, or by different shapes of obstacles. At most speeds, a smooth, curved obstacle will produce less turbulence than a squarer one.

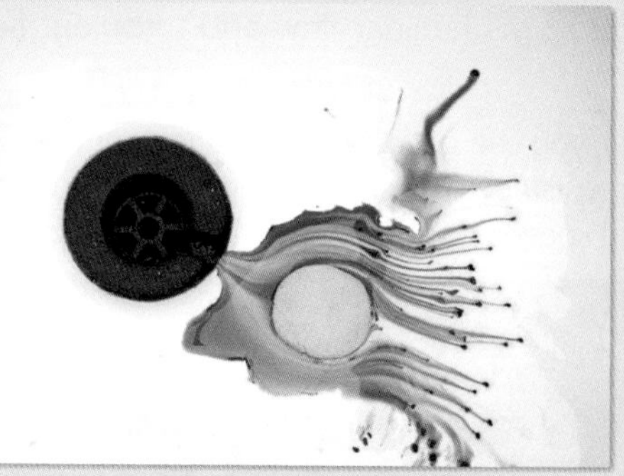

▲ **fig E** A few crystals of potassium manganate(VII) will produce purple trails in the water flow which can then be made to pass around objects made from clay. You can alter the flow rate and the obstacle shapes in order to see how the flow changes.

Safety Note: Avoid skin contact with the potassium manganate (vii), and the clay after it has been used.

CHECKPOINT

SKILLS ADAPTIVE LEARNING

1. Give three examples of objects that are designed to reduce the amount of turbulent flow of air or water over them.
2. Sketch diagrams to illustrate the basic definitions of streamline flow and turbulent flow. Explain how your diagrams show each type of flow.
3. The text on streamlining above says the water flowing at a particular point in a pipe 'will always move in the same direction and at the same speed'.

   Explain why the smoke in the wind tunnel can change direction and move up and over the car.
4. Describe and explain the differences in the water surfaces in the two pictures in **fig D**.

SUBJECT VOCABULARY

laminar flow/streamline flow a fluid moves with uniform lines in which the velocity is constant over time

turbulent flow fluid velocity in a particular place changes over time, often in an unpredictable manner

streamlines lines of laminar flow in which the velocity is constant over time

SPECIFICATION REFERENCE **1.4.25b**

LEARNING OBJECTIVES

- Understand the concept of viscosity.
- Know how viscosity is related to temperature.

VISCOSITY

fig A The 'sharkskin' suit helped break many swimming world records at the Beijing Olympics – it was later banned by swimming's governing body as 'technology doping'. Despite the manufacturer's claim that this was due to the material's extremely low viscous drag, scientists discovered this to be false and put the success down to its physiological benefits for athletes.

When you walk in a swimming pool, you find it much harder than walking through air. The friction acting against you is greater in water than it is in air. This frictional force in fluids is due to **viscosity**. If the frictional force caused by movement through the fluid is small, we say the viscosity is low.

Newton developed a formula for the friction in liquids that includes several factors. One of these factors relates to the particular liquid in question. It would be even harder to wade through a swimming pool of oil than one full of water. This fluid-dependent factor is called the **coefficient of viscosity** and has the symbol η, the Greek letter eta.

As viscosity determines the friction force acting within a fluid, it has a direct effect on the rate of flow of the fluid. Consider the differing rates of flow of a river of lava (see **table A**) compared with the viscosity of lava.

LAVA TYPE	SILICA CONTENT	VISCOSITY	APPROXIMATE FLOW RATE / km h^{-1}
basaltic	least	least	30–60
andesitic	in between	in between	10
rhyolitic	most	most	1

table A How is the rate of flow related to the viscosity of the fluid?

The rate of flow of a fluid through a pipe is inversely proportional to the viscosity of the fluid. In 1838, Jean Poiseuille, a French doctor and physiologist, investigated the flow of fluids in pipes and proved the connection between flow rate and viscosity. Poiseuille was interested in blood flow through the body, but his law is very important in industrial design. For example, the rate of flow of liquid chocolate through pipes in the manufacture of sweets will vary with the chocolate's viscosity, which will vary depending on the exact recipe used to produce it. More sugar may mean greater viscosity and thus slower flow through the pipes, and so less chocolate per sweet.

PRACTICAL SKILLS

Investigating flow rates

You can investigate how fluid flow rate depends on the fluid's viscosity by doing an experiment very similar to those carried out by Poiseuille in the mid-nineteenth century. Using a constant pressure, water forced through a narrow pipe will flow at a certain rate, inversely proportional to its viscosity. By varying the height of the water tank, you can record measurements of this 'head of pressure', h, against the flow rate. The gradient of the best-fit line will allow you to calculate the viscosity of the water.

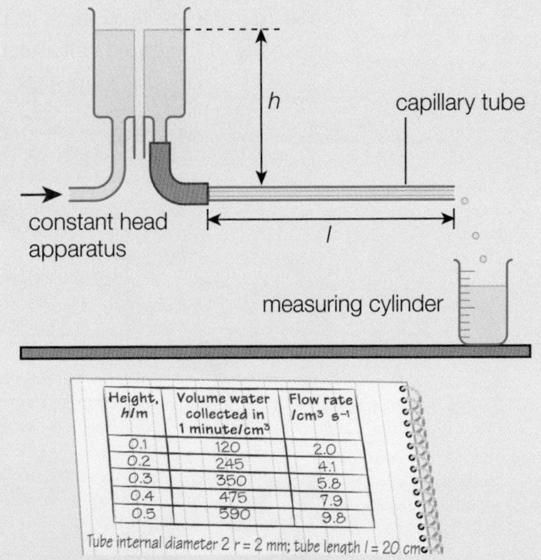

Height, h/m	Volume water collected in 1 minute/cm^3	Flow rate /cm^3 s^{-1}
0.1	120	2.0
0.2	245	4.1
0.3	350	5.8
0.4	475	7.9
0.5	590	9.8

Tube internal diameter $2r = 2$ mm; tube length $l = 20$ cm

fig B Experimental set up and sample results for an investigation into Poiseuille flow.

You can plot a graph of the flow rate (F) against height (h) and hence calculate the viscosity of water (η).

Poiseuille's equations tells us that the gradient of the graph = $\frac{\pi \rho g r^4}{8 \eta l}$

where r is the internal radius of the capillary tube, ρ is the density of water, and $g = 9.81$ N kg^{-1}.

Safety Note: Secure the constant head apparatus to a stand so that it cannot fall over even when filled with water.

An even greater variation in viscosity of liquid chocolate is caused by changes in its temperature. The sweet manufacturer can account for variation in a recipe (which might come from something as minor as a change in supplier of cocoa beans) by adjusting the flow rate by altering the temperature. Viscosity is directly related to fluid temperature. In general, liquids have a lower coefficient of viscosity at higher temperatures. For gases, viscosity increases with temperature.

FLUID	TEMPERATURE / °C	VISCOSITY / Pa S
air	0	0.000017
air	20	0.000018
air	100	0.000022
water	0	0.0018
water	20	0.0010
water	100	0.0003
glycerine	−40	6700
glycerine	20	1.5
glycerine	30	0.63
chocolate	30	100
chocolate	50	60

table B The viscosities of different fluids at different temperatures.

PRACTICAL SKILLS

Investigating how viscosity changes with temperature

You can investigate how the viscosity of a liquid changes with temperature using a re-sealable tin or bottle half-full of a test fluid (such as syrup). The temperature of the liquid is varied using a water bath. The viscosity of the liquid will affect the rate at which the tin or bottle rolls down a fixed ramp.

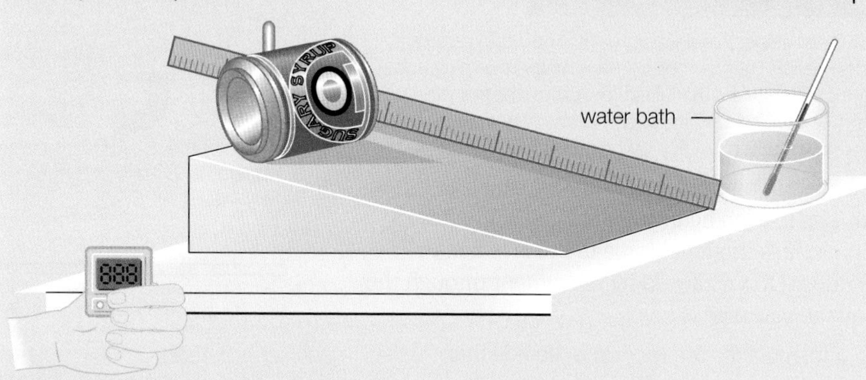

▲ **fig C** How does the viscosity of syrup change with its temperature?

Safety Note: If edible syrup or oils are used, do not taste. Avoid skin contact with mineral oil or motor oil. If the liquid is very hot it will stick to skin and cause severe burns.

SKILLS ANALYSIS

CHECKPOINT

1. Why is the world record for 100 m swimming a longer time than that for 100 m sprinting?
2. Describe how temperature affects viscosity for liquids and gases.
3. How and why would holding a swimming competition in a warmer pool affect the times achieved by swimmers?
4. Why might a chocolate manufacturer alter their machinery so it functioned at a higher temperature?
5. Draw a graph of the experimental results shown in **fig B** in order to find the viscosity of water. How does your value compare with the figures in **table B**?

SUBJECT VOCABULARY

viscosity how resistant a fluid is to flowing
coefficient of viscosity a numerical value given to a fluid to indicate how much it resists flow

2A 4 TERMINAL VELOCITY

LEARNING OBJECTIVES

- Use the equation for viscous drag.
- Use a falling ball method to determine the viscosity of a liquid.

You have previously learned that acceleration due to gravity near the surface of the Earth is about 9.81 m s^{-2}. An object falling in a vacuum does accelerate at this rate. However, it is unusual for objects to be dropped near the surface of the Earth in a vacuum (in nearly all such cases a physics teacher is likely to be demonstrating to a class). In reality, in order to calculate an object's actual acceleration when falling, we need to take account of all the forces acting on it, combine these to find a resultant force, and then use Newton's second law ($\boldsymbol{a} = \frac{\Sigma \boldsymbol{F}}{m}$) to calculate the resulting acceleration.

▲ **fig A** A skydiver will fall at a constant speed if the forces acting on him are balanced.

For a falling object such as a skydiver, this means we need to include the weight, the upthrust caused by the object being in the fluid air, and the viscous drag force caused by the movement. The difficult part is that the viscous drag varies with speed through the fluid, and speed is constantly changing as a result of the acceleration. Usually, we consider the equilibrium situation, in which the weight exactly balances the sum of upthrust and drag, which means that the falling velocity remains constant. This constant velocity is the **terminal velocity**.

EXAM HINT

The phrase 'terminal velocity' is only defined to be for objects falling under gravity with a constant weight. For a similar situation horizontally, for example a car using a constant thrust force, an alternative phrase such as 'maximum velocity' should be used.

STOKES' LAW

In the mid-nineteenth century, Sir George Gabriel Stokes, an Irish mathematician and physicist at Cambridge University, investigated fluid dynamics and came up with an equation for the viscous drag ($\boldsymbol{F}$) on a small sphere at low speeds. This formula is now called Stokes' law:

$$\boldsymbol{F} = 6\pi r \eta \boldsymbol{v}$$

where r is the radius of sphere (m), $\boldsymbol{v}$ is the velocity of sphere (m s^{-1}), and η is the coefficient of viscosity of the fluid (Pa s).

Thus, in such a simple situation, the drag force is directly proportional to the radius of the sphere, and directly proportional to the velocity.

▲ **fig B** Along with Lord Kelvin and James Clerk Maxwell, Sir George Gabriel Stokes helped to build the reputation of Cambridge University in many areas of mathematical physics.

For simplicity, we will only consider simple situations, such as a solid sphere moving slowly in a fluid. Imagine a ball bearing dropping through a column of oil, for example. If you consider the terminal velocity of the sphere in terms of the forces in detail, then:

weight = upthrust + Stokes' force

$m_s\boldsymbol{g}$ = weight of fluid displaced + $6\pi r \eta \boldsymbol{v}_{\text{term}}$

where m_s is the mass of the sphere and $\boldsymbol{v}_{\text{term}}$ is its terminal velocity.

For the sphere, the mass m_s is given by:

m_s = volume × density of sphere = $\frac{4}{3}\pi r^3 \times \rho_s$

so the weight of the sphere $\boldsymbol{W}_s$ is given by:

$$\boldsymbol{W}_s = m_s\boldsymbol{g} = \tfrac{4}{3}\pi r^3 \rho_s \boldsymbol{g}$$

For the sphere, the upthrust is equal to the weight of fluid displaced. The mass m_f of fluid displaced is given by:

m_f = volume × density of fluid = $\frac{4}{3}\pi r^3 \times \rho_f$

so the weight of fluid displaced $\boldsymbol{W}_f$ is given by:

$$\boldsymbol{W}_f = m_f\boldsymbol{g} = \tfrac{4}{3}\pi r^3 \rho_f \boldsymbol{g}$$

Overall then:

$$\tfrac{4}{3}\pi r^3 \rho_s \boldsymbol{g} = \tfrac{4}{3}\pi r^3 \rho_f \boldsymbol{g} + 6\pi r \eta \boldsymbol{v}_{\text{term}}$$

We can rearrange the equation to find the terminal velocity:

$$\boldsymbol{v}_{\text{term}} = \frac{\frac{4}{3}\pi r^3 \boldsymbol{g}(\rho_s - \rho_f)}{6\pi r \eta}$$

Cancelling the π and the radius term:

$$\boldsymbol{v}_{\text{term}} = \frac{2r^2 \boldsymbol{g}(\rho_s - \rho_f)}{9\eta}$$

So terminal velocity is proportional to the square of the radius. This means that a larger sphere falls faster. Furthermore, because the radius is squared, it falls much faster.

EXAM HINT

At a fixed temperature, viscosity is constant.

'Viscous drag' increases with velocity, but 'viscosity' does not change.

WORKED EXAMPLE

Find the terminal velocity of (a) a steel ball bearing of radius 1 mm and (b) a steel ball bearing of radius 2 mm falling through glycerine in a measuring cylinder.

The viscosity of glycerine is highly temperature dependent: at 20 °C we can take η = 1.5 Pa s

density of steel = 7800 kg m^{-3}

density of glycerine = 1200 kg m^{-3}

g = 9.81 m s^{-2}

(a) For a 1 mm radius ball bearing:

$$v_{term} = \frac{2r^2 g(\rho_s - \rho_f)}{9\eta}$$

$$v_{term} = \frac{2\,(1 \times 10^{-3})^2 \times 9.81 \times (7800 - 1200)}{9 \times 1.5}$$

$$v_{term} = 9.6 \times 10^{-3}\ \mathrm{m\,s^{-1}}$$

(b) For a 2 mm radius ball bearing:

$$v_{term} = \frac{2r^2 g(\rho_s - \rho_f)}{9\eta}$$

$$v_{term} = \frac{2\,(2 \times 10^{-3})^2 \times 9.81 \times (7800 - 1200)}{9 \times 1.5}$$

$$v_{term} = 3.8 \times 10^{-2}\ \mathrm{m\,s^{-1}}$$

Comparing the values, you can see that doubling the radius of the ball makes its terminal velocity four times as great.

EXAM HINT

Note that the steps and layout of the solution in this worked example are suitable for terminal velocity questions in the exam.

PRACTICAL SKILLS CP2

Investigating terminal velocity

You can investigate the viscosity of a liquid by allowing various differently sized spheres to fall through it and then measure their terminal velocity and radii.

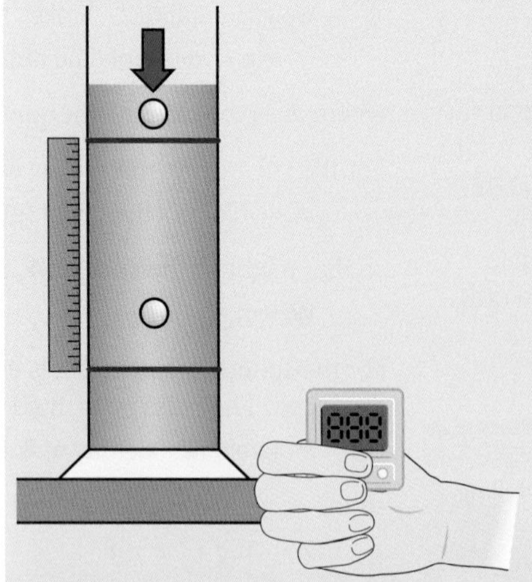

fig C How does the terminal velocity of a falling sphere depend on its radius and the viscosity of the fluid?

You can plot a graph of the terminal velocity, v_{term}, against the square of the sphere radius, r^2, and hence calculate the viscosity of water, η.

Stokes' law tells us that the gradient of the graph = $\frac{2g(\rho_s - \rho_f)}{9\eta}$

where ρ_s is the density of the material of the spheres and ρ_f is the density of the fluid they fall through, and g = 9.81 N kg^{-1}. You may need to do an additional experiment to find the two densities.

Water is usually not viscous enough to give measurably different terminal velocities in this experiment. However, if you do use water, you can then compare the answer for its viscosity with that found from the Poiseuille flow experiment in **Section 2A.3**.

VISCOUS DRAG

You would find it difficult to wade through a swimming pool filled with oil because of the oil's viscous drag. This is the friction force between a solid and a fluid. Calculating this fluid friction force can be relatively simple. On the other hand, it can be very complicated for large objects, fast objects and irregularly shaped objects, as the turbulent flow creates an unpredictable situation.

It must be remembered that the simple slow-falling sphere of Stokes' law is not a common situation and in most real applications the terminal velocity value is a result of more complex calculations. However, the principle that larger objects generally fall faster holds true for most objects without a parachute.

FALLING OBJECT	TERMINAL VELOCITY / $m\,s^{-1}$
skydiver	60
golf ball	32
hail stone (0.5 cm radius)	14
raindrop (0.2 cm radius)	9

table A The terminal velocities of various objects falling in air. Note that the skydiver value varies greatly with the shape in which the body is held when falling.

CHECKPOINT

SKILLS CRITICAL THINKING

1. Use Stokes' law to calculate the viscous drag on a ball bearing with a radius of 1 mm, falling at 1 mm s^{-1} through liquid chocolate at 30 °C.
2. Why is it difficult to calculate the terminal velocity for a cat falling from a high rooftop?
3. A spherical meteorite, of radius 2 m and made of pure iron, falls towards Earth.
   (a) For its fall through the air, use Stokes' law to calculate the meteorite's terminal velocity.
   (b) The meteorite lands in a tropical freshwater lake that is at 20 °C and continues sinking underwater. Use Stokes' law to calculate its new terminal velocity.
   (c) What assumptions have you made in order to make these calculations?
   (See tables of density data in **Section 2A.1** and viscosities in **Section 2A.3**.)
4. Use **table A** to estimate the terminal velocity of the cat in question 2.
5. The experiment shown in **fig C** was carried out using glycerine as the liquid through which the sphere was dropped, but the experiment was repeated at different temperatures from 10 °C up to 50 °C.
   (a) (i) In what way would the density of the glycerine change as the temperature increased for each experiment.
   (ii) In what way would the density of the ball bearing change as the temperature increased for each experiment.
   (b) In what way would the upthrust on the ball bearing change as the temperature of the glycerine was increased for each experiment.
   (c) Explain how the viscosity of glycerine changes as the temperature increases (see **table B** in **Section 2A.3**).
   (d) At each temperature, the student drew a graph of terminal velocity against the square of the radius of the ball bearing. Explain how the gradient of the graph would change with the temperature changes, and why.
   (e) Why is this experiment likely to be inconclusive if water were used instead of glycerine?

EXAM HINT

Students often confuse the terms 'drag' and 'upthrust'. Make sure you are clear on their definitions, and use the right word for the right upward force, depending on what is causing the force.

SUBJECT VOCABULARY

terminal velocity the velocity of a falling object when its weight is balanced by the sum of the drag and upthrust acting on it

2A THINKING BIGGER

THE PLIMSOLL LINE

SKILLS CRITICAL THINKING, PROBLEM SOLVING, REASONING/ARGUMENTATION, INTERPRETATION, ADAPTIVE LEARNING, INITIATIVE, PRODUCTIVITY, ETHICS, COMMUNICATION, ASSERTIVE COMMUNICATION

Load lines are painted on the side of ships to show how low they may safely sit in the water. Although usually associated with the British Member of Parliament Samuel Plimsoll (1824–1898), such lines have been in use for hundreds of years.

In this activity, we will look at how the 'Plimsoll line' came to be required by British law.

OCEAN SERVICE WEBSITE

WHAT IS A PLIMSOLL LINE?

A commercial ship is properly loaded when the ship's waterline is at the ship's Plimsoll line.

The Plimsoll line is a reference mark located on a ship's hull that indicates the maximum depth to which the vessel may be safely immersed when loaded with cargo. This depth varies with a ship's dimensions, type of cargo, time of year, and the water densities encountered in port and at sea. Once these factors have been accounted for, a ship's captain can determine the appropriate Plimsoll line needed for the voyage (see **fig B**).

Samuel Plimsoll (1824–1898) was a member of the British Parliament and he was concerned with the loss of ships and crews due to overloading. In 1876, he persuaded Parliament to pass the Unseaworthy Ships Bill, which required marking a ship's sides with a line that would disappear below the waterline if the ship was overloaded. The line, also known as the Plimsoll mark, is found in the middle of the ship on both the port and starboard sides of cargo ships and is still used worldwide by the shipping industry.

▲ **fig A** Samuel Plimsoll.

LTF – Timber Tropical Fresh Water	TF – Tropical Fresh Water
LF – Timber Fresh Water	F – Fresh Water
LT – Timber Tropical Seawater	T – Tropical Seawater
LS – Timber Summer Seawater	S – Summer Temperate Seawater
LW – Timber Winter Seawater	W – Winter Temperate Seawater
LWNA – Timber Winter North Atlantic	WNA – Winter North Atlantic

▲ **fig B** Plimsoll mark on the hull of a floating ship, and a chart indicating variables such as water density.

From the website of the National Ocean Service, an office of the U.S. National Oceanic and Atmospheric Administration, http://oceanservice.noaa.gov/facts/plimsoll-line.html

SCIENCE COMMUNICATION

1 The text opposite is from the website of the National Ocean Service, an office of the U.S. National Oceanic and Atmospheric Administration. Consider the text and comment on the type of writing used. Try and answer the following questions:

(a) How has the author attempted to maintain the reader's interest?

(b) Discuss the level of the science presented, in relation to the intended audience.

INTERPRETATION NOTE

Think about the type of reader who is likely to visit the webpage. Also, think about why the National Ocean Service would include this webpage if it is 'a dull subject'.

PHYSICS IN DETAIL

Now we will look at the physics in detail. Some of these questions link to topics earlier in this book, so you may need to combine concepts from different areas of physics to work out the answers.

2 Explain the information in the first paragraph of the text, with relation to the heights of the marks in **fig B** and the table of the codes shown on a typical load line.

3 The highest load lines are for timber. These allow for timber to be loaded on to the deck of the ship. This can easily be thrown overboard if needed. Explain why the marks are higher, and how and when it could help if this extra deck cargo were thrown off the ship.

4 Explain why the ships in **fig B** have the shape that they do. Why are they designed differently from large square barges that do not travel on the ocean? Justify your answer.

THINKING BIGGER TIP

Think about how the density of the water leads to the depth that the ship sinks into the water. Would saltier water be more or less dense than freshwater? And how would the temperature of the water affect its density?

ACTIVITY

Write a comment piece for a magazine. Argue that although Plimsoll's campaign was 150 years ago, the capitalist approach to business has not changed very much. Include examples of at least one current business practice, anywhere in the world, where safety technology is not used enough because of its cost. Your article should include a paragraph explaining to a general audience the science behind the safety technology mentioned, and information on your sources of evidence.

DID YOU KNOW?

In the year 1873–74, 411 ships sank off the coast of the United Kingdom, with the loss of 506 lives. Overloading and poor repair made some ships so dangerous that they became known as 'coffin ships'.

2A EXAM PRACTICE

1 Which is the correct expression for calculating density?

A Mass × volume

B Mass ÷ volume

C Upthrust × volume

D Weight ÷ volume [1]

(Total for Question 1 = 1 mark)

2 What is the mass of a spherical stone with a diameter 25 mm, if the density of the stone is 2900 kg m^{-3}?

A 2.82×10^{-9} kg

B 24 g

C 190 g

D 3.55×10^{8} kg [1]

(Total for Question 2 = 1 mark)

3 Which is the correct definition of 'laminar flow'?

A At a given point in a flowing fluid, the velocity of flow varies uniformly over time.

B At a given point in a flowing fluid, the velocity of turbulent flow is proportional to the viscosity.

C At a given point in a flowing fluid, the velocity of flow does not vary over time.

D At a given point in a flowing fluid, the velocity of flow does not vary over the distance from the centre of flow. [1]

(Total for Question 3 = 1 mark)

4 Which row in the table is correct?

	Gas viscosity	Liquid viscosity
A	increases at higher temperature	increases at higher temperature
B	increases at higher temperature	decreases at higher temperature
C	decreases at higher temperature	increases at higher temperature
D	decreases at higher temperature	decreases at higher temperature

[1]

(Total for Question 4 = 1 mark)

5 A stone is dropped into water. It has a mass of 288 g and there is an upthrust force of 0.58 N when it is released in the water. What is the resultant force on the stone at that moment of release, when it is initially not moving?

A 287.4 N

B 2.83 N

C 2.25 N

D zero [1]

(Total for Question 5 = 1 mark)

6 Blood clots can lead to heart attacks. Blood flow through arteries is normally laminar, but an obstruction may cause the blood flow to become turbulent. This can lead to the formation of blood clots.

(a) The diagram shows an artery containing an obstruction.

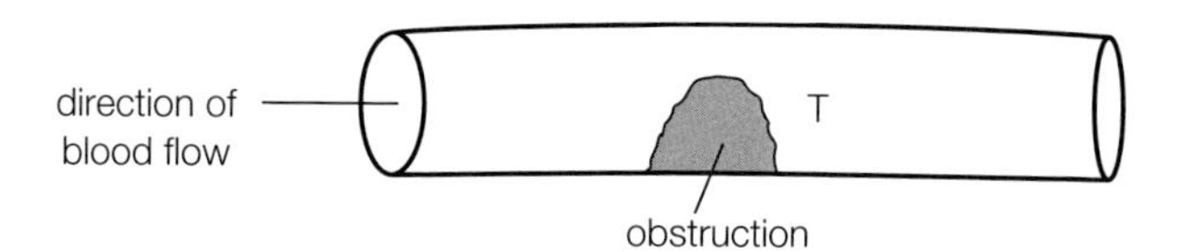

After passing the obstruction the laminar flow becomes turbulent in the area marked T.

(i) Add flow lines to the diagram to show laminar flow changing to turbulent flow after passing the obstruction. [2]

(ii) Explain what is meant by laminar flow and turbulent flow. [2]

(b) In one experiment on blood flow, the viscosity of the blood and the velocity of blood flow were measured.

(i) Describe how you would expect the velocity of blood flow to vary with the viscosity. [1]

(ii) Suggest and explain how a rise in the temperature of the blood would affect the velocity of flow. [2]

(Total for Question 6 = 7 marks)

7 The photograph shows oil being poured into a cold frying pan and spreading out.

Explain the difference that using a hot pan would make to how the oil spreads. [2]

(Total for Question 7 = 2 marks)

8 In the game of table tennis, a ball is hit from one end of the table to the other over a small net.

(a) Making a table tennis ball spin when it is hit can affect its flight. The diagram shows the path of air around a spinning ball. It contains regions of laminar flow and turbulent flow. The flow changes from one to the other at points A and B.

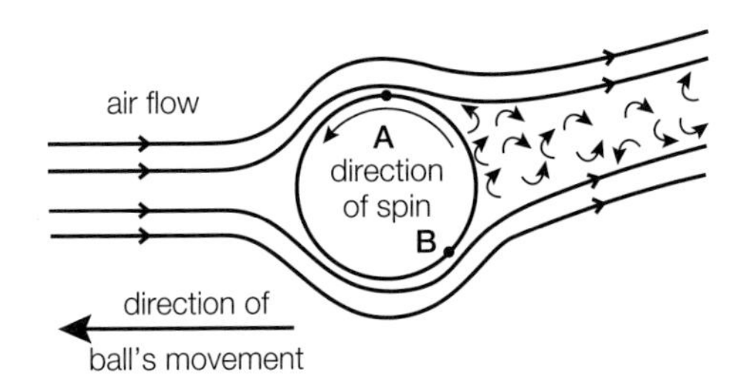

(i) With reference to the diagram, explain what is meant by laminar flow and turbulent flow. [2]

(ii) The ball is spinning in the direction shown in the diagram.
Suggest why there is a larger region of turbulent flow on the top of the ball than the bottom. [1]

(b) The diagram shows that the air is deflected upwards after passing the ball.
Explain why this means there must be a downwards component of force on the ball in addition to its weight. [2]

(c) Spinning a table tennis ball allows it to be hit harder and still hit the table on the other side of the net.

(i) A table tennis ball is hit, without any spin, from one end of a table so that it leaves the bat horizontally with a speed of $31\,\text{m s}^{-1}$. The length of the table is 2.7 m.
Show that the ball falls a vertical distance of about 4 cm as it travels the length of the table. (3)

(ii) The net is 15 cm high. Explain how the spin helps the ball hit the table on the other side of the net. (3)

(Total for Question 8 = 11 marks)

9 Soil is usually made up of a variety of particles of different sizes. The photograph shows what happens when soil is mixed up with water and the particles are allowed to settle.

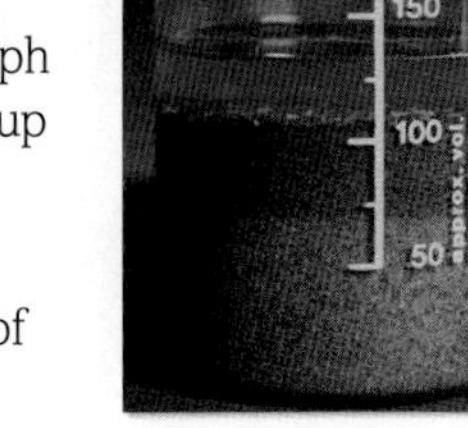

(a) The dot below represents a particle of the soil falling through water.

(i) Add labelled arrows to show the three forces acting on the particle as it falls through the water. [2]

●

(ii) Explain why a particle held stationary in water and then released accelerates downwards at first but then reaches a steady downwards speed. [4]

(iii) Write an expression showing the relationship for these forces when the particle is falling at a steady speed. [1]

(b) A typical particle of sand in the sample has the following properties:

diameter = $1.6 \times 10^{-3}\,\text{m}$

volume = $2.1 \times 10^{-9}\,\text{m}^3$

density = $2.7 \times 10^{3}\,\text{kg m}^{-3}$

weight = $5.7 \times 10^{-5}\,\text{N}$

(i) Show that the upthrust acting on the particle is about $2 \times 10^{-5}\,\text{N}$.
density of water = $1.0 \times 10^{3}\,\text{kg m}^{-3}$ [2]

(ii) Calculate the steady downwards speed this particle would achieve if allowed to fall through water.
viscosity of water = $1.2 \times 10^{-3}\,\text{Pa s}$ [3]

(c) The different types of particles in soil can be defined according to their diameters, as in the following table.

Soil particle	Particle diameter
clay	less than 0.002 mm
silt	0.002 mm – 0.05 mm
sand	0.05 mm – 2.00 mm
fine pebbles	2.00 mm – 5.00 mm
medium pebbles	5.00 mm – 20.00 mm
coarse pebbles	20.00 mm – 75.00 mm

The photograph shows that when soil is allowed to settle in water, the pebbles tend to be found towards the bottom, followed by sand, silt and clay in succession.
Explain why this happens. Assume that all particles have the same density. [3]

(Total for Question 9 = 15 marks)

10 Explain how a stone dropped into a lake, from the surface, will reach a maximum velocity as it falls to the lake bottom, and how this will be different in summer and winter. [6]

(Total for Question 10 = 6 marks)

TOPIC 2 MATERIALS

CHAPTER 2B SOLID MATERIAL PROPERTIES

You probably rarely consider the likelihood that a bridge will fail as you walk or drive over it. However, when designing the bridge, and selecting the materials to use in it, an engineer has to ensure that this likelihood is minimal.

Design processes for any new project need to include specifications indicating the extremes of conditions that may apply. A bridge needs to withstand an overloaded lorry that is not complying with weight restrictions; a snowboarding helmet might be used by a person on a snowmobile; or a motorist might try to use a climbing rope to tow their car. Some of these uses might be classed as unreasonable, and the designer would not expect to have to make their product strong enough to stand up to such extreme forces.

If you were designing a new harbour wall, what extreme weather conditions would you ensure that the wall could withstand? Would it be good enough to make sure it could withstand a storm with such strong winds and waves that only occurs once every 50 years? 100 years? 1000 years? And how do you achieve the strength required within budget?

In this chapter we will see how the strength of materials can be measured, and thus how to choose the right material for any given job.

MATHS SKILLS FOR THIS CHAPTER

- **Units of measurement** (*e.g. the pascal, Pa*)
- **Calculating areas of circles** (*e.g. finding the cross-sectional area of a wire in order to find stress*)
- **Use of standard form and ordinary form** (*e.g calculating Young modulus*)
- **Making order of magnitude calculations** (*e.g. comparing Young modulus for different materials*)
- **Changing the subject of an equation** (*e.g. re-arranging the stress equation*)
- **Substituting numerical values into algebraic equations** (*e.g. calculating the strain*)
- **Determining the slope of a linear graph** (*e.g. finding the Young modulus from experimental results*)
- **Estimating, by graphical methods as appropriate, the area between a curve and the x-axis, and realising the physical significance of the area that has been determined** (*e.g. finding the energy stored in a stretched material from graphical data*)

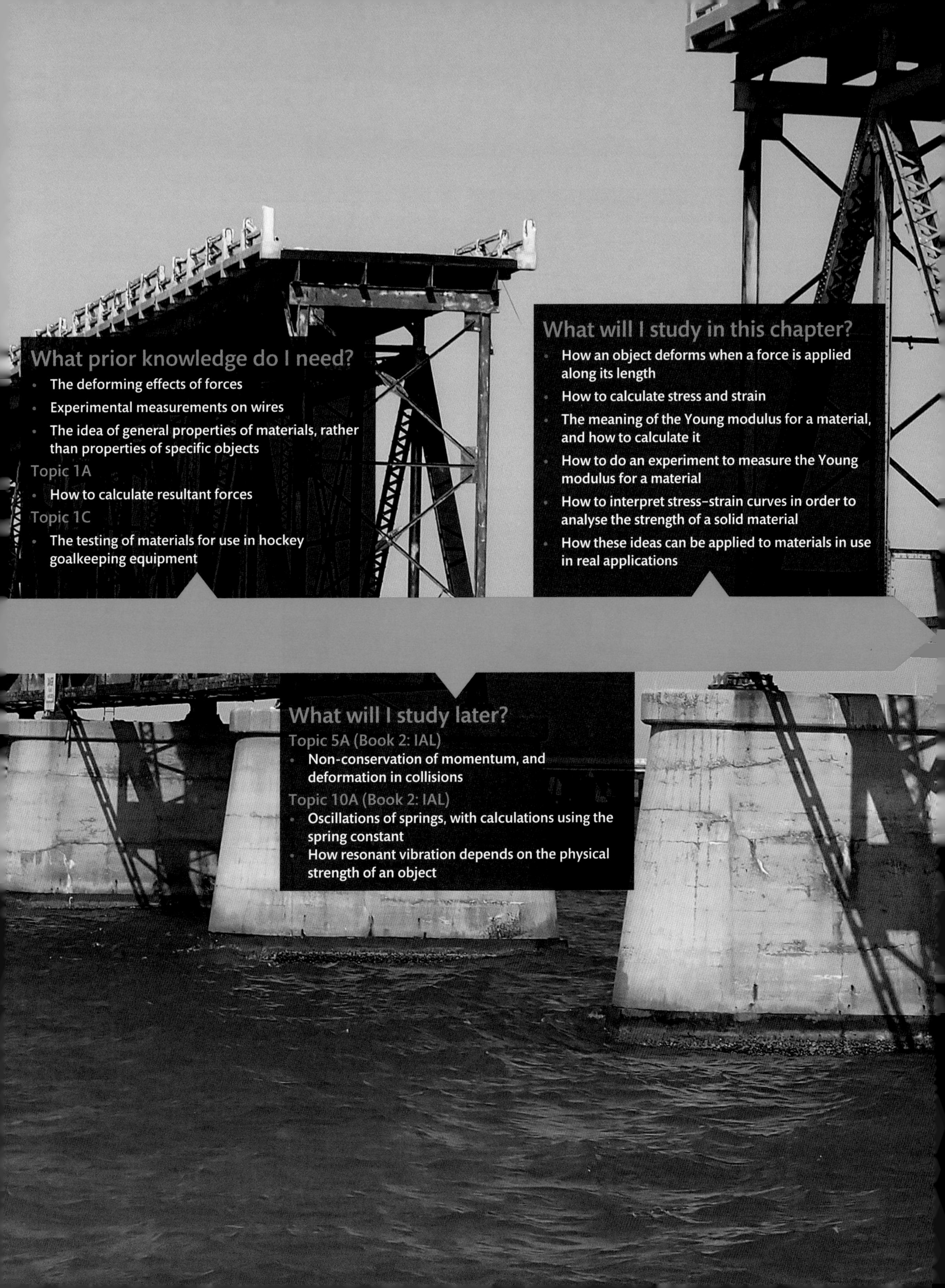

What prior knowledge do I need?

- The deforming effects of forces
- Experimental measurements on wires
- The idea of general properties of materials, rather than properties of specific objects

Topic 1A

- How to calculate resultant forces

Topic 1C

- The testing of materials for use in hockey goalkeeping equipment

What will I study in this chapter?

- How an object deforms when a force is applied along its length
- How to calculate stress and strain
- The meaning of the Young modulus for a material, and how to calculate it
- How to do an experiment to measure the Young modulus for a material
- How to interpret stress–strain curves in order to analyse the strength of a solid material
- How these ideas can be applied to materials in use in real applications

What will I study later?

Topic 5A (Book 2: IAL)

- Non-conservation of momentum, and deformation in collisions

Topic 10A (Book 2: IAL)

- Oscillations of springs, with calculations using the spring constant
- How resonant vibration depends on the physical strength of an object

2B 1 HOOKE'S LAW

SPECIFICATION REFERENCE 1.4.27 1.4.32

LEARNING OBJECTIVES

- Understand Hooke's law and be able to make calculations using it.
- Calculate the elastic strain energy stored in a deformed material sample.
- Estimate the elastic strain energy stored from a force–extension graph for a sample.

Whenever a force acts on a material sample, the sample will be deformed to a different size or shape. If it is made longer, the force is referred to as **tension**, and the extra length is known as the **extension**. For a material being squashed to a smaller size, both the force and the decrease in size are called **compression**, although the decrease in size could be referred to as a negative extension.

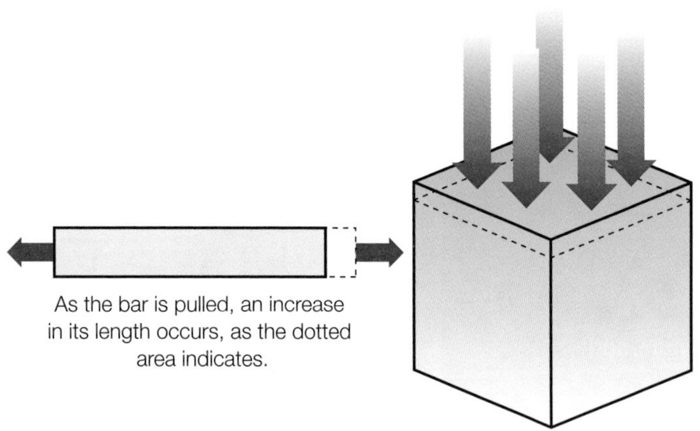

▲ **fig A** A tensile force causes extension. A compressive force causes a negative extension.

HOOKE'S LAW

Robert Hooke, a rival of Isaac Newton, was an exceptional experimental scientist. His interests were very diverse and he discovered many things that are more important than the rule called Hooke's law.

Hooke's law states that the force needed to extend a spring is proportional to the extension of the spring. A material only obeys Hooke's law if it has not passed what is called the **limit of proportionality**.

Up to a certain limit of proportionality, the force needed to extend a spring is proportional to the extension of the spring. If an object is subject to only a small force, it will deform elastically. When the force is removed, it returns to its original size and shape, as long as the **elastic limit** was not passed. We will look in much more detail at elastic and plastic **deformation** in **Section 2B.3**.

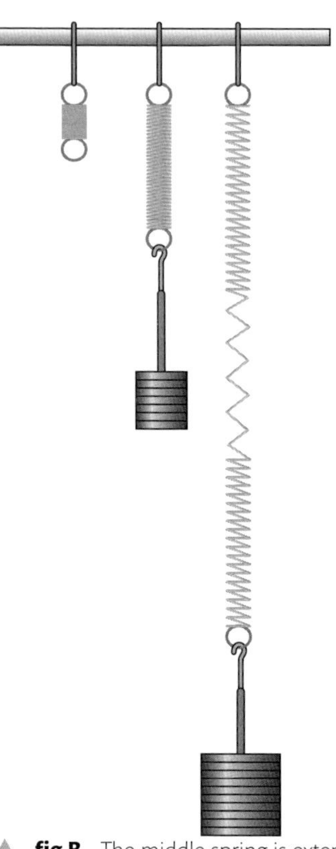

▲ **fig B** The middle spring is extended in proportion to the applied load, obeying Hooke's law. The right-hand spring has been stretched beyond its limit of proportionality, so no longer obeys Hooke's law.

Hooke's law is best described mathematically with the equation:

force applied (N) = stiffness constant ($N\,m^{-1}$) × extension (m)

$$\Delta F = k\Delta x$$

EXAM HINT

The stiffness constant for a spring is usually referred to as the **spring constant**. Either phrase refers to k in the Hooke's law equation.

WORKED EXAMPLE

A spring has a stiffness constant of $50\,N\,m^{-1}$ and is 3.0 cm long. How long would it be if a 200 g mass were hung from it?

Weight force:

$$W = mg = 0.200 \times 9.81$$
$$= 1.962\,N$$
$$\Delta F = k\Delta x$$
$$\therefore \quad \Delta x = \frac{\Delta F}{k} = \frac{1.962}{50} = 0.03924\,m$$
$$= 3.924\,cm$$

final length = original length + extension

$$L = L_0 + \Delta x = 3.0 + 3.9$$
$$L = 6.9\,cm$$

EXAM HINT

Note that the steps and layout of the solution in this worked example are suitable for questions about Hooke's law in the exam.

PRACTICAL SKILLS

Investigating Hooke's law

You can perform a simple experiment to measure the stiffness constant for a spring. By hanging various masses on a spring and measuring the corresponding extensions, you can gather a set of results for ΔF and Δx. In each case, you could calculate the spring constant from a pair of readings, but you will get a more accurate final answer for k if you plot a graph of the results and find its gradient.

MASS / g	WEIGHT / N	ORIGINAL LENGTH / cm	LOADED LENGTH / cm	EXTENSION / cm
0	0.000	2.4	2.4	0.0
100	0.981	2.4	2.8	0.4
200	1.962	2.4	3.1	0.7
300	2.943	2.4	3.6	1.2
400	3.924	2.4	3.9	1.5
500	4.905	2.4	4.3	1.9
600	5.886	2.4	4.8	2.4
800	7.848	2.4	5.8	3.4
1000	9.810	2.4	6.5	4.1

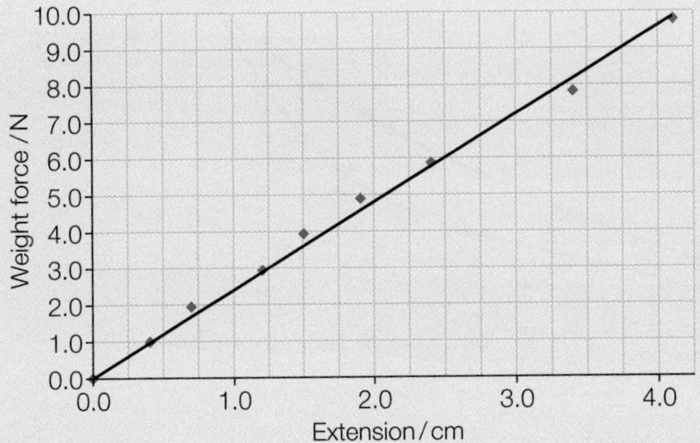

fig C The results of a Hooke's law investigation of a spring before it reaches its elastic limit. The fact that the best-fit line is straight shows that the spring obeys Hooke's law.

The graph in **fig C** has been plotted with the dependent variable on the x-axis, so that when the gradient of the line is calculated, this will give you the value for the spring constant:

$$\text{Gradient, } m = \frac{\Delta y}{\Delta x} = \frac{\Delta F}{\Delta x} = k$$

Safety Note: Use eye protection in case the spring becomes detached and flies back. Keep hands and feet clear of the 'drop zone' if large masses are used.

ELASTIC STRAIN ENERGY

The work done in deforming a material sample before it reaches its elastic limit will be stored within the material as elastic strain energy, E_{el}. We have previously seen that work done can be calculated by multiplying the force by the distance moved in the direction of the force. This is true in deforming materials too, but Hooke's law means that the force value varies for different extensions. If we plot the extension of a spring with increasingly large masses hanging on it, as in **fig C**, the graph will follow Hooke's law. To find the work done to extend the spring a certain amount, we must calculate using the average force over the distance of the extension.

$$\Delta E_{el} = \tfrac{1}{2}F\Delta x$$

WORKED EXAMPLE

In **fig C**, how much elastic strain energy is stored in the spring when it is extended by 2.5 cm?

At Δx = 2.5 cm, the corresponding force value is F = 6.0 N.

$$\Delta E_{el} = \tfrac{1}{2}F\Delta x$$

$$\Delta E_{el} = \tfrac{1}{2} \times 6.0 \times 0.025$$

$$\Delta E_{el} = 0.075\,\text{J}$$

WORK FROM FORCE–EXTENSION AND FORCE–COMPRESSION GRAPHS

The work done in deforming a material is calculated by multiplying the extension or compression by an appropriate average force value. If the force is varying in a non-linear way, which is common for some materials, it might not be a straightforward process to find the average force. However, the area between the line on a force–extension (F–Δx) graph and the extension axis will represent the work done. Finding the area under the line is easy with the linear type of relationship, shown in **fig C**. As the area under the line on **fig C** is a triangle, the formula for the area of a triangle gives us the same equation as was used above:

$$\Delta E_{el} = \tfrac{1}{2}F\Delta x$$

If the relationship is non-linear, as shown in the example of **fig D**, the work done can still be found from the area under the line.

WORKED EXAMPLE

What is the elastic strain energy stored in the material shown in **fig D** if it is extended by 12 cm?

The area under the line, up to the value Δx = 0.12 m, can be split into three sections:

- Triangle from origin to x = 0.04 m:
  area = $\tfrac{1}{2}$ × base × height = $\tfrac{1}{2}$ × 0.04 × 4 = 0.08 J
- Rectangle from x = 0.04 m to x = 0.08 m:
  area = base × height = 0.04 × 4 = 0.16 J
- Trapezium from x = 0.08 m to x = 0.12 m:
  area = base × average height = 0.04 × (4 + 6)/2 = 0.20 J
  total work = 0.08 + 0.16 + 0.20
  elastic potential energy = 0.44 J

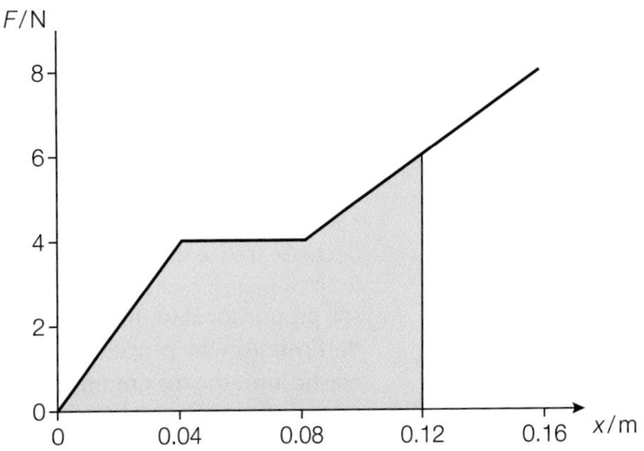

fig D The elastic potential energy stored in a material can be found from a non-linear force–extension graph by working out the area under the line up to the required extension.

If a non-linear force–extension or force–compression graph has a curved line, finding the area may involve estimating or counting the squares on the graph paper under the line. In this case, you will also need to multiply the number of squares by the elastic strain energy value ($F \times \Delta x$) for each individual square.

EXAM HINT

Be careful to distinguish between 'length' and 'extension'. You may be given a measurement for either length or extension. You should check carefully to see if you need to calculate the extension from information you have about lengths.

CHECKPOINT

1. What is the spring constant for a spring that starts at a length of 25 cm and extends to a length of 32 cm when a mass of 50 g is added?

SKILLS REASONING / INTERPRETATION

2. In **fig C**, how would the line appear different if it were for a spring of constant $k = 280\ \text{N m}^{-1}$?
3. Explain how the elastic strain energy stored in a stretched spring could be calculated from the formula $\Delta E_{el} = \frac{1}{2}k(\Delta x)^2$.

SKILLS ANALYSIS

4. **Fig E** shows the results of a Hooke's law experiment loading and unloading a rubber band to an extension of 24 cm.

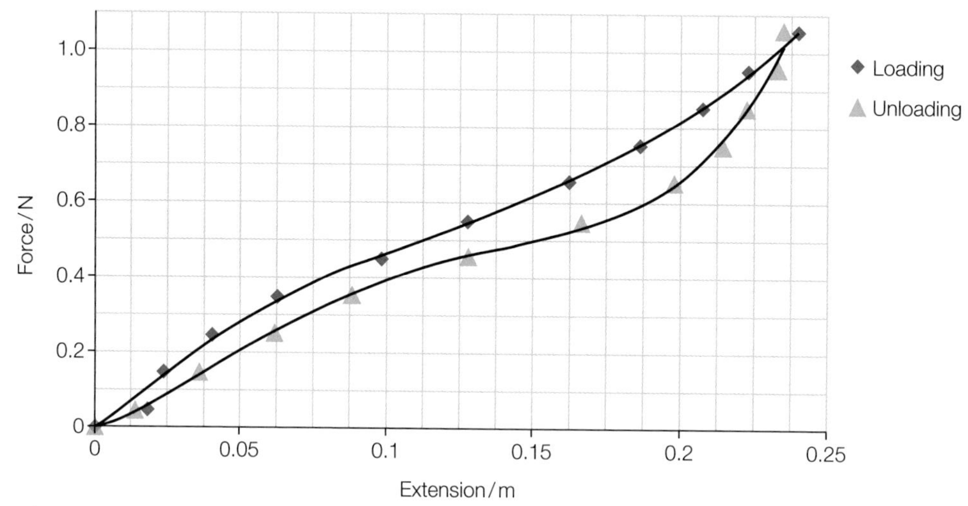

fig E Different extensions during loading and unloading is an example of **hysteresis**.

(a) From the graph, estimate the work done in loading the rubber band.

(b) When unloading, the rubber band releases the stored elastic strain energy. Estimate, from the graph, the area underneath the unloading curve, which will be the elastic strain energy released.

(c) The difference and the answers to parts (a) and (b) represents energy lost in this process, which is mostly used in internal heating of the rubber band. How much **thermal** energy does the rubber gain from the complete cycle of loading and unloading?

SUBJECT VOCABULARY

tension a force acting within a material in a direction that would extend the material
extension an increase in size of a material sample caused by a tension force
compression a force acting within a material in a direction that would squash the material. Also the decrease in size of a material sample under a compressive force
limit of proportionality the maximum extension (or strain) that an object (or sample) can have, which is still proportional to the load (or stress) applied
deformation the process of alteration of form or shape
elastic limit the maximum extension or compression that a material can undergo and still return to its original dimensions when the force is removed
spring constant the Hooke's law constant of proportionality, k, for a spring under tension
hysteresis where the extension under a certain load will be different depending on its history of past loads and extensions
thermal connected with heat

2B 2 STRESS, STRAIN AND THE YOUNG MODULUS

SPECIFICATION REFERENCE 1.4.28

LEARNING OBJECTIVES

- Calculate tensile/compressive stress.
- Calculate tensile/compressive strain.
- Calculate the Young modulus.

If you pull on the two ends of a metal bar, you are unlikely to deform it at all. However, pulling with the same force on the two ends of a very thin piece of wire made from the same metal may have very different results: it will probably extend elastically, and may pass its elastic limit and start to experience permanent deformation. Depending on the specific metal and the exact dimensions, it is possible that your force may be strong enough to break the wire. This example demonstrates that engineers need more information about the forces that may be encountered within the structures they are building than just the numbers of newtons. You used the same force and the same metal, but in the case of the metal bar you made no impression on it, whilst the thin wire broke. To make fair comparisons between samples, we need to consider the forces on them and their sizes as well as the materials from which they are made.

STRESS

Tensile (or compressive) **stress** is a measure of the force within a material sample, but it takes account of the cross-sectional area across the sample. This allows force comparisons to be made between samples of different sizes, so that they are measured under comparable conditions.

$$\text{stress (pascals, Pa, or N m}^{-2}) = \frac{\text{force (N)}}{\text{cross-sectional area (m}^2)}$$

$$\sigma = \frac{F}{A}$$

WORKED EXAMPLE

A cylindrical stone column has a diameter of 60 cm and supports a weight of 2500 N (**fig A**). What is the compressive stress in the column?

$$\text{Area} = \pi r^2 = 3.14 \times 0.30^2 = 0.2826 \text{ m}^2$$

$$\sigma = \frac{F}{A} = \frac{2500}{0.2826}$$

$$\sigma = 8850 \text{ Pa}$$

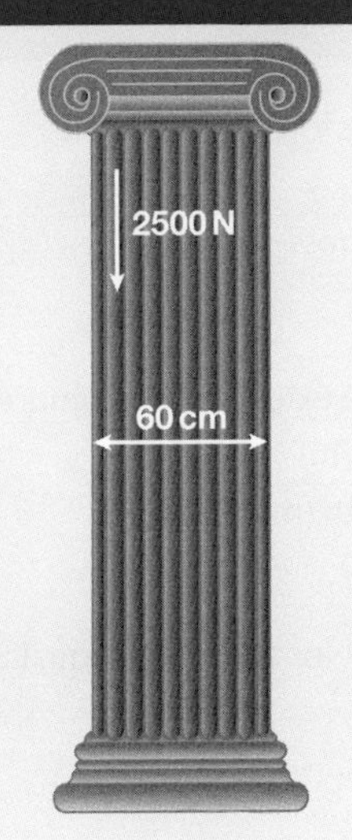

EXAM HINT

Note that the steps and layout of the solution in this worked example are suitable for stress questions in the exam.

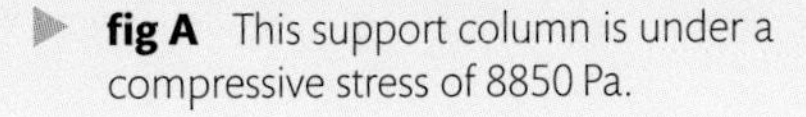

fig A This support column is under a compressive stress of 8850 Pa.

STRAIN

Tensile (or compressive) **strain** is a measure of the extension (or compression) of a material sample, but it takes account of the original length of the sample. This allows extension comparisons to be made between samples of different sizes, so that they are measured under comparable conditions.

$$\text{strain (no units)} = \frac{\text{extension (m)}}{\text{original length (m)}}$$

$$\varepsilon = \frac{\Delta x}{x}$$

As strain is a ratio, it has no units. However, it is often expressed as a percentage by multiplying the ratio by 100%.

WORKED EXAMPLE

A copper wire of length 1.76 m is stretched by a force to a length of 1.80 m. What is the tensile strain in the wire?

$$\varepsilon = \frac{\Delta x}{x} = \frac{(1.80 - 1.76)}{1.76} = \frac{0.04}{1.76}$$

$$\varepsilon = 0.023 = 2.3\%$$

EXAM HINT

Note that the steps and layout of the solution in this worked example are suitable for strain questions in the exam.

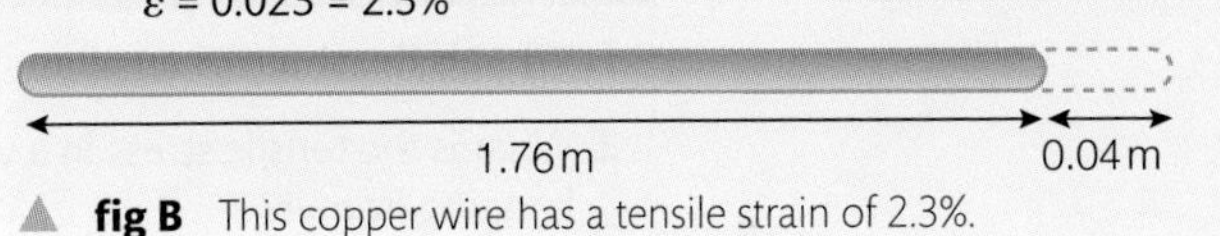

fig B This copper wire has a tensile strain of 2.3%.

YOUNG MODULUS

If a material is deformed elastically, stress will be proportional to strain, with a constant of proportionality that is a measure of the stiffness of the material – how much it deforms under a certain stress. The stiffness constant is called the **Young modulus**. So, the Young modulus is a measure of the stiffness of a material, which takes account of the shape and size of the sample, so that different samples of the same material will all have the same value for the Young modulus. The idea of stiffness is a measure of how much the material deforms when forces are applied to it.

$$\text{Young modulus (Pa)} = \frac{\text{stress (Pa)}}{\text{strain (no units)}}$$

$$E = \frac{\sigma}{\varepsilon}$$

LEARNING TIP

The stiffness constant, k, from Hooke's law relates to a particular object, such as a spring. For a material, the Young modulus, E, is the stiffness constant for the material in general, regardless of sample size.

The definition for the Young modulus also includes the fact that the material must be undergoing elastic deformation. Beyond the limit of proportionality, this equation will no longer work to calculate the stiffness of the material.

LEARNING TIP

From the original definitions of stress and strain, the Young modulus can also be calculated from:

$$E = \frac{Fx}{A\Delta x}$$

WORKED EXAMPLE

The copper wire from above has a diameter of 0.22 mm and was stretched using a force of 100 N. What is the Young modulus of copper?

$$A = \pi r^2 = 3.14 \times (1.1 \times 10^{-4})^2 = 3.80 \times 10^{-8}\,\text{m}^2$$

$$\sigma = \frac{F}{A} = \frac{100}{3.8 \times 10^{-8}}$$

$$\therefore \quad \sigma = 2.63 \times 10^9\,\text{Pa}$$

$$E = \frac{\sigma}{\varepsilon} = \frac{2.63 \times 10^9}{0.023}$$

$$E = 1.16 \times 10^{11}\,\text{Pa}$$

(Note that the stress has been calculated from the original data and the unrounded value has been used in this calculation to give the value shown here.)

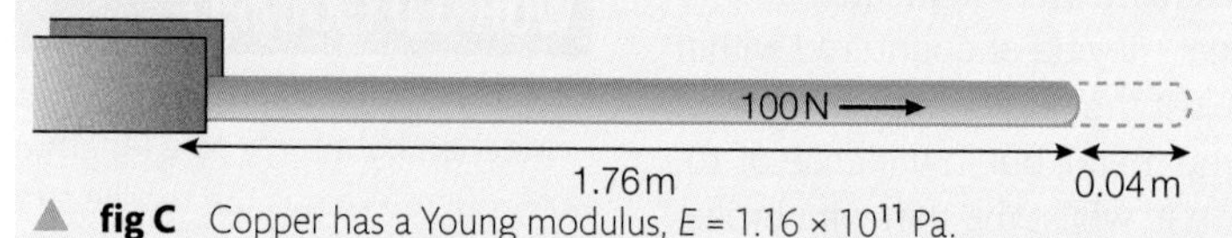

fig C Copper has a Young modulus, $E = 1.16 \times 10^{11}$ Pa.

EXAM HINT

Note that the steps and layout of the solution in this worked example are suitable for Young modulus questions in the exam.

CHECKPOINT

1. What is the strain of an aluminium wire if it extends from 97 cm to 1.04 m?
2. What is the tensile stress in a vertical steel wire that has a diameter of 0.40 mm and a 1.0 kg mass hanging from it?
3. Use these data to find the Young modulus of human hair: d = 0.1 mm; x = 12 cm; F = 0.60 N; Δx = 1.8 mm.
4. (a) What is the compressive stress in each upper leg bone of a 7 tonne elephant if the bone is a vertical cylinder with a diameter 25 cm and the elephant is standing normally?
   (b) If the Young modulus of elephant bone is 19 GPa, and the bones are originally 95 cm long, how much would they reduce in length if the elephant stood up on its back legs? What assumption do you need to make?

SUBJECT VOCABULARY

stress a proportionate measure of the force on a sample:

$$\text{stress (pascals Pa, or N m}^{-2}) = \frac{\text{force (N)}}{\text{cross-sectional area (m}^2)}$$

$$\sigma = \frac{F}{A}$$

strain a proportionate measure of the extension (or compression) of a sample:

$$\text{strain (no units)} = \frac{\text{extension (m)}}{\text{original length (m)}}$$

$$\varepsilon = \frac{\Delta x}{x}$$

Young modulus the stiffness constant for a material, equal to the stress divided by its corresponding strain

3 STRESS–STRAIN GRAPHS

SPECIFICATION REFERENCE

1.4.29 1.4.30 1.4.31 CP3

CP3 LAB BOOK PAGE 14

LEARNING OBJECTIVES

- Interpret stress–strain graphs.
- Understand and apply the terms limit of proportionality, elastic limit, yield point, breaking stress, elastic deformation and plastic deformation in relation to stress–strain graphs.

STRESS–STRAIN ANALYSIS

From the definition of the Young modulus, the stress should be proportional to the strain if a material is undergoing elastic deformation. Therefore we should get a straight-line graph if we plot stress against strain. When this is done, we do find a straight-line relationship for small stresses. Once the limit of proportionality is passed, the internal structure of the material starts to behave differently. This means that the graph starts to curve. Depending on the material under test, the graph will go through various phases as the molecular structure of each material determines its response to increasing stress. Eventually, the stress will become too great, and the material will fracture. At this point, the line on the graph must end, as no further data can be obtained about the material. It cannot withstand higher stresses, so the graph cannot be plotted any further.

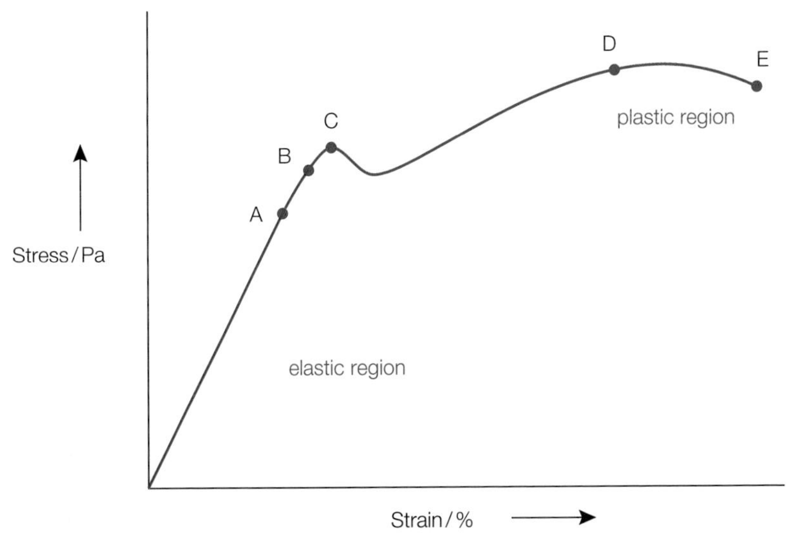

▲ **fig A** The stress–strain graph for a metal gives detailed information about how the material behaves under different levels of stress. The gradient of the straight-line portion of the graph will be equal to the value of the Young modulus.

LEARNING TIP

It can be confusing to see the sections of a stress–strain graph where the line goes down. In reality, these graphs are generated from results of materials testing in which the strain is continuously increased on a machine that can measure the stress within a material. This stress can go down if there is a change in the arrangement of the molecular structure of the material.

Fig A is a generalised stress–strain graph for a metal, such as copper. Most metals will follow the shape shown here, and there are various areas of interest on the graph.

In the straight-line portion from the origin to **point A**, the metal extends elastically, and will return to its original size and shape when the force is removed. The gradient of the straight-line portion of the graph is equal to the Young modulus for the metal.

Point A is the limit of proportionality. Slightly beyond this point, the metal may still behave elastically, but it cannot be relied upon to increase strain in proportion to the stress.

Point B is the elastic limit. Beyond this point, the material is permanently deformed and will not return to its original size and shape, even when the stress is completely released.

Point C is the **yield point**, beyond which the material undergoes a sudden increase in extension as its atomic substructure is significantly re-organised. The metal 'gives' just beyond its yield point as the metal's atoms slip past each other to new positions where the stress is reduced.

LEARNING TIP

It is quite common for the limit of proportionality, the elastic limit and the yield point to be in the same place on the graph. They have been separated here to explain the different ideas behind each.

Point D represents the highest possible stress within this material. It is called the Ultimate Tensile Stress, or UTS, σ_U.

Point E is the fracture stress, or breaking stress. It is the value that the stress will be in the material when the sample breaks.

WORKED EXAMPLE

The European Space Agency (ESA) have tested a material known as 'IMPRESS intermetallic alloy'. **Fig B** illustrates a stress–strain graph for this material at two different temperatures, to see how it would stand up under different conditions.

Considering first the Young modulus, you can see that the ESA have drawn measurements on the graph to make the calculation of the gradient of the straight-line portions of the graphs for each temperature plot. For example, at 15 °C:

$$\text{gradient} = E = \frac{\Delta y}{\Delta x} = \frac{\sigma}{\varepsilon} = \frac{400\text{ MPa}}{0.3\%}$$

$$\therefore \quad E = \frac{400 \times 10^6}{0.003}$$

$$E = 1.33 \times 10^{11}\text{ Pa}$$

The values labelled as $\sigma_{0.2}$ are the stress values at each temperature, which will result in a permanent deformation strain of 0.2%. These are in the plastic region, and the permanent deformations are indicated by the dashed lines connected back to 0.2% on the x-axis.

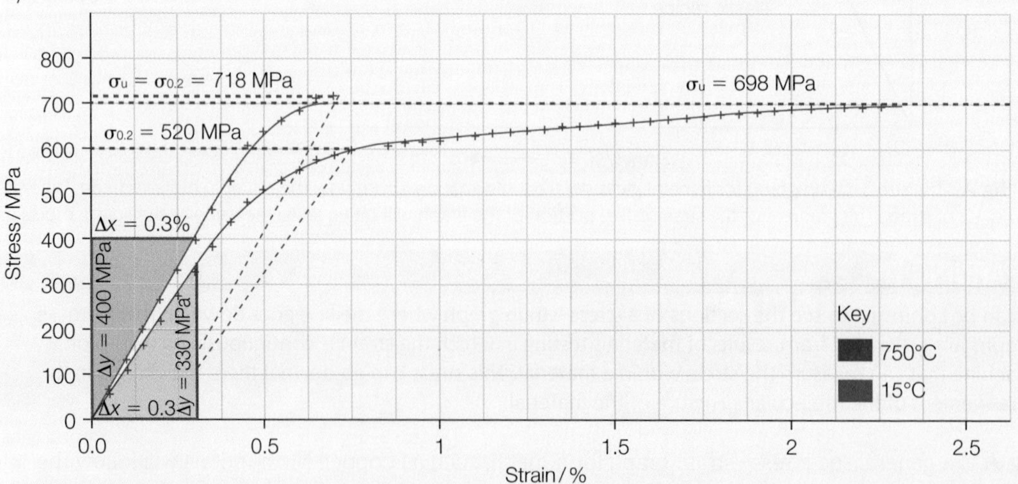

▲ **fig B** Stress–strain graphs, at two temperatures, for the IMPRESS intermetallic alloy, for use in the European space programme.

PRACTICAL SKILLS CP3

Investigating stress–strain relationships for metals

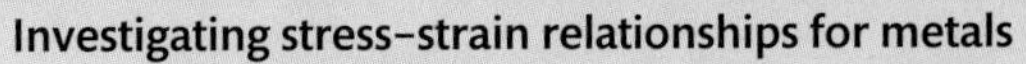

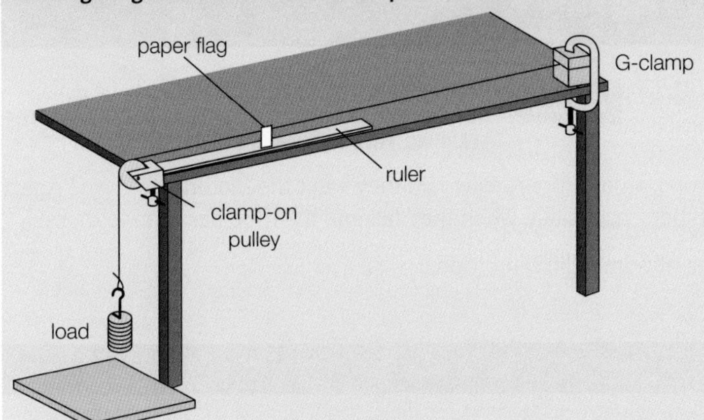

▲ **fig C** An experiment to find the stress–strain relationships when stretching a metal wire. With detailed analysis of the stress–strain graph of the metal, you can find the Young modulus.

You can perform a simple experiment to measure the stiffness constant – the Young modulus – for a metal by stretching a thin wire. The original length and diameter of the wire must be measured first. Then, using increasing forces as the independent variable, you will need to take measurements of the extension corresponding to each force. There will be some readings during the wire's elastic deformation region which will increase uniformly. Beyond the elastic limit, it will increase with greater extensions for each increase in the load force until, eventually, the fracture stress will be reached and the wire will snap. Safety goggles must be worn during this experiment, as wires that snap under tension are a hazard to the eyes.

The graph in **fig B** has been plotted with the strain on the x-axis, so that when the gradient of the line is calculated this will give you the value for the stiffness constant, or Young modulus:

$$\text{gradient, } m = \frac{\Delta y}{\Delta x} = \frac{\sigma}{\varepsilon} = E$$

Normal practice is to plot experimental results with the independent variable on the x-axis. Although your experiment will have the stress as the independent variable, it is usual to draw the curve of these results with stress on the y-axis. Plot this graph and find the Young modulus of the metal you used, by finding the gradient of the straight-line portion of the graph.

Safety Note: Use eye protection to prevent injury if the wire snaps. Use a 'drop box' on the landing pad to keep hands and feet out of the 'drop zone'.

EXAM HINT

You need to be able to describe all the steps in an experimental method for finding the Young modulus, including safety points, and how to use the measurements to calculate the Young modulus.

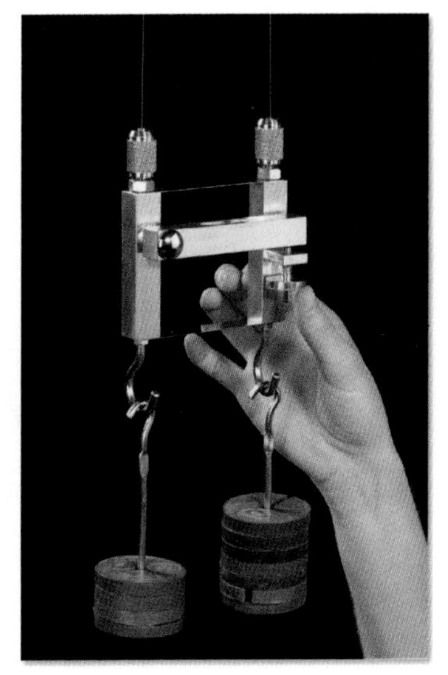

▲ **fig D** Searle's apparatus.

CHECKPOINT

SKILLS PROBLEM SOLVING, CREATIVITY

1. Estimate the limit of proportionality for the IMPRESS alloy at 750 °C, as shown in **fig B**.
2. Searle's apparatus, as shown in **fig D** is is used in experiments to find the Young modulus for a metal wire. With this apparatus, the test wire hangs vertically, parallel to an identical control wire. Weights are added to load the test wire only, and its extension, measured as an excess over the length of the control wire rather than from its original length, is calculated. Explain how this setup will allow the experimenter to avoid possible error caused by variations in the room temperature.

SUBJECT VOCABULARY

yield point a strain value beyond which a material undergoes a sudden and large plastic deformation

2B THINKING BIGGER

GET ROPED IN

SKILLS CRITICAL THINKING, PROBLEM SOLVING, ANALYSIS, REASONING/ARGUMENTATION, INTERPRETATION, ADAPTIVE LEARNING, INITIATIVE, PRODUCTIVITY, COMMUNICATION

If a manufacturer of climbing ropes is to choose a material for a new rope, what properties would they want the material to have? Apart from low cost, they will need to consider whether it will safely hold climbers, especially when they fall and the rope has to save them.

In this activity, we will look at some important aspects of the properties of some climbing ropes.

VARIOUS SOURCES

PRODUCT	COLOUR	DIAMETER	LENGTHS (m)	WEIGHT (per m)	NO OF FALLS (UIAA)	IMPACT FORCE
PITCH	NAVY BLUE	8.5 mm	50, 60	49 g	13	6.5 kN
PITCH	ORANGE	8.5 mm	50, 60	49 g	13	6.5 kN
ZONE	SUNSET RED	9.8 mm	50, 60, 70, 80	63 g	6	8.7 kN
ORBIT	COBALT BLUE	9.6 mm	50, 60, 70, 80	61 g	8	8.7 kN
COULOIR	RED	8.0 mm	50, 60	43 g	7	6.6 kN
COULOIR	BLUE	8.0 mm	50, 60	43 g	7	6.6 kN

From the website of rope manufacturer DMM International, http://dmmclimbing.com/products/ropes-&-cord/

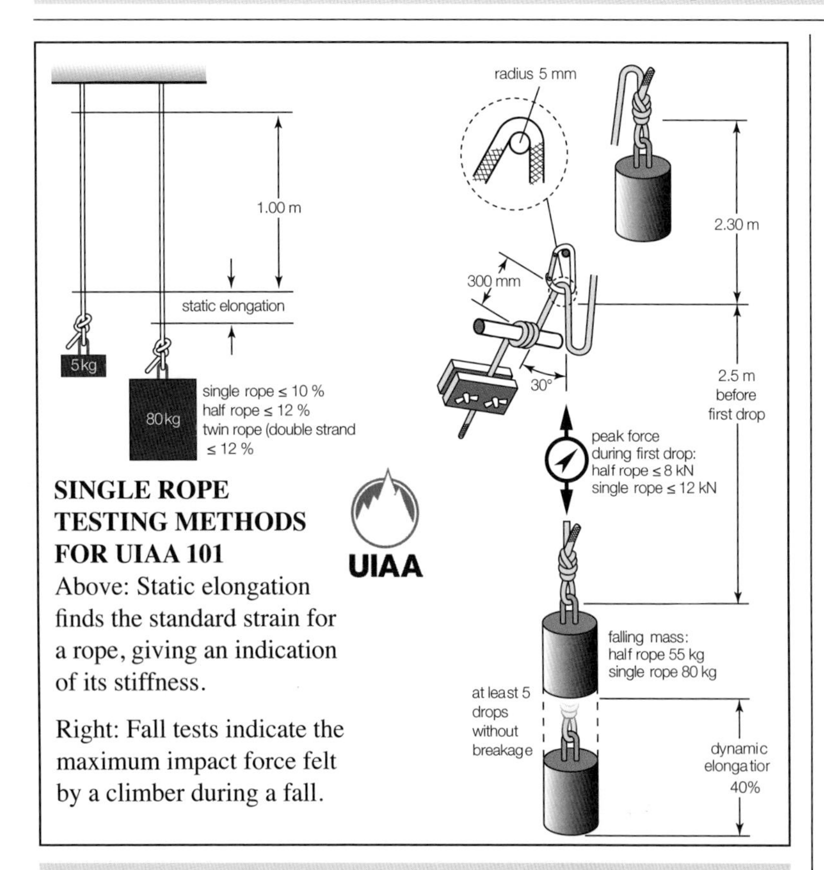

SINGLE ROPE TESTING METHODS FOR UIAA 101

Above: Static elongation finds the standard strain for a rope, giving an indication of its stiffness.

Right: Fall tests indicate the maximum impact force felt by a climber during a fall.

From the Safety Standards pages of UIAA, the International Climbing and Mountaineering Federation

ROPE ANCHORING

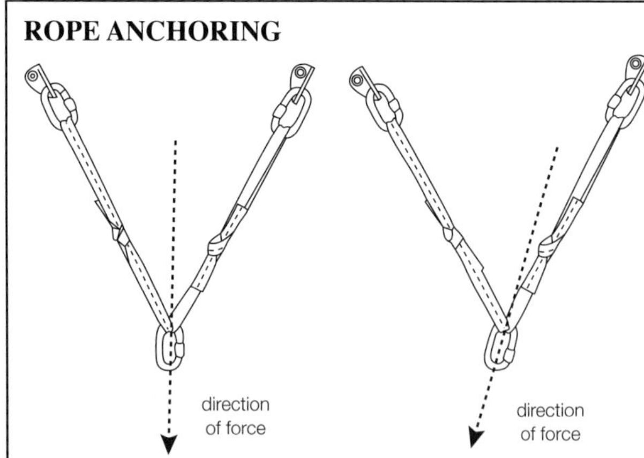

Fig 7–19 a, line representing direction of force **bisects** angle, thus load on the two anchors is equal; **b**, direction of force is to one side, thus load on right-hand anchor is greater than on left-hand anchor.

Angle (in degrees)	Force on each anchor
0	50%
60	60%
90	70%
120	100%
150	190%
170	580%

From the 6th edition of *Mountaineering: The Freedom of the Hills*, a climbing book

SCIENCE COMMUNICATION

1 The extracts opposite are from three different sources relating to the use of ropes for rock climbing.

(a) Compare and contrast the style of presentation from each source.

(b) Discuss the safety considerations that the authors (and their insurers) will have had to think about before publishing the information in each case.

PHYSICS IN DETAIL

Now we will look at the physics in detail. Some of these questions link to topics earlier in this book, so you may need to combine concepts from different areas of physics to work out the answers.

2 (a) Calculate the weight of a 50 m length of DMM's ***Couloir*** rope.

(b) Calculate the density of DMM's ***Orbit*** rope.

3 (a) Explain the physics behind the caption in **fig 7-19** from the ***Mountaineering*** book.

(b) The table of angles from the ***Mountaineering*** book refers to the total angle between the slings (short support ropes) in the symmetrical situation shown in **fig 7-19a**. Explain why the force on each **anchor** would increase as suggested, and calculate how accurately the force percentages have been reported.

4 (a) Looking at the fall tests, explain why lower impact force ropes will stretch more.

(b) Considering part (a), explain why a larger percentage dynamic elongation, compared with the static elongation, minimises injuries from the rope stopping a fall.

5 A recent update of the UIAA rope testing standards includes a method for calculating the energy absorbed by a rope before failure. This is given by the equation:

$$E_{rupt} = \int_{t_{tens}}^{t_{rupt}} F_{(S)} S dS$$

This equation is an integration which means it calculates the sum of all the $F \times S$ values over the range of tensions up to rupture. How could such calculations produce results for ropes as presented in the suggested picture below?

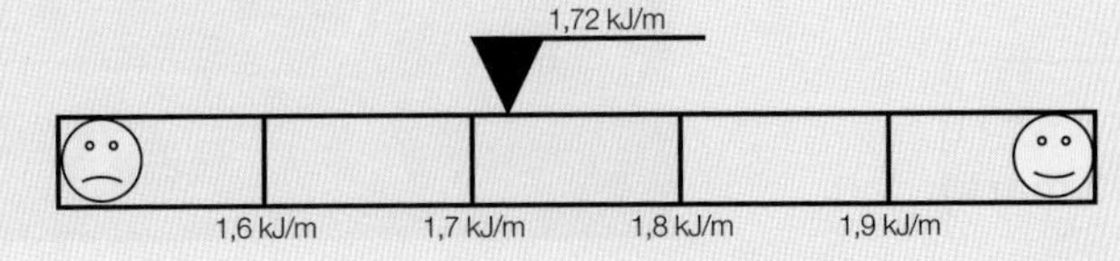

THINKING BIGGER TIP

Include an explanation of what physics term should be used instead of the word 'elongation'. Note that elongation is quoted as a percentage.

ACTIVITY

Write a review comparing the DMM ropes listed in terms of physics properties. Include at least:

- calculations of density, and a commentary about which ropes would float in water.
- calculations of stress in each rope under static elongation, when it must support an 80 kg mass, and assuming the maximum 8% elongation.
- comparison of the Young modulus values for each rope, and compare these with their impact forces transferred in the UIAA fall tests. Is there a relationship?

SUBJECT VOCABULARY

bisects to divide something into two, usually equal, parts
anchor an object, usually on a rope or chain, designed to give support

2B EXAM PRACTICE

1 The pascal is the unit for the Young modulus. Which unit is equivalent?

A mm/m

B %

C $N\,m^{-1}$

D $N\,m^{-2}$ [1]

(Total for Question 1 = 1 mark)

2 A 200 g mass hung on a spring stretches it from 8.2 cm to 9.0 cm. What is the spring constant for this spring?

A $2.5\,N\,m^{-1}$

B $25\,N\,m^{-1}$

C $250\,N\,m^{-1}$

D $250\,000\,N\,m^{-1}$ [1]

(Total for Question 2 = 1 mark)

3 The diagram shows the stress–strain curve for a metal. Which label corresponds to the Ultimate Tensile Stress for this metal?

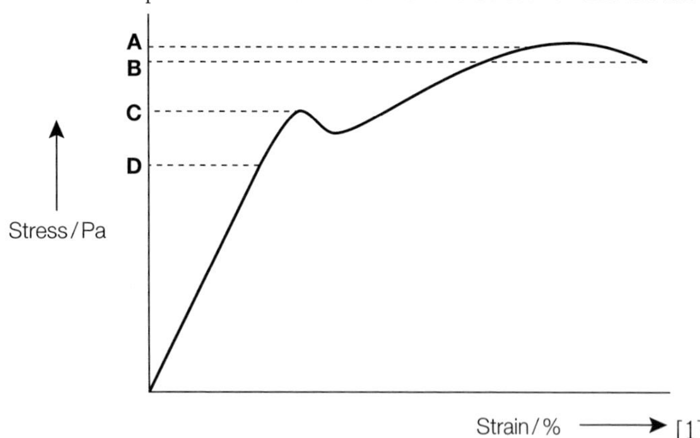

[1]

(Total for Question 3 = 1 mark)

4 Which is the correct definition for the 'elastic limit' for a metal wire?

A The maximum extension that the wire can have, which is still proportional to the load applied.

B The highest value that the stress can ever reach within this wire.

C The value that the stress will be in the wire when it breaks.

D The maximum extension or compression that the wire can undergo and still return to its original dimensions when the force is removed. [1]

(Total for Question 4 = 1 mark)

5 When a spring is extended by Δx, elastic strain energy, E_{el}, is stored in the spring. If the extension changes, then the elastic strain energy also changes.

Which row in the table correctly gives the new elastic strain energy for the change in extension given?

	New extension	New elastic strain energy
A	$2\Delta x$	$2E_{el}$
B	$\frac{\Delta x}{2}$	$2E_{el}$
C	$2\Delta x$	$4E_{el}$
D	$\frac{\Delta x}{2}$	$4E_{el}$

[1]

(Total for Question 5 = 1 mark)

6 You are asked to find the Young modulus for a metal using a sample of wire.

(a) Describe the apparatus you would use, the measurements you would take and explain how you would use them to determine the Young modulus for the metal. [8]

(b) State **one** safety precaution you would take. [1]

(c) Explain **one** experimental precaution you would take to ensure you obtain accurate results. [2]

(Total for Question 6 = 11 marks)

7 The photograph shows a tin bought from a joke shop. When the lid is removed, a long spring, covered in fabric to resemble a snake, flies out of the tin.

The spring on its own is shown here.

The graph shows length against force for the spring.

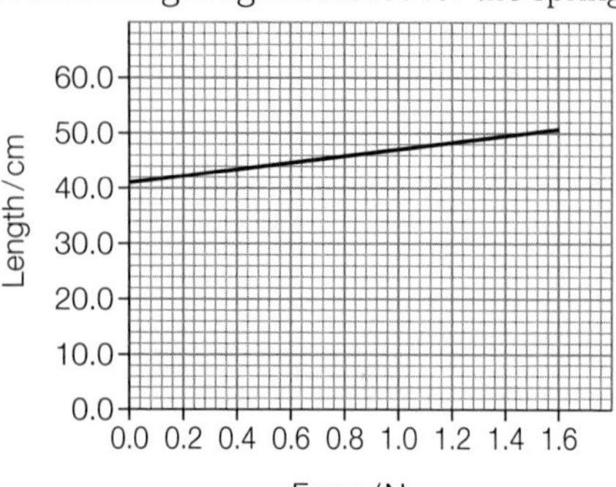

(a) Explain whether the spring obeys Hooke's law. [2]

(b) Show that the spring constant k of the spring is about $20\,\mathrm{N\,m^{-1}}$. [3]

(c) The original length of the spring is 41.0 cm and the length of the tin is 9.0 cm.

(i) Calculate the force that must be applied to the spring to get it into the tin. [2]

(ii) Calculate the energy stored in the spring when it is compressed to fit into the tin. [2]

(d) In fact, the bottom of the tin contains a device that makes a squeak when the spring is released, making the internal length of the tin less than 9.0 cm.
Explain the effect this has on the speed at which the spring leaves the tin. [3]

(Total for Question 7 = 12 marks)

8 The diagram shows a submarine and one of the forces acting on it. The submarine moves at a constant depth and speed in the direction shown.

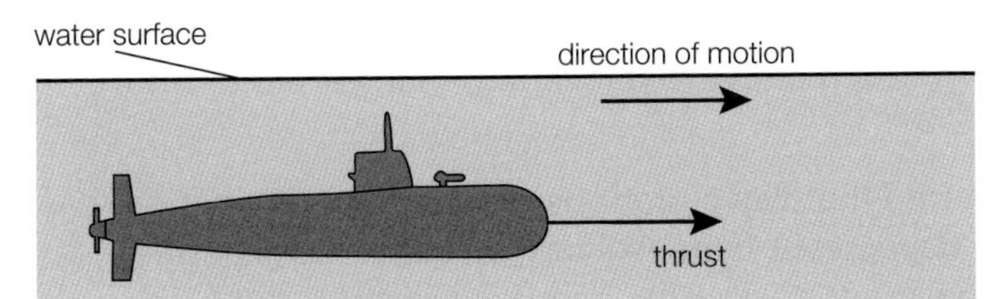

(a) State two equations that show the relationship between the forces acting on the submarine. [2]

(b) The submarine has a volume of $7100\,\mathrm{m^3}$.

Show that the weight of the submarine is about $7 \times 10^7\,\mathrm{N}$.

Density of seawater = $1030\,\mathrm{kg\,m^{-3}}$ [2]

(c) The submarine can control its depth by changing its weight. This is done by adjusting the amount of water held in ballast tanks.
As the submarine dives to greater depths the increased pressure of the surrounding water produces a compressive strain.

(i) Explain what is meant by compressive strain. [1]

(ii) This decreases the volume of the submarine. Explain the action that should be taken to maintain a constant depth as the volume of the submarine is decreased. [2]

(iii) The submarine is made from steel. Suggest why a material, such as fibreglass, which has a much smaller Young modulus than steel would be unsuitable at greater depths. [2]

(Total for Question 8 = 9 marks)

9 The photograph shows a device for swatting flies.

The device consists of a handle, a spring and a disc as shown in the photograph below.

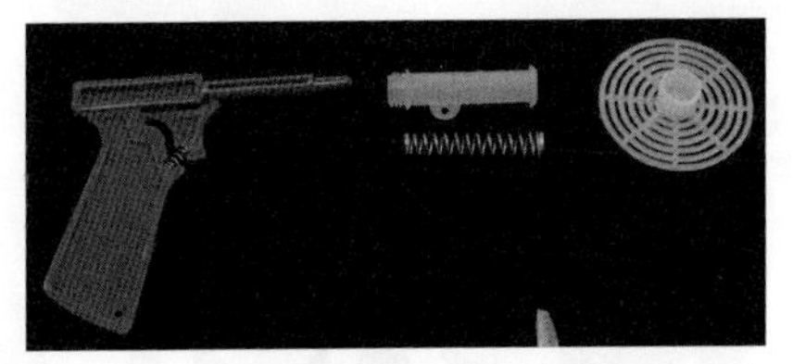

When the button is pushed, the compressed spring is released, launching the disc at the fly.

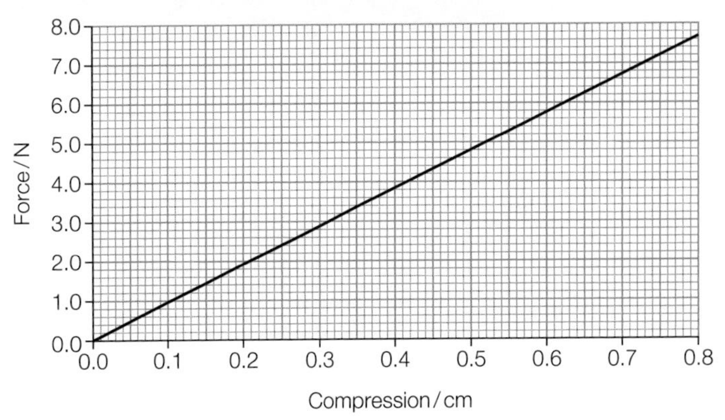

(a) Show that the force constant for the spring is about $1000\,\mathrm{N\,m^{-1}}$. [2]

(b) The spring is 6.3 cm long. When it is compressed in the device, the length of the spring is reduced to 1.6 cm.

Assuming that the spring obeys Hooke's law throughout the compression, show that the energy stored in the spring before releasing the disc is about 1 J. [2]

(c) The disc and spring have a combined mass of 9.4 g.

(i) Show that the maximum speed at which they can be launched is about $15\,\mathrm{m\,s^{-1}}$. [2]

(ii) State an assumption that you have made. [1]

(d) The disc is launched horizontally at a fly on the wall 3.0 m away.

(i) Calculate the velocity of the disc as it hits the wall. Ignore the effects of air resistance. [4]

(ii) The fly is 20 cm below the horizontal level at which the disc is launched. Show that the disc is close enough to hit the fly if it does not move. The disc has a radius of 3 cm. [3]

(e) Suggest an advantage of the disc used over a solid disc. [1]

(Total for Question 9 = 15 marks)

TOPIC 3 WAVES AND THE PARTICLE NATURE OF LIGHT

CHAPTER 3A BASIC WAVES

As waves are a means to transfer energy, they are very important in the world around us. From the destruction caused by earthquakes, to the determination of the chemical composition of distant stars, to whether or not it's worth taking your surfboard to the beach at the weekend, we study waves in virtually every part of human experience.

Topic 3 covers many aspects of the science behind waves, and in this chapter we will begin this study with some basic definitions and the wave equation. As waves and wave mathematics are relevant to many areas of future study, it is important to understand these basic definitions.

Even when scientists consider new areas of study, they often find that wave motions are the basis of what they are trying to understand. The fundamental causes of gravity have never been well understood. For example, experimental observations of gravitational waves in order to expand that understanding are currently being done by a number of universities around the world.

MATHS SKILLS FOR THIS CHAPTER

- **Units of measurement** (*e.g. the hertz, Hz*)
- **Use of standard form and ordinary form** (*e.g. using the speed of light*)
- **Changing the subject of an equation** (*e.g. using the wave equation*)
- **Solving algebraic equations** (*e.g. comparing the two equations for wave speed*)
- **Substituting numerical values into algebraic equations** (*e.g. a radar calculation to find the distance to an aeroplane*)

What prior knowledge do I need?

- Using the terms frequency, wavelength, amplitude and speed to describe waves
- Detailed knowledge of the parts of the electromagnetic spectrum – their uses, dangers and properties
- Knowledge of sound waves and light waves
- Models of waves and wave processes
- Calculations of wave speed
- Examples of transverse and longitudinal waves

What will I study in this chapter?

- Definitions of the properties of waves, such as frequency, wavelength, amplitude and speed
- The definition of wave phase, and its relationship to frequency and time period
- The difference between transverse and longitudinal waves
- The transmission of waves, and graphical representations of them
- The wave equation
- The use of wave calculations, in pulse-echo techniques such as ultrasound scanning, and in seismology

What will I study later?

Topic 3B

- How waves can combine when they meet
- The development of standing waves, and what affects their properties
- Applications of standing wave patterns, such as the interference or diffraction of light, and the development of musical instruments

Topic 3C

- Effects on the properties and movement of a wave due to changes in its medium

Topic 3D

- The wave–particle duality of light and electrons

Topic 10A (Book 2: IAL)

- Connections between simple harmonic motion and wave motion

Topic 11B (Book 2: IAL)

- The implications of the Doppler effect in astronomy

1 WAVE BASICS

SPECIFICATION REFERENCE 2.3.33 2.3.34 2.3.38 CP4

CP4 LAB BOOK PAGE 18

LEARNING OBJECTIVES

- Understand the terms amplitude, frequency, period and wavelength.
- Define wave speeds, measure the speed of sound in air, and use the wave equation $v = f\lambda$
- Explain how pulse-echo techniques allow the measurement of distances.

ENERGY TRANSFER

A **wave** is a means for transferring energy via oscillations. Whilst energy moves from one place to another, the waves cause no net movement of any matter.

A mechanical wave is one in which there needs to be some sort of material medium – a substance that oscillates to allow the transfer of the energy. For example, a sound wave of a human voice transfers energy from one person's vocal cords to another's eardrum by the repeated vibrations of air molecules.

Electromagnetic waves can transfer energy through repeated oscillations of electric and magnetic fields, but these fields do not need matter to support them. Indeed, the interaction between electromagnetic waves and matter generally slows their transfer of energy. For example, light travels more slowly in water than it does in a vacuum.

GRAPHING WAVES

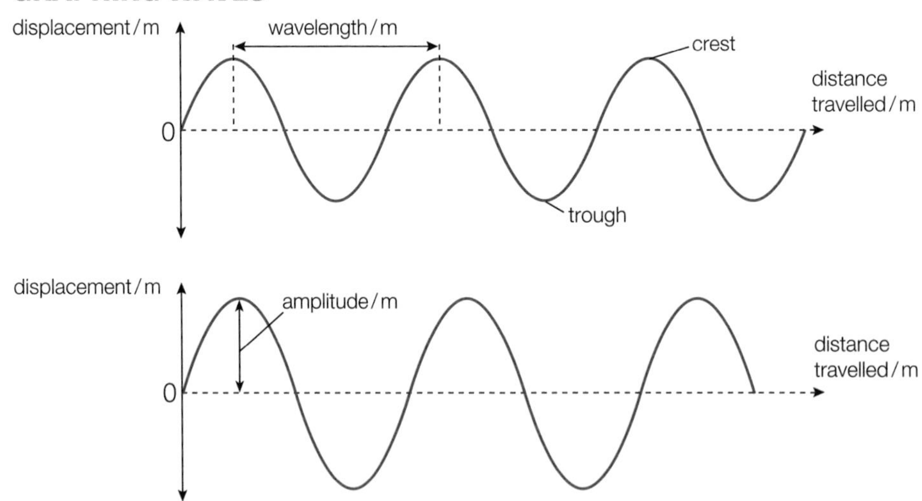

▲ **fig A** The vibration of a wave over a certain distance, as if frozen at an instant in time.

Wave motions can be plotted on graphs. A plot of the displacement against distance travelled for a wave, as in **fig A**, shows the physical scale of the oscillations and the movement of the energy.

Alternatively, a plot of the displacement versus time, as in **fig B**, shows how the vibrations occur over time.

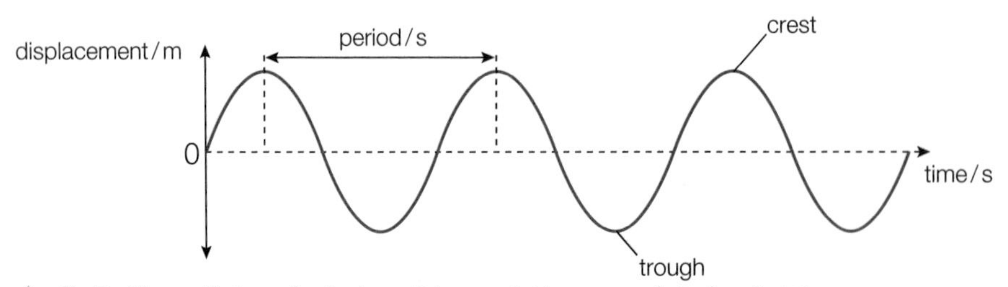

▲ **fig B** The oscillation of a single particle or point in a wave plotted against time.

WAVE MEASUREMENTS

There are a number of important properties of waves that scientists measure for various purposes. In most cases, you can determine these from the graphs shown in **fig A** and **fig B**.

Displacement: the position of a particular point on a wave, at a particular instant in time, measured from the mean (equilibrium) position. (Symbol: various, often x; SI units: m.)

Amplitude: the magnitude of the maximum displacement reached by an oscillation in the wave. (Symbol: A; SI units: m.)

Frequency: the number of complete wave cycles per second. This may sometimes be measured as the number of complete waves passing a point per second. (Symbol: f; SI units: hertz, Hz.)

Wavelength: the distance between a point on a wave and the same point on the next cycle of the wave, for example, the distance between adjacent wave peaks. (Symbol: λ; SI units: m.)

Period: the time taken for one complete oscillation at one point on the wave. This will also be the time taken for the wave to travel one wavelength. (Symbol: T; SI units: s.)

Phase: the stage a given point on a wave is through a complete cycle. Phase is measured in angle units, as a complete wave cycle is considered to be the same as travelling around a complete circle, that is 360° or 2π radians. (No standard symbol; SI units: rad.)

Wave speed: the rate of movement of the wave – the same as speed in general. (Symbol: v, or c for speed of electromagnetic waves; SI units: m s^{-1}.)

You can find the speed of a wave from distance divided by time, as with the speed of anything. However, as each wave has a certain wavelength, and the frequency tells us how many of these wavelengths pass per second, you can also find the speed by multiplying frequency and wavelength together. This is often referred to as the **wave equation**:

$$\text{wave speed (m s}^{-1}\text{)} = \text{frequency (Hz)} \times \text{wavelength (m)}$$

$$v = f\lambda$$

PRACTICAL SKILLS CP4

Investigating the speed of sound

Using the simple definition of wave speed that is distance divided by time, we can use a **twin beam oscilloscope** to find the extra time a sound takes to travel a short extra distance. One beam trace shows a sound picked up by a microphone held 50 cm from the loudspeaker. The other trace shows the same sound, picked up by a second microphone held further from the loudspeaker. The difference in positions of the peaks on the two oscilloscope traces shows the time taken, t, for the sound to travel the extra distance, d. If we measure this carefully, then the speed will be given by:

$$v = \frac{d}{t}$$

It can be difficult to make accurate measurements from the screen of an oscilloscope, so we need to synchronise the traces to minimise the effect of random error in taking such a measurement. Firstly, with both microphones at the same distance from the loudspeaker, the two traces appear in identical phase positions (*in phase*). If we slide the second microphone slowly away from the loudspeaker, we will move the traces out of phase with each other until eventually they come back to exact synchronisation. At this point, the distance between the two microphones is exactly one wavelength, λ. We set the frequency on the signal generator, and so the wave equation can be used to find the speed: $v = f\lambda$.

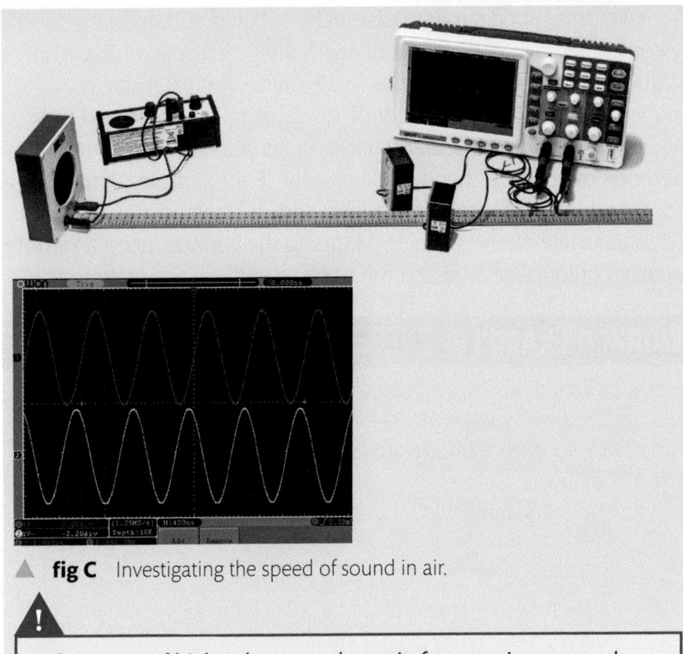

▲ **fig C** Investigating the speed of sound in air.

Safety Note: If high volumes and certain frequencies are used, some may find them distressing or experience mild nausea. Anyone using a hearing aid may experience unpleasant interactions with the noise from the loudspeakers.

PULSE-ECHO MEASUREMENTS

The natural habitat for some bats is woodland, and they fly through the trees at 10–15 km h^{-1}, depending on species. In order to catch insects, the bats need to be able to sense their location precisely. Their well-known echolocation system, using very high frequency (50–100 kHz – ultrasound) sound pulses, gives a detailed perception of the world at distances of less than 5 metres. At greater distances, the echo is too quiet for the bat to use. But how does it work?

The bat will make a 'chirp' through its nose. This sound pulse will typically last 3 milliseconds. When the sound hits nearby objects, it will be reflected back to the bat's sensitive ears and its brain can accurately measure the time between making the sound and hearing the echo. The bat's brain has also evolved to calculate the distance to the reflecting object using the equation:

distance = speed × time.

Experiments with dolphins have shown that they use a similar echolocation system, but for them it is so good that dolphins can build up an image of the shapes of nearby objects. They can 'see' with sound. This is probably also true for bats, to ensure that they eat an insect and not an insect-sized leaf, but dolphins can respond to experimental scientists better than bats.

▲ **fig D** Dolphin echolocation is good enough to make out the shapes of objects in murky waters.

We have developed similar pulse-echo ranging and imaging systems in a very wide range of technological applications, from sonar on ships and submarines to air traffic control radar, medical imaging and the measurement of distance to asteroids and to the Moon. This is particularly useful in situations where other methods might be difficult or dangerous. For example, Venus has a very hot, high pressure, atmosphere with sulfuric acid clouds that block our view of the surface. Mapping the surface using a remote radar technique has avoided the need to land on the planet.

WORKED EXAMPLE

An air traffic control system sends out a pulse of radio waves that are reflected by a jumbo jet. The reflection is picked up by the radio dish 0.007 seconds after the emission of the pulse. How far away is the plane?

$$\text{speed} = \frac{\text{distance}}{\text{time}}$$

distance = speed × time

$s = v \times t \qquad v = \text{speed of light} = 3.0 \times 10^8\ \text{m s}^{-1}$

$s = 3.0 \times 10^8 \times 0.007 = 2.1 \times 10^6\ \text{m}$

$s = 2100\ \text{km}$

In that 0.007 seconds, the radio pulse has travelled to the plane and back again, so the actual distance to the plane is half that calculated:

distance = 0.5 × 2100 = 1050 km

CHECKPOINT

SKILLS PROBLEM SOLVING, CRITICAL THINKING

1. What are the amplitudes of the three waves shown in **fig E**?

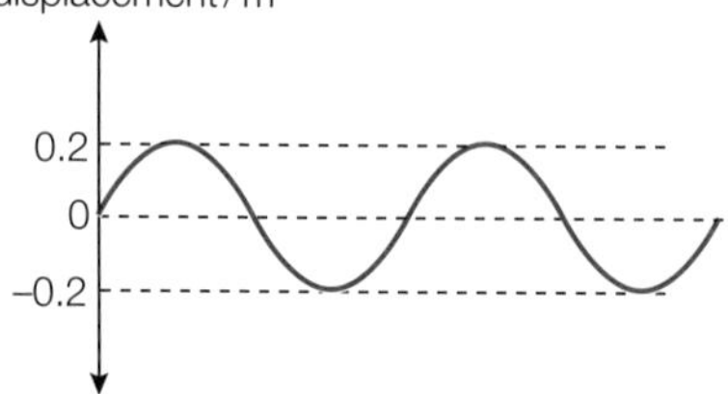

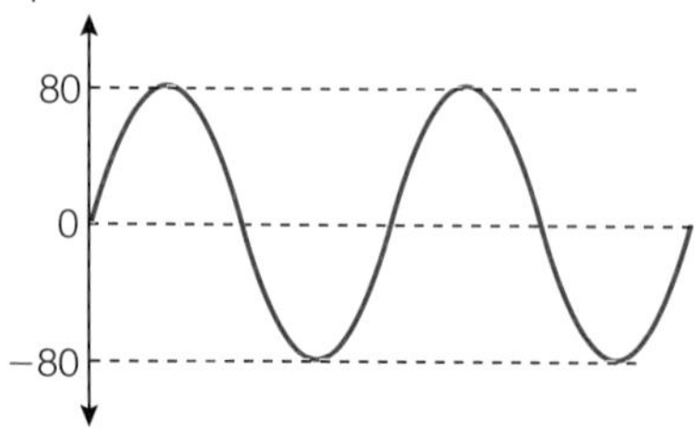

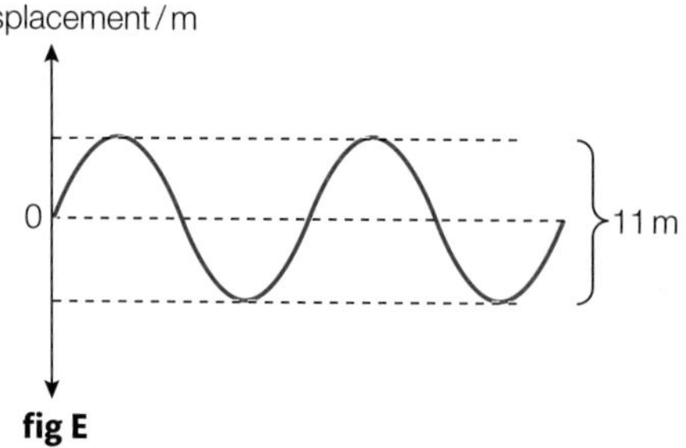

▲ **fig E**

2. Sound travels at 330 m s^{-1} in air. If the sound of thunder arrives 3.75 s after the lightning flash is seen, how far away was the lightning strike?
3. Ultraviolet light has a wavelength of 368 nm. What is the frequency of this light?
4. Explain why the two equations for wave speed are actually the same calculation.
5. Estimate the speed of the waves the last time you saw the sea.

EXAM HINT

If you are asked to calculate wave speed, make sure the equation you use matches the information you have: frequency and wavelength are multiplied in the wave equation BUT distance is divided by time if distance and time are the measurements you are given.

SUBJECT VOCABULARY

wave a means for transferring energy via oscillations
displacement the position of a particular point on a wave, at a particular instant in time, measured from the mean (equilibrium) position
amplitude the magnitude of the maximum displacement reached by an oscillation in the wave
frequency the number of complete wave cycles per second:

$$\text{frequency (Hz)} = \frac{1}{\text{time period (s)}}$$

$$f = \frac{1}{T}$$

wavelength the distance between a point on a wave and the same point on the next cycle of the wave
period (also **time period**) the time taken for one complete oscillation at one point on the wave:

$$\text{time period (s)} = \frac{1}{\text{frequency (Hz)}}$$

$$T = \frac{1}{f}$$

phase the stage a given point on a wave is through a complete cycle, measured in angle units, rad
wave speed the rate of movement of the wave (not the rate of movement within oscillations)
wave equation:

$$\text{wave speed (m s}^{-1}) = \text{frequency (Hz)} \times \text{wavelength (m)}$$

$$v = f\lambda$$

twin-beam oscilloscope an oscilloscope with two inputs. It displays each as a line on the screen, and both are shown at the same time to compare the inputs

3A 2 WAVE TYPES

SPECIFICATION REFERENCE 2.3.35 2.3.36 2.3.37

LEARNING OBJECTIVES

- Explain the difference between longitudinal and transverse waves.
- Describe longitudinal waves, particularly in terms of pressure variation and the displacement of molecules.
- Describe transverse waves.

TRANSVERSE WAVES

A **transverse wave** is one where the movements of the particles, or fields in an electromagnetic wave, are up and down, or left and right, whilst the energy travels forwards. This is illustrated in **fig A**, where one student is vibrating the rope up and down, but the energy travels along the rope towards the other student. As the first student moves the rope up and down, the particles pull their neighbours up and down. These then pass the vibration on to their neighbours, through their intermolecular forces, and the wave moves along the rope.

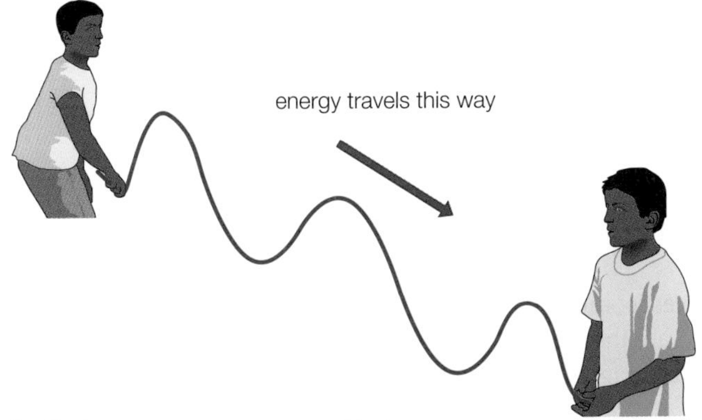

fig A A transverse wave on a skipping rope.

LONGITUDINAL WAVES

A **longitudinal wave** in a fluid, such as air, is generated by squashing particles together and then stretching them apart from each other, repeatedly – thus vibrating them 'longitudinally'. The areas of higher pressure cause the particles to push apart from each other, but this makes the particles move and squash their neighbours. This higher pressure – a **compression** – then pushes them away to squash their neighbours. Similarly, the areas where there are too few particles (compared with the uniform spread of the particles when the wave is not present) cause particles to move into the vacant space – this is known as the **rarefaction** – filling the vacant space up, but causing a vacancy behind these particles, and the wave moves along. An example of a longitudinal wave in a solid material can be seen by squashing and releasing a spring, as shown in **fig B**.

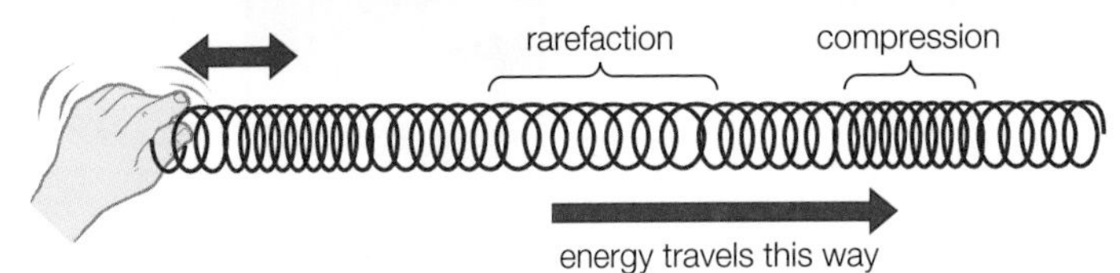

fig B A longitudinal wave generated in a long spring by repeatedly squashing and stretching one end. Areas of higher pressure are called compressions, and areas of lower pressure are called rarefactions.

GRAPHING WAVES

In **fig A** of **Section 3A.1** we saw how waves can be shown on a graph of displacement versus distance. This is relatively easy with a transverse wave, as the graph appears like a picture of the wave, so it is easy to understand. The wavelength can be found by measuring along the graph's x-axis from one point on a wave cycle to the same point on the next wave cycle. Amplitude is measured from the x-axis vertically to a maximum displacement point.

Longitudinal waves are less easy to visualise from their graph, and making measurements from them can be difficult. **Fig C** illustrates how a longitudinal wave can be represented on a graph of displacement versus distance along the wave. When a longitudinal wave is drawn like this, you can make measurements more easily. The wavelength and amplitude can be read from the graph just like for a transverse wave.

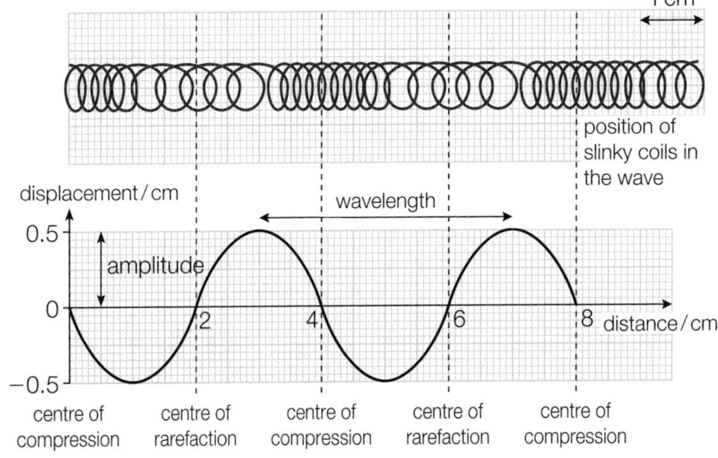

fig C A longitudinal wave can be easier to measure if it is drawn on a displacement–distance graph.

LONGITUDINAL WAVES IN ACTION

Sound waves are caused by oscillations of particles of the medium (in **fig D** the medium is air), causing compressions and rarefactions along the line of movement of the wave. Areas of higher pressure and lower pressure in the air continue the movement of the vibrations. An area of higher pressure, a compression, will occur when the particles on either side are displaced towards it. This means particles are displaced in opposite directions towards each other to squash together and increase the pressure, as shown in **fig E**.

fig D A loudspeaker produces sound waves by moving its cone back and forth to set up vibrations of air molecules in line with the direction of movement of the sound wave energy – the oscilloscope shows the vibrations over time.

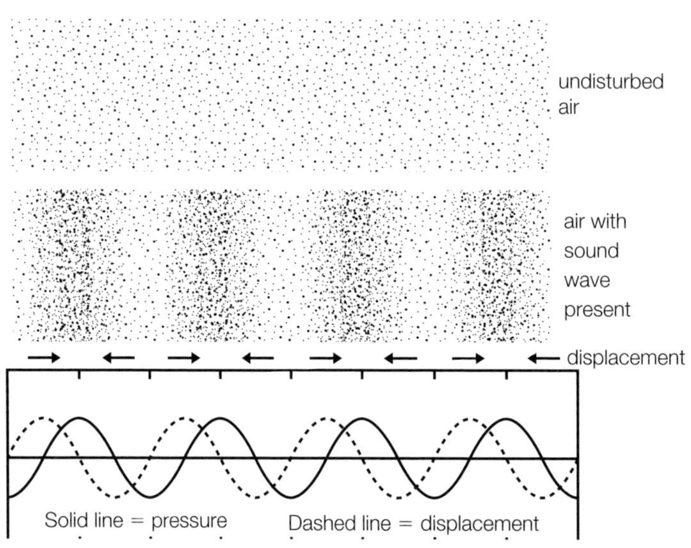

fig E Sound waves are the result of areas of increased pressure moving through a body of particles by vibrating the particles back and forth.

Earthquake waves, also called seismic waves, come in different types. **Fig F** shows Primary, or 'P', waves and Secondary, or 'S', waves. These are the standard longitudinal and transverse seismic waves, with P-waves travelling faster, and arriving first, leading to their name.

EXAM HINT

The details of seismic wave types are beyond the specification, but are good examples of longitudinal and transverse waves.

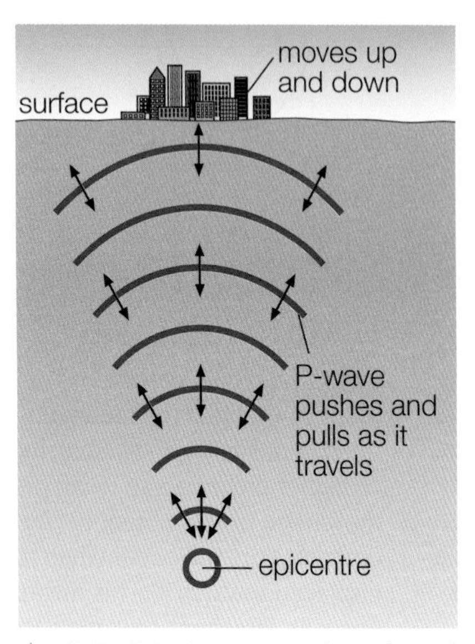

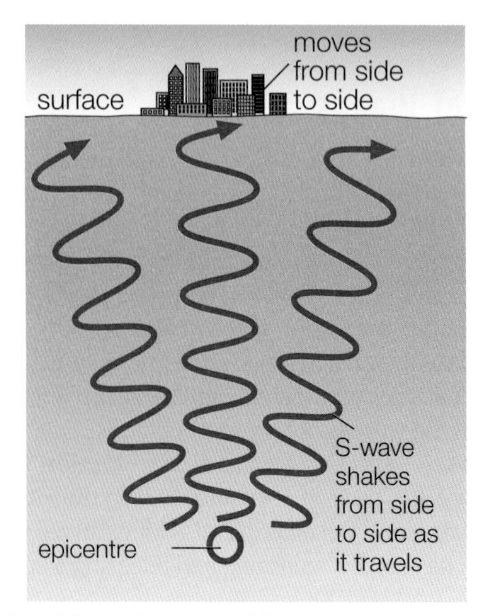

fig F Seismic waves can vibrate the rock particles of the Earth's crust in different ways.

CHECKPOINT

SKILLS ANALYSIS, ADAPTIVE LEARNING

1. What are the amplitudes and the wavelengths of the spring waves shown in **fig C**?
2. (a) Explain why a seismic S-wave should be classified as a transverse wave.
   (b) Describe a seismic P-wave in terms of the pressure variations and displacement of the rock particles.
3. Referring to **fig G**, explain why sounds with a greater amplitude sound louder.

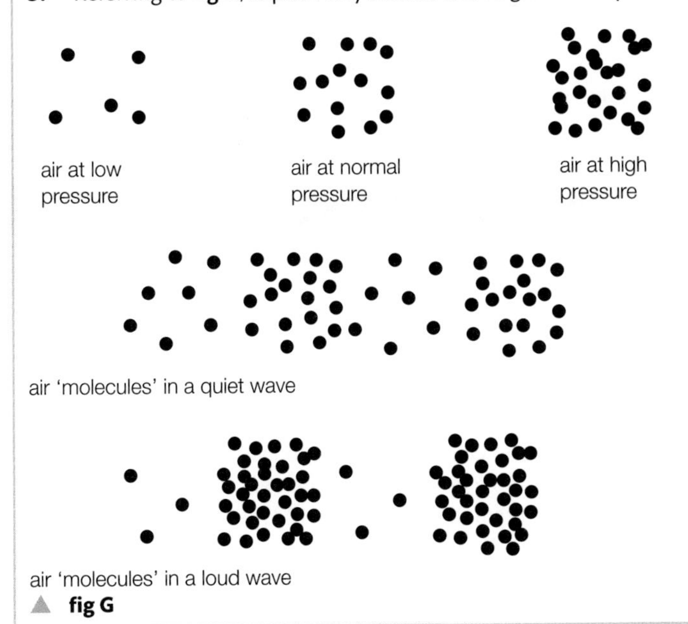

▲ **fig G**

EXAM HINT

When defining transverse and longitudinal waves, you must write both parts of the idea: oscillations direction *and* direction of wave travel.

SUBJECT VOCABULARY

transverse wave a wave in which the oscillations occur perpendicular to the direction of movement of the wave energy
longitudinal wave a wave in which the oscillations occur parallel to the direction of movement of the wave energy
compression an area in a longitudinal wave in which the particle oscillations put them closer to each other than their equilibrium state
rarefaction an area in a longitudinal wave in which the particle oscillations put them further apart from each other than their equilibrium state

3A THINKING BIGGER

EARTHQUAKE

SKILLS CRITICAL THINKING, PROBLEM SOLVING, ANALYSIS, INTERPRETATION, ADAPTIVE LEARNING, PRODUCTIVITY, COMMUNICATION

Earthquakes are among the most deadly natural hazards. There are around 100 earthquakes each year of a size that could cause serious damage.

In this activity, we will look at some of the detail about the detection of earthquake waves.

USGS WEBSITE

EARTHQUAKE FAQs

What is an earthquake and what causes them to happen?

An earthquake is caused by a sudden slip on a fault. The tectonic plates are always slowly moving, but they get stuck at their edges due to friction. When the stress on the edge overcomes the friction, there is an earthquake that releases energy in waves that travel through the Earth's crust and cause the shaking that we feel.

▲ **fig A**

What are P-waves? What are S-waves?

When an earthquake occurs, it releases energy in the form of seismic waves that radiate from the earthquake source in all directions. The different types of energy waves shake the ground in different ways and also travel through the earth at different velocities. The fastest wave, and therefore the first to arrive at a given location, is called the P-wave. The P-wave, or compressional wave, alternately compresses and expands material in the same direction it is travelling. The S-wave is slower than the P-wave and arrives next, shaking the ground up and down and back and forth perpendicular to the direction it is travelling. Surface waves follow the P- and S-waves.

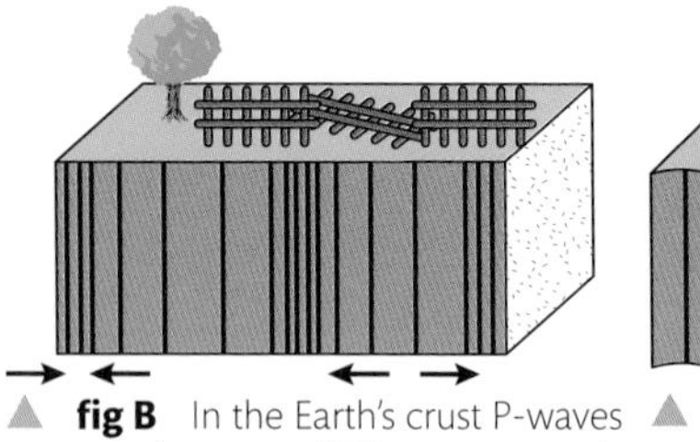

▲ **fig B** In the Earth's crust P-waves travel at around 7 km per second.

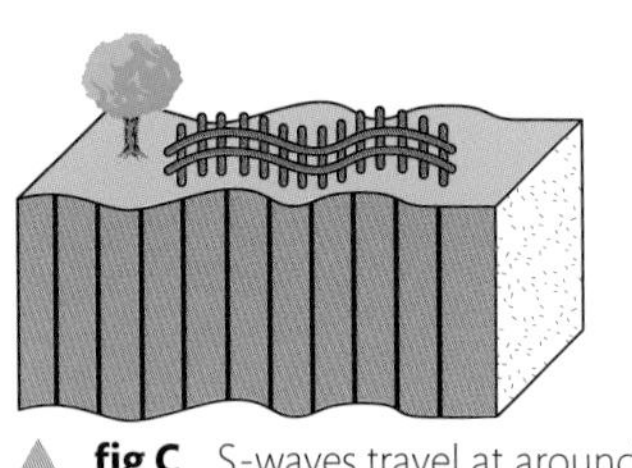

▲ **fig C** S-waves travel at around 4 km per second.

What are seismograms? How do you read them?

Seismograms are the records (paper copy) produced by seismographs used to calculate the location and magnitude of an earthquake. They show how the ground moves with the passage of time. On a seismogram, the HORIZONTAL axis = time (measured in seconds) and the VERTICAL axis = ground displacement (usually measured in millimetres). When there is NO EQ reading there is just a straight line except for small movements caused by local disturbance or 'noise' and the time markers.

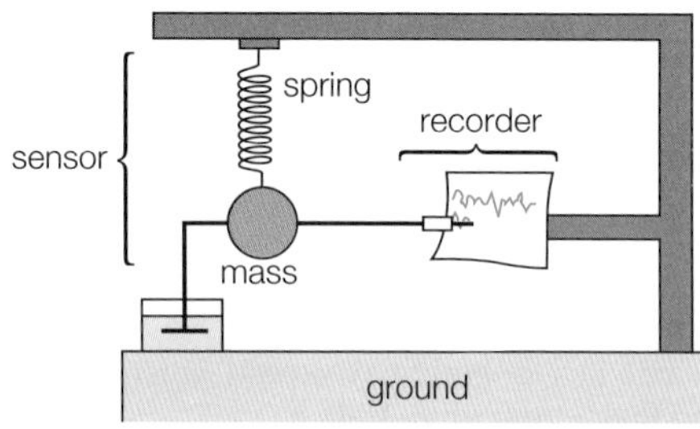

▲ **fig D** There are many different types of seismometers, but they all are based on a fundamental principle – that the differential motion between a free mass (which tends to remain at rest) and a supporting structure anchored in the ground (which moves with the vibrating Earth) can be used to record seismic waves.

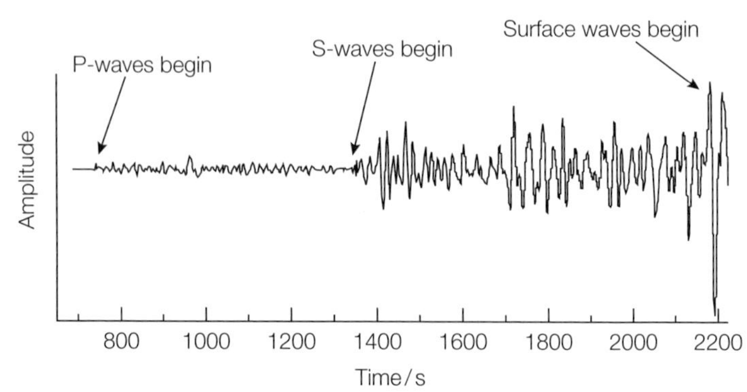

▲ **fig E** A seismogram recorded in Göttingen, Germany.

From the website of the U.S. Geological Survey, http://earthquake.usgs.gov

SCIENCE COMMUNICATION

1 The extract opposite is from educational webpages produced by the U.S. Geological Survey. The USGS is a United States government funded organisation that researches all geological matters. Consider the extract and comment on the type of writing that is used. Then answer the following questions:

(a) Why would the USGS spend time and money producing these and other webpages?

(b) Discuss the audience you think the USGS is aiming at. You need to consider the scientific depth in this extract, along with the style of presentation of the science.

INTERPRETATION NOTE

On this page, we only have an extract: it is useful to visit the webpages and see all the different sections that it covers.

PHYSICS IN DETAIL

Now we will look at the physics in detail. Some of these questions link to topics earlier in this book, so you may need to combine concepts from different areas of physics to work out the answers.

2 Explain which wave type (transverse or longitudinal) is shown by each of the earthquake waves.

3 Why can't S-waves travel through air or liquids, including the liquid iron outer core inside the Earth?

4 Why would the waves travel faster the deeper they are in the Earth?

(a) Explain how the information on the seismogram in **fig E** could be used to work out the distance to the earthquake. Calculate this distance.

(b) Then explain why this distance alone cannot determine where the earthquake occurred.

THINKING BIGGER TIP

Although you don't know when the earthquake occurred, think about how the difference in the times they arrive could help find the time the earthquake occurred, and then answer this question.

ACTIVITY

Write a similar webpage, including any ideas you have for diagrams. Explain how three seismograms of the same earthquake, produced by seismometers in different places on Earth, would allow a seismologist to find the original location of the earthquake.

THINKING BIGGER TIP

Consider your answers to question 4 – these should give you an idea as to how this is done.

3A EXAM PRACTICE

1 Which of the following gives the correct values of the amplitude and wavelength for the wave shown?

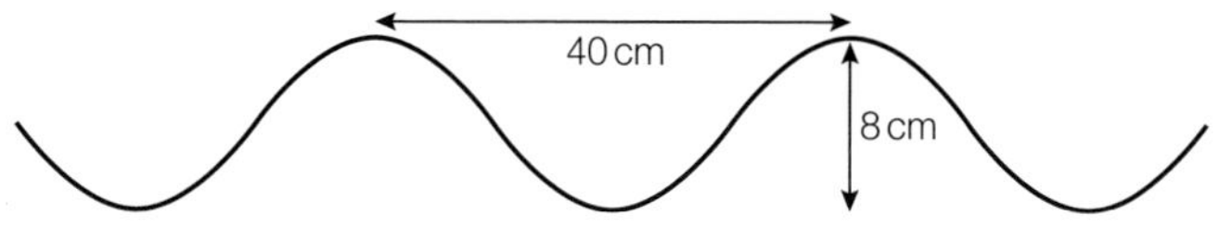

A $A = 4$ cm; $\lambda = 20$ cm

B $A = 4$ cm; $\lambda = 40$ cm

C $A = 8$ cm; $\lambda = 20$ cm

D $A = 8$ cm; $\lambda = 40$ cm [1]

(Total for Question 1 = 1 mark)

2 Which of the following has the lowest frequency?

A 256 Hz

B Ultrasound

C Red light ($\lambda = 700$ nm)

D Sea waves with a speed of 2.8 m s^{-1} and a wavelength of 4.4 m [1]

(Total for Question 2 = 1 mark)

3 Which of these is a correct equation connecting wave properties?

A $v = f\lambda$

B $v = \frac{f}{\lambda}$

C $\lambda = \frac{1}{f}$

D $f = \frac{\lambda}{v}$ [1]

(Total for Question 3 = 1 mark)

4 Sonar is a pulse-echo technique in which ships use sound pulses to measure the depth of the sea. In seawater, the speed of sound is 1500 m s^{-1}. If the echo of a pulse is detected 0.87 s after it is emitted, what is the depth of the sea at that point?

A 650 m

B 850 m

C 1300 m

D 1700 m [1]

(Total for Question 4 = 1 mark)

5 A sound wave pulse-echo technique can be used to find the thickness of old railway lines. A probe emitted a sound pulse into a steel rail and detected the echo. The pulse and its echo are shown on the oscilloscope trace below.

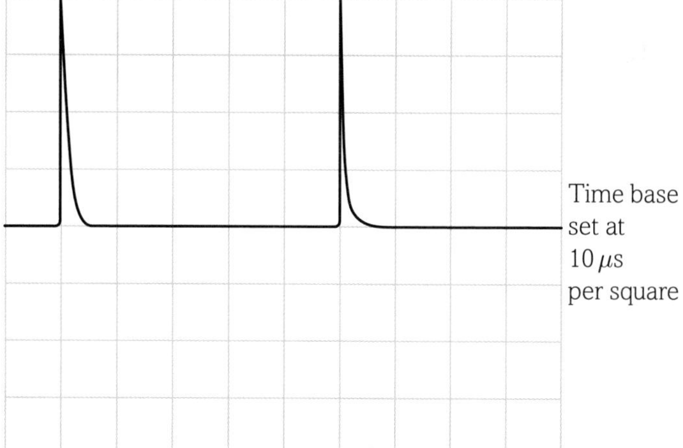

The sound waves used have a frequency of 3.8 MHz and the speed of sound in this steel is 5900 m s^{-1}.

(a) Explain why the echo has a smaller amplitude than the emitted pulse. [2]

(b) Calculate the thickness of this steel rail. [3]

(Total for Question 5 = 5 marks)

6 Frequencies below the audible range for humans are called infrasound. Infrasound is produced by earthquakes.

(a) Describe how sound waves travel through air. [3]

(b) State what is meant by frequency. [1]

(c) An infrasound wave has a wavelength of 1500 m and a frequency of 2.0 Hz.
Calculate the speed of infrasound in the ground. [2]

(d) In 2004, a huge earthquake produced a very large tidal wave which swept across the Indian Ocean towards Sri Lanka. Many large animals in Sri Lanka moved away from the coast before the tidal wave hit.
Suggest a reason for the animals behaving in this way. (2)

(Total for Question 5 = 8 marks)

7 A London radio station broadcasts at a frequency of 95.8 MHz. Calculate the wavelength in the air of these radio waves. [3]

(Total for Question 7 = 3 marks)

8 (a) Ultrasound has a frequency above the limit of human hearing. Ultrasound scanning can be used by doctors to obtain information about the internal structures of the human body without the need for surgery. Pulses of ultrasound are sent into the body from a transmitter placed on the skin.

(i) The ultrasound used has a frequency of 4.5 MHz. State why waves of this frequency are called ultrasound. [1]

(ii) A pulse of ultrasound enters the body and its reflection returns to the transmitter after a total time of 1.6×10^{-4} s.
Calculate how far the reflecting surface is below the skin.
Average speed of ultrasound in the body = $1500\,\text{m}\,\text{s}^{-1}$ [3]

(iii) State why the ultrasound is transmitted in pulses. [1]

(b) Another way of obtaining information about the internal structures of the human body is by the use of X-rays.

(i) Give one property of X-rays which makes them more hazardous to use than ultrasound. [1]

(ii) State two other differences between X-rays and ultrasound. [2]

(Total for Question 8 = 8 marks)

9 The diagram shows an experiment with sound waves.

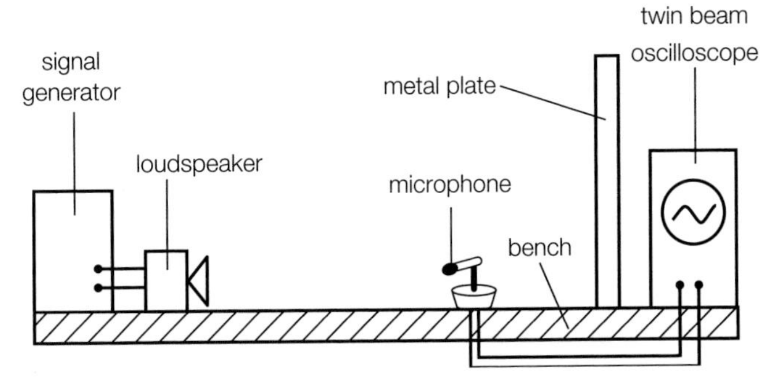

A loudspeaker is connected to a signal generator. A microphone is connected to a twin beam oscilloscope. Sound waves reach the microphone directly from the loudspeaker and after reflection from the metal plate.

A student set up the apparatus as shown in an attempt to measure the speed of sound by comparing the two traces on the oscilloscope screen. However, only one of the traces showed a sound wave whilst the other was a flat line.

(a) Describe and explain the difference in amplitude of the sound waves reaching the microphone directly, and those reaching it after reflection from the metal plate. [2]

(b) (i) Why does only one trace on the oscilloscope show the sound wave? [1]

(ii) Describe how the student should change the experimental set up in order to be able to measure the speed of sound. [1]

(iii) Explain how the student could use the new set up to find the speed of sound in air. Include measurements, experimental adjustments, how to analyse the results to find the speed of sound, and why they should also remove the metal plate. [6]

(c) With the frequency set at 3.7 kHz, the wavelength is measured as 9.0 cm. Calculate the speed of sound in air. [2]

(Total for Question 9 = 12 marks)

TOPIC 3 WAVES AND THE PARTICLE NATURE OF LIGHT

CHAPTER 3B THE BEHAVIOUR OF WAVES

Some of the most interesting and beautiful phenomena are the product of two or more waves combining their effects. It is surprising that waves can pass through each other, creating some spectacular effects, and then continue on their way as if they had never met.

We will look at the ways in which different waves combine, along with the reasons why they might not combine, and learn how to predict and measure the outcomes of these combinations. The production of stationary wave patterns can be unexpected, but it is the basis for the functioning of musical instruments. The variety and beauty of musical instruments, along with the variation in the sounds they produce, is a result of their need to generate a stationary wave by combining multiple wave movements together.

Diffraction and interference can be explained with some simple mathematical ideas. When these relationships are repeated over and over again, in different places, it can lead to striking sights, such as the rainbow reflected from an oily puddle. Also it leads to the limits of scientific knowledge, such as the discovery of gravitational waves.

MATHS SKILLS FOR THIS CHAPTER

- **Units of measurement** (*e.g. the radian, rad*)
- **Translating information between graphical, numerical and algebraic forms** (*e.g. superposing wave displacements to discover a standing wave*)
- **Understanding the relationship between degrees and radians and translating from one to the other** (*e.g. comparing the phase relationship between waves*)
- **Plotting two variables from experimental data, and understanding that $y = mx + c$ represents a linear relationship** (*e.g. using a graph of experimental results to verify the standing wave properties in a string under tension*)
- **Solving algebraic equations** (*e.g. comparing the positions of interference maxima and minima*)
- **Substituting numerical values into algebraic equations** (*e.g. finding the speed of waves in a string under tension*)

What prior knowledge do I need?

Topic 3A

- Definitions of the properties of waves, such as frequency, amplitude, wavelength and speed
- The definition of wave phase, and its relationship to frequency and time period
- The difference between, and examples of, transverse and longitudinal waves
- The transmission of waves, and graphical representations of them
- Calculations of wave properties

What will I study in this chapter?

- How waves can combine when they meet
- The development of standing waves, and what affects their properties
- Diffraction and interference
- Applications of standing wave patterns, such as the interference or diffraction of light, and the development of musical instruments

What will I study later?

Topic 3C

- Effects on the properties and movement of a wave due to changes in its medium

Topic 3D

- Interference and diffraction of fast-moving particles such as electrons

Topic 10A (Book 2: IAL)

- The connections between waves and simple harmonic motion

Topic 11B (Book 2: IAL)

- The use of diffraction in studying the spectra of stars

3B 1 WAVE PHASE AND SUPERPOSITION

SPECIFICATION REFERENCE 2.3.39

LEARNING OBJECTIVES

- Understand what is meant by the terms wavefront, superposition and phase.
- Explain examples of wave superposition.

PHASE

All points in a wave will be at some position through their cycle of oscillation. Some of these positions in the cycle have their own names, such as peak (or crest), trough, compression or rarefaction. The remaining points through a cycle are important, but there are too many to give them all names. We describe any position through a cycle by a number, referred to as the phase. A complete cycle is considered to be equivalent to rotating through a complete circle, so the phase position will be an angle measurement, where a complete cycle is equivalent to 360°.

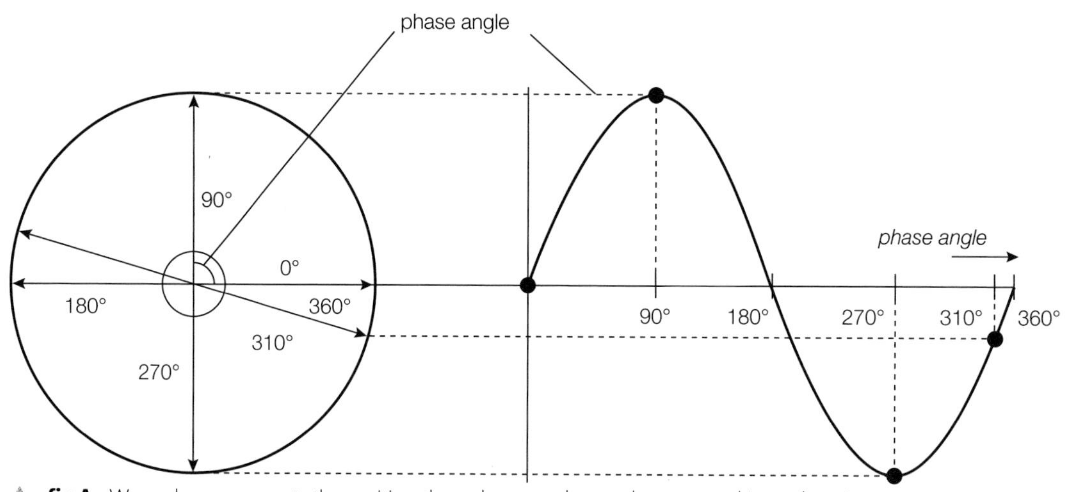

▲ **fig A** Wave phase represents the position through a complete cycle, measured in angle units.

It is quite common to measure phase in the alternative angle units of radians, where one complete cycle is 2π rad. **Table A** shows common phase positions in the two equivalent angle units, along with the fraction of a complete wavelength that they represent. As waves are, by definition, a repetitive cyclic process, phases of more than 360° could be reduced to their equivalent value within a first cycle. For example, the wave peak after 1.25 cycles has exactly the same phase as the wave peak in the first cycle, so its phase could be quoted as either 450° or 90° $\left(\frac{5\pi}{2} \text{ or } \frac{\pi}{2} \text{rad}\right)$.

EXAM HINT

If a question asks for a phase value, just give one answer. If you try to express the same answer in radians and in degrees, you must get it right both times or it is marked wrong.

WAVE CYCLE POSITION	START	$\frac{1}{4}$ OF A CYCLE	$\frac{1}{2}$ OF A CYCLE	$\frac{3}{4}$ OF A CYCLE	A WHOLE CYCLE	1.5 CYCLES	2 COMPLETE CYCLES
PHASE / °	0	90	180	270	360	540	720
PHASE / RAD	0	$\frac{\pi}{2}$	π	$\frac{3\pi}{2}$	2π	3π	4π
NUMBER OF WAVELENGTHS	0	$\frac{\lambda}{4}$	$\frac{\lambda}{2}$	$\frac{3\lambda}{4}$	λ	1.5λ	2λ

table A Phase angles in degrees and radians, and also compared to wavelengths.

WAVEFRONTS

Diagrams of waves are often drawn as lines, as in **fig B**, where all the points on a line represent points on the wave that are at exactly the same phase position, perhaps a wave crest. These lines are called **wavefronts**. They are how the sea might look observed from a helicopter, where the troughs are all in shadow and so appear darker.

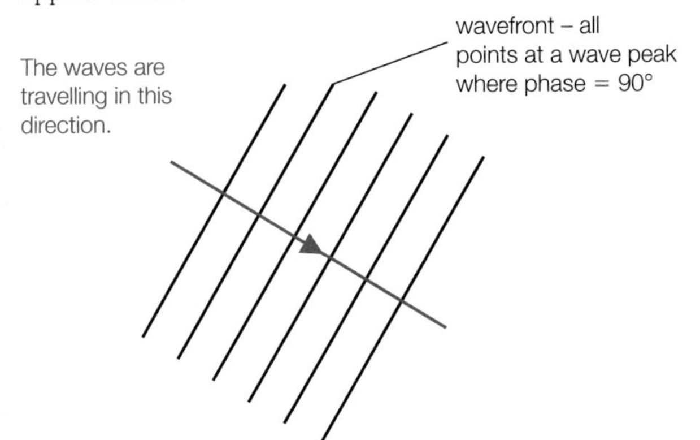

▲ **fig B** All points on each wavefront (black lines) are in the same phase position: 90°, or $\pi/2$. The green line showing direction of travel is called a ray. Rays must be perpendicular to wavefronts.

WAVE SUPERPOSITION

When waves meet, each wave will be trying to cause a wave displacement according to its phase at that location. The overall displacement will be the vector sum of the displacements caused by the individual waves. This is called wave **superposition**. Afterwards, each wave will continue past each other, as the energy progresses in the same direction it was originally travelling. This can be most simply illustrated if we consider just wave pulses passing each other, as in **fig D**.

▲ **fig C** When waves meet, the displacements would add to each other to give the overall displacement for each location.

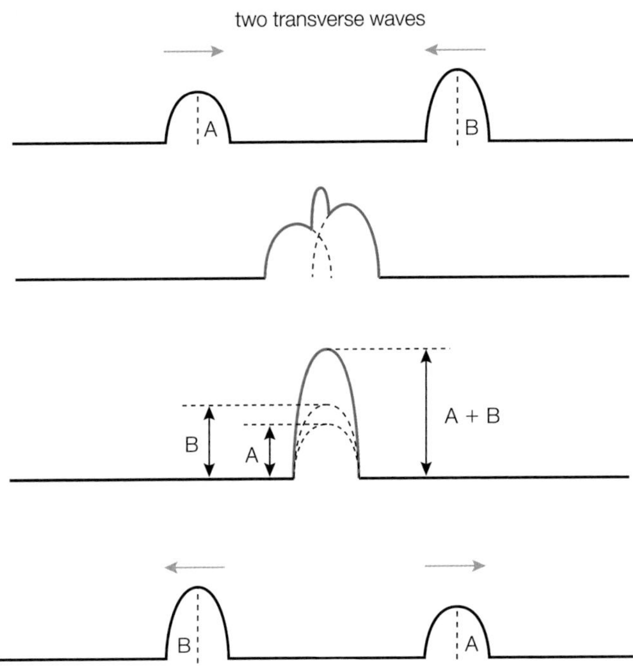

▲ **fig D** Wave pulse displacements add when they meet in the same location.

If the wave pulses in **fig D** are travelling on the surface of water, then each would be trying to displace the water molecules upwards briefly as it moves. When both meet at the same location, wave pulse A lifts the water a little, and then pulse B is also lifting up, so the water raises even higher – to the level of the two pulse displacements added together. This effect is thought to explain the tales from sailors that they saw a sudden very, very large wave which then disappeared as quickly as it had appeared. With many waves travelling in many different directions, they might be at the same point at the same time. Adding these together in wave superposition could cause a huge displacement at that moment. It then drops again as the waves continue past each other.

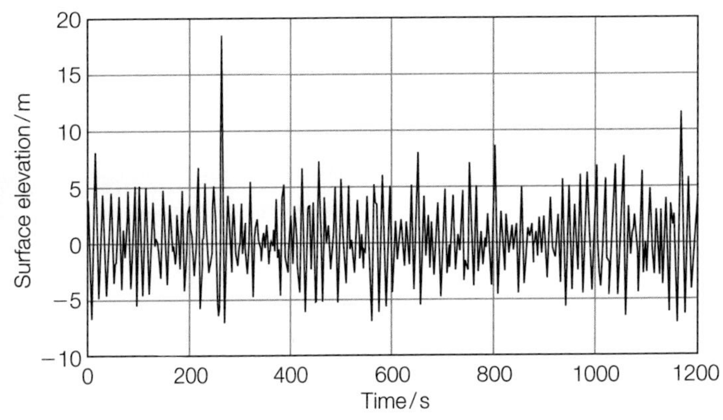

▲ **fig E** Multiple small waves coinciding in the same location can lead to a brief very large displacement. This graph is of sea surface height at the Draupner North Sea Oil Platform on 1st January 1995, showing a freak 18 m wave.

SUPERPOSITION OF CONTINUOUS WAVES

If, rather than a single point along the path of the waves, we consider waves superposing over a large space, the outcome is a continuous wave that is the sum of the displacements over time in each location. If the two waves are in phase, their effect will be to produce a larger-amplitude resultant wave. This is known as **constructive interference**. If identical waves meet and are exactly out of phase – if their phase difference is 180° or π radians – then the resultant is a zero-amplitude wave. This idea of complete **destructive interference** can be confusing: imagine shining two beams of light to the same place, and at that point you see blackness. In the right circumstances, as explained in **Section 3B.4**, this is exactly what happens.

▲ **fig F** Constructive and destructive interference.

It is important to know the phase difference between two waves, in order to determine what will happen as a result of their superposition. Waves that constructively interfere, as with the top pair in **fig F**, must be in phase, which means they have a phase difference of zero. As phase repeats every complete cycle, this could also be said to be a phase difference of 360° or 2π radians.

CHECKPOINT

1. What is the difference between a ray and a wavefront?
2. What is the phase difference between waves that completely destructively interfere?
3. **SKILLS** PROBLEM SOLVING — Imagine waves on the surface of the sea. What are the phase differences between:
   (a) two wave crests
   (b) a crest and the next trough
   (c) a trough and the crest three waves in front?
4. On graph paper, draw three diagrams of the wave pulses shown in **fig G**, at time = 1.0 s, 2.0 s and 3.0 s if **fig G** is at time = 0. Superpose the wave pulses as necessary to draw the correct overall displacements of the string.

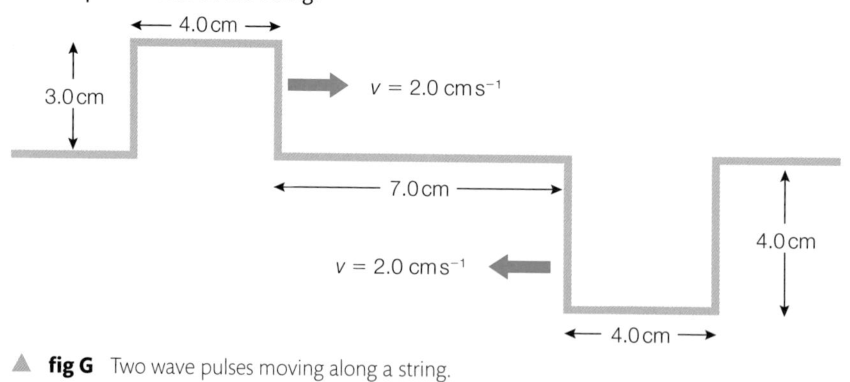

▲ **fig G** Two wave pulses moving along a string.

SUBJECT VOCABULARY

wavefronts lines connecting points on the wave that are at exactly the same phase position
superposition when more than one wave is in the same location, the overall effect is the vector sum of their individual displacements at each point where they meet
constructive interference the superposition effect of two waves that are in phase, producing a larger amplitude resultant wave
destructive interference the superposition effect of two waves that are out of phase, producing a smaller amplitude resultant wave

LEARNING OBJECTIVES

- Explain what is meant by coherent waves.
- Explain how a standing/stationary wave forms and identify nodes and antinodes.
- Use the equation for the speed of transverse waves on a string.
- Verify experimentally what factors affect the frequency of standing waves on a string.

STANDING WAVES

Continuous waves travelling in opposite directions will superpose continuously, and this can set up a **stationary wave** pattern, also known as a **standing wave**. The waves need to be of the same speed and frequency, with similar amplitudes, and have a constant phase relationship. Waves with the same frequency and a constant phase relationship are said to be **coherent**.

Stationary waves have this name because the profile of the wave does not move along, it only oscillates. This also means that wave energy does not pass along a standing wave, so they do not meet our strict definition of waves, which do transfer energy and are more precisely called **progressive waves**. A situation such as that in **fig B** can be set up when waves travel along a string and are reflected from a fixed end, which is how the wave in **fig A** was produced. In this situation, such as when a guitar string is played, the first wave will meet its own reflection. These two coherent waves set up a standing wave on the string. There are points where the resultant displacement is always zero. These points never move, and are called **nodes**. The points of maximum amplitude are called **antinodes**. Note that all points between one node and the next are in the same phase at all times, although their amplitude of vibration varies up to the antinode and back to zero at the next node.

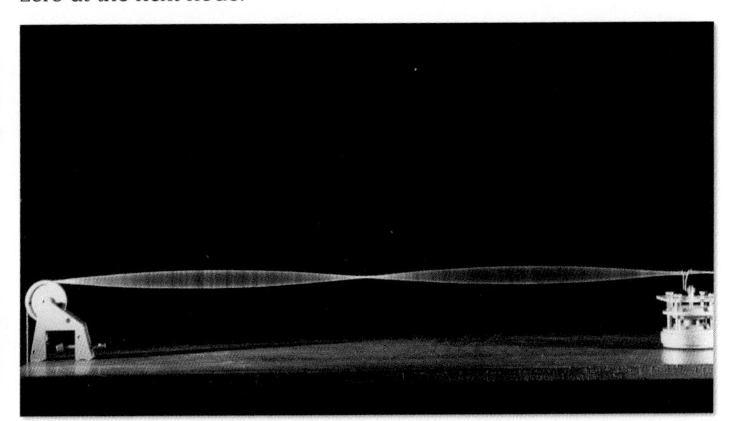

fig A When coherent waves meet, a stationary wave is set up with points of zero amplitude (nodes) and points of maximum amplitude (antinodes). This can be demonstrated in the school laboratory with a vibration generator and a rubber cord.

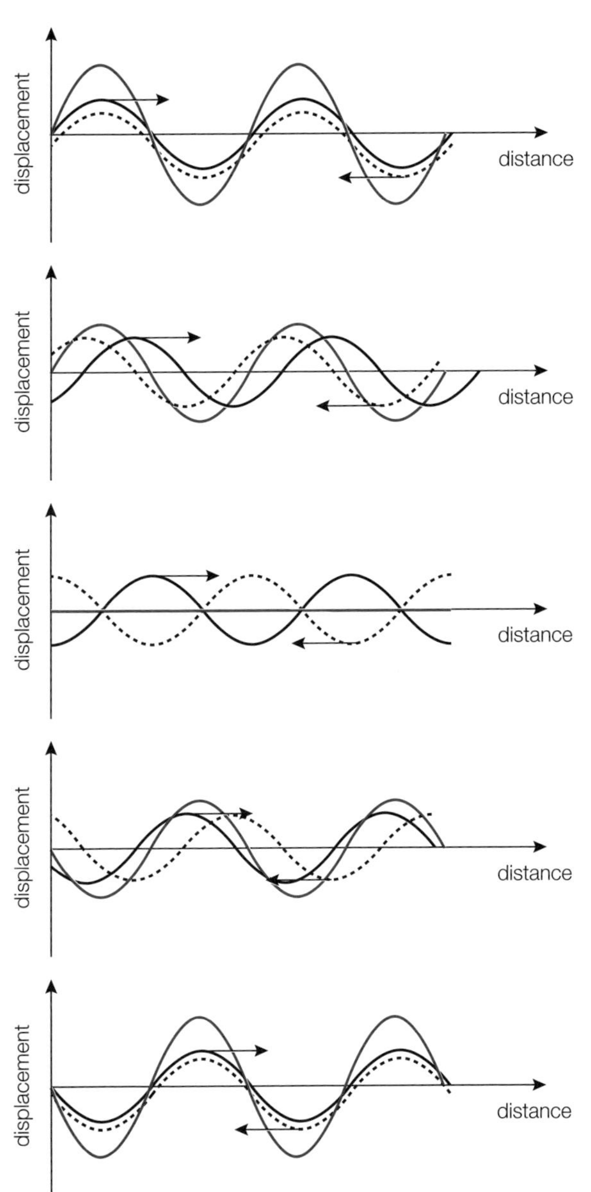

fig B As coherent waves pass through each other, the resultant wave (pink) will be stationary.

Depending on the frequency of the wave that produces the stationary wave by constructively interacting with its own reflection, the standing wave pattern that can be set up will have a corresponding wavelength, as the wave equation must still be satisfied. For example, in **fig C**, the first overtone exactly fits one complete wave in the length of the string: $\lambda = L$. In the third overtone, there are two complete waves in length L, so the wavelength is halved. The speed of waves on the string is the same in both situations, so the frequency in the third overtone must be double that of the first overtone. If a string is plucked and allowed to vibrate freely, all modes of vibration will be present at the same time, but the fundamental will be the most significant.

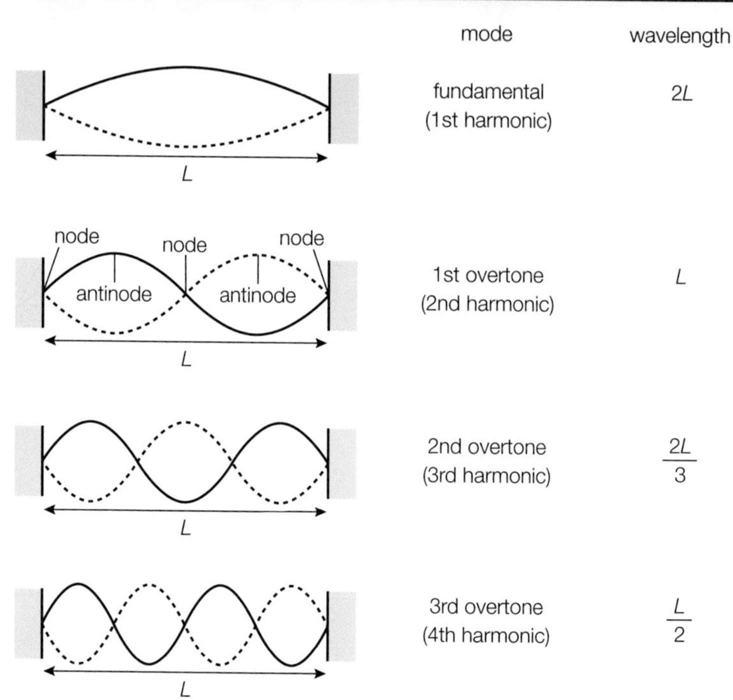

▲ **fig C** As the string is fixed at each end, these end points must always be nodes. The standing wave can only occur if its wavelength exactly allows a node at each end.

WORKED EXAMPLE

A guitar string of length 65 cm vibrates in its fundamental mode. The speed of waves in the string is 362 m s^{-1}. What are the wavelength and frequency of this standing wave?

$$\lambda = 2L = 2 \times 0.65$$
$$\lambda = 1.30\text{ m}$$
$$f = \frac{v}{\lambda} = \frac{v}{2L}$$
$$f = \frac{362}{1.30}$$
$$f = 278\text{ Hz}$$

EXAM HINT

Note that the steps and layout of the solution in this worked example are suitable for wave questions in the exam.

STRING WAVE SPEEDS

Waves on a stretched spring travel at a speed that is affected by the tension in the string, T (in newtons) and the mass per unit length of the string, μ (in kg m^{-1}). The equation for the speed of a wave in a string is:

$$v = \sqrt{\frac{T}{\mu}}$$

If this equation is combined with the wave equation, we get an equation that tells us how the frequency of string vibrations is affected by other factors:

$$v = \sqrt{\frac{T}{\mu}} \quad \text{and} \quad v = f \times \lambda$$

$$\therefore \quad f \times \lambda = \sqrt{\frac{T}{\mu}}$$

$$f = \frac{1}{\lambda}\sqrt{\frac{T}{\mu}}$$

In the fundamental mode of vibration, this means the fundamental frequency, f_0, depends on the length of the string, its tension and its mass per unit length from:

$$f_0 = \frac{1}{2L}\sqrt{\frac{T}{\mu}}$$

WORKED EXAMPLE

How fast do waves travel in a string that is held under 75 N tension and has a mass per unit length of 5.0×10^{-4} kg m^{-1}? What would be its fundamental frequency if the string was 75.5 cm long?

$$v = \sqrt{\frac{T}{\mu}}$$

$$v = \sqrt{\frac{75}{5 \times 10^{-4}}}$$

$$v = 387 \text{ ms}^{-1}$$

$$f_0 = \frac{1}{2L}\sqrt{\frac{T}{\mu}}$$

$$f_0 = \frac{1}{2 \times 0.755}\sqrt{\frac{75}{5 \times 10^{-4}}}$$

$$f_0 = 256 \text{ Hz}$$

EXAM HINT

Note that the steps and layout of the solution in this worked example are suitable for fundamental frequency questions in the exam.

PRACTICAL SKILLS CP5

Investigating the factors affecting the fundamental frequency of a string

We can verify the equation $f_0 = \frac{1}{2L}\sqrt{\frac{T}{\mu}}$ experimentally. In order to confirm it, we need to do three separate investigations to verify each part of the relationship, whilst keeping the other variables as control variables. So, we need to verify: $f_0 \propto \frac{1}{L}$; $f_0 \propto \sqrt{T}$; and $f_0 \propto \sqrt{\frac{1}{\mu}}$.

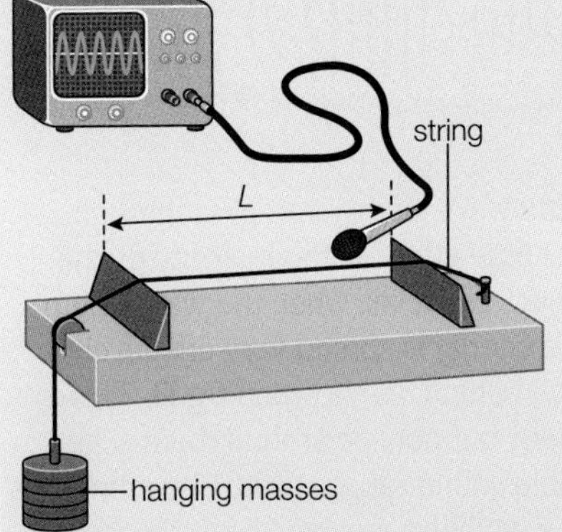

fig D Sonometer experiments to verify the factors affecting the fundamental frequency of a stretched string.

We can use a microphone connected to an oscilloscope to monitor the sounds produced by a **sonometer** string and to measure its frequency of vibration. This can be easier if a datalogging computer is used instead of an oscilloscope, so the screen can be frozen to look at in detail.

1 The string supports (called bridges) on a sonometer are moveable, so that we can find the frequency with varying lengths, L, whilst the same string (constant μ) and the same hanging mass (constant T) keep the other variables controlled. We can plot a graph to verify f_0 on the y-axis is proportional to $1/L$ on the x-axis.
2 We can find the fundamental frequency using a fixed length of the same string throughout (constant L and constant μ) for varying masses hung over the pulley, i.e. varying T. We can plot a graph to verify f_0 on the y-axis is proportional to $\sqrt{T}$ on the x-axis.
3 The final experiment requires a set of different strings (varying diameter metal wires could be used). Maintaining the same length and the same hanging mass (constant L and T) keeps the other variables controlled. Measure the mass of each wire using a digital balance, and its full length, in order to calculate the mass per unit length for each wire or string used. We can plot a graph to verify f_0 on the y-axis is proportional to $\sqrt{\frac{1}{\mu}}$ on the x-axis.

Safety Note: Use eye protection if working close to the wire in case it snaps. Place a 'drop box' below the hanging masses if they are very large.

CHECKPOINT

SKILLS ANALYSIS

1. A guitar string of length 75 cm vibrates in its second harmonic. The speed of waves in the string is 420 m s^{-1}. What are the wavelength and frequency of this standing wave?
2. (a) How fast do waves travel in a string that is held under 90 N tension and has a mass per unit length of 4.8×10^{-4} kg m^{-1}?
   (b) If the string is 74.0 cm long, calculate the fundamental frequency.
3. A sonometer experiment produced the following data (**table A**). Draw a suitable graph to find out what the mass per unit length of the vibrating string was.

LENGTH / M	FREQUENCY / HZ
0.60	190
0.65	180
0.70	165
0.75	150
0.80	140
0.85	135
0.90	130
0.95	120
1.00	115

table A Data from a sonometer experiment.

The hanging mass used was 2.4 kg ∴ T = 23.5 N.

4. Estimate the mass per unit length of piano wire.

SUBJECT VOCABULARY

stationary or **standing wave** a wave which has oscillations in a fixed space, with regions of significant oscillation and regions with zero oscillation, which remain in the same locations at all times
coherence waves which must have the same frequency and a constant phase relationship. Coherent waves are needed to form a stable standing wave
progressive wave a means for transferring energy via oscillations
nodes regions on a stationary wave where the amplitude of oscillation is zero
antinodes regions on a stationary wave where the amplitude of oscillation is at its maximum
sonometer an apparatus for experimenting with the frequency relationships of a string under tension, usually consisting of a horizontal wooden sounding box and a metal wire stretched along the top of the box

LEARNING OBJECTIVES

- Understand what is meant by the term diffraction.
- Understand the factors that affect the amount of diffraction.
- Describe an experiment to observe diffraction effects.

▲ **fig A** **Diffraction** is a spreading of wave energy through a gap or around an obstacle.

DIFFRACTION

When a wave passes the edge of an obstacle, the wave energy spreads into the space behind the obstacle. If the obstacle is relatively small, this can mean that the wave energy will pass around both sides of it and continue travelling past the obstacle, with no shadow, as if the obstacle were not there, as shown in **fig B**.

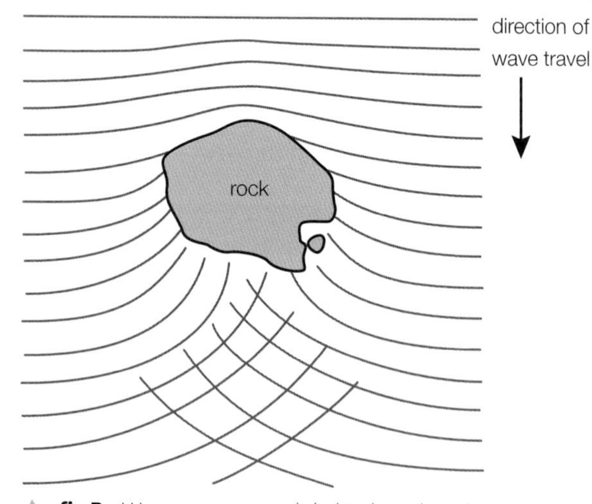

▲ **fig B** Wave energy spreads behind an obstacle.

If the obstacle is larger, then there may be a shadow region behind it, but there will still be diffraction around each edge. If there are two close obstacles forming a gap, then there will be diffraction from each edge of the gap, causing the waves to spread out through the gap. Also see **Section 3D.1** for an explanation of why diffraction occurs.

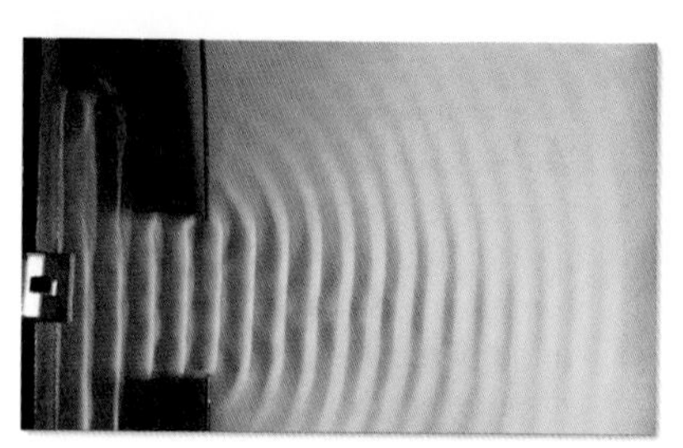

▲ **fig C** Water waves in a ripple tank diffract when passed through a gap.

FACTORS AFFECTING DIFFRACTION

The amount of diffraction around an obstacle depends on the size comparison between the obstacle and the wavelength of the wave. As the resulting diffraction through a gap is caused by the diffraction at each end of the gap; there is an optimum size for maximum diffraction, or spreading, through the gap. Diffraction happens most when the gap size is the same size as the wavelength.

▲ **fig D** The amount of diffraction depends on the size of the gap relative to the wavelength.

If the gap is too small, very little wave energy can pass through. If the gap is very large, there is little effect as the majority of the wave passes through undisturbed. However, when the wavelength matches the gap size, the wave energy is spread very effectively through the gap to fill the space behind, as shown in **fig D**. This is why we can hear around corners, but not see around them. The typical wavelength for an audible sound ranges from 15 centimetres to 15 metres, depending on the exact frequency. These are the same sort of size as common diffracting obstacles. For example, a doorway might be a metre wide, and a sound of this wavelength would have a frequency around 330 Hz, which is a typical speech frequency for humans. This means that a conversation in a room could be heard in the corridor outside the door, as the waves diffract from their original direction through the doorway to be heard out of line of sight. However, with visible frequencies of light being of the order of 10^{-7} metres, these will not diffract through the doorway enough to see. So light from the speakers will only be visible in direct line of sight through the gap.

DIFFRACTION PATTERNS

The diffraction pattern observed when light passes through a narrow slit, such as in **fig E**, shows a central **maximum** and then areas of darkness and then **maxima** of decreasing intensity. It can be confusing to see the areas where there is no light at all, when the diffraction through a gap suggests that the wave energy spreads out behind the gap. You might expect to see the screen behind the slit illuminated everywhere. We will look more carefully at interference in the next section, but the diffraction patterns we have seen with a laser are examples of an interference pattern. Considering the waves being diffracted from each end of the slit, gives us two waves that will meet at the screen. We saw in **Section 3B.1** that if waves meet in phase they will superpose to give a larger amplitude resultant wave. This would appear as a light spot on the screen. At a slightly different place on the screen, these two waves will be completely out of phase and the sum of their displacements will always be zero, so it appears as a dark spot. The diffraction pattern is an example of a standing wave on the screen, where the dark spots are nodes and the light spots are antinodes.

LEARNING TIP

If waves come to a boundary or obstacle we can say the waves are 'incident' on it.

SINGLE SLIT DIFFRACTION

If we change the width of the diffraction slit, or the wavelength of the light, we will alter the diffraction pattern that we observe. A narrower slit widens the central maximum, as well as the further maxima and **minima**.

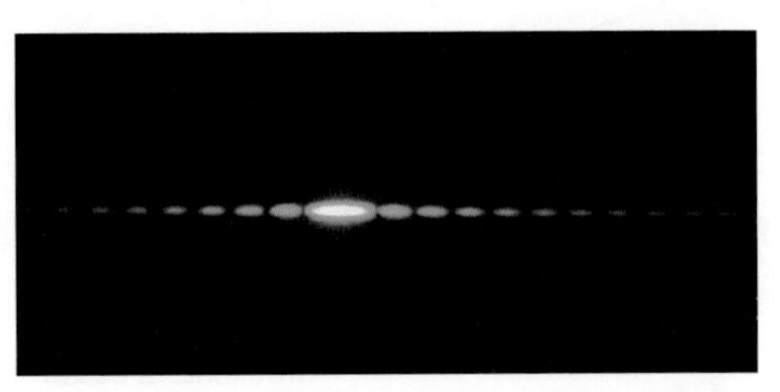

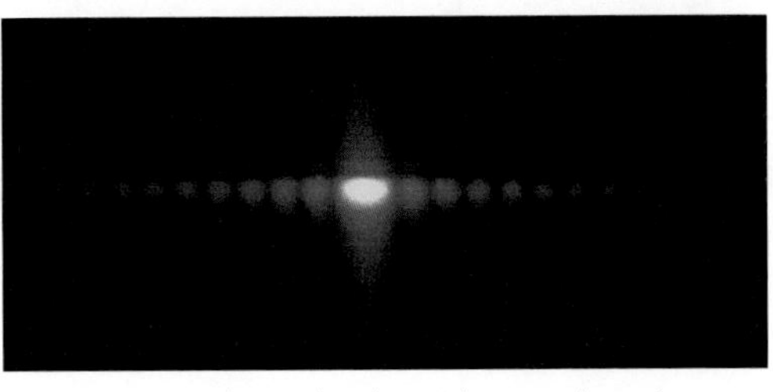

▲ **fig E** Diffraction patterns of light observed on a screen show a bright, wide central maximum, surrounded by areas of darkness, and further bright spots that have decreasing brightness from the centre.

LEARNING TIP

'**Monochromatic**' light has only one colour, so only one wavelength.

THE DIFFRACTION GRATING

A diffraction grating is a device that will cause multiple diffraction patterns, which then overlap. This creates an interference pattern with a mathematically well-defined spacing between bright and dark spots. It is a collection of a very large number of slits through which the waves can pass. These slits are parallel and have a fixed distance between each slit.

▲ **fig F** Diffraction gratings can be used to examine light spectra.

The pattern produced by each colour passing through a diffraction grating follows the equation:

$$n\lambda = d\sin\theta$$

where θ is the angle between the original direction of the waves and the direction of a bright spot, λ is the wavelength of the light used, d is the spacing between the slits on the grating, and n is called the 'order'. The order is the bright spot number from the central maximum (which is $n = 0$).

LEARNING TIP

The grating spacing is often quoted as a number of lines per metre (or per millimetre), which means that to find d you will need to do an initial calculation in which:

$$d = \frac{1}{\text{number per metre}}$$

OR:

$$d = \frac{1 \times 10^{-3}}{\text{number per millimetre}}$$

PRACTICAL SKILLS CP6

Investigating diffraction with a laser
You can use a laser pointer or laboratory experimental laser to demonstrate diffraction. A diffraction grating investigation allows careful study of the light making up the spectrum from any light source. Astronomers use this to study the light spectra from stars.

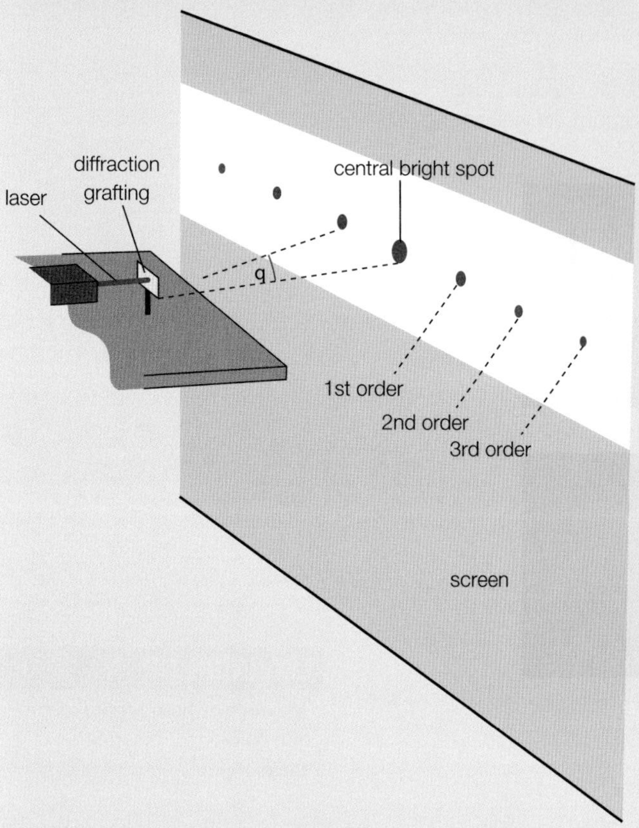

▲ **fig G** A diffraction grating will produce a series of bright spots at increasing angles from the centre.

Diffraction gratings are manufactured to have a fixed spacing, *d*, between lines on the grating, and this will be printed on the grating. By measuring the angle to each maximum brightness spot created by a diffraction grating, you can calculate the wavelength of the light used from the diffraction grating equation.

▲ **fig H** CDs have a series of very close lines marked on them and act as a diffraction grating that reflects. This causes the spectrum of colours that can be seen on them from white light that hits the surface.

CHECKPOINT

1. In stormy seas, a sailor thought it would be a good idea to position his small boat behind a large oil tanker to avoid being hit by the waves that were coming from the other side of the tanker. Explain why this plan might not help.

2. (a) Explain why the radio waves from the two dishes in **fig I** spread out differently in each case.
   (b) Suggest examples in which each situation in **fig I** might be useful for different purposes.

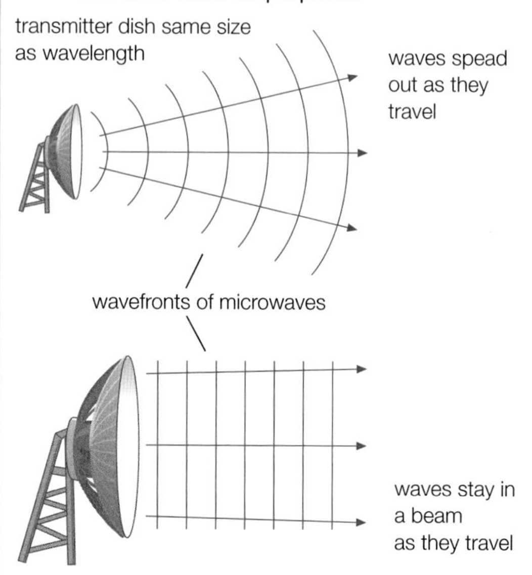

▲ **fig I** The size of a radio transmitter can be important depending on its purpose.

3. In an experiment to examine the colours of light produced by hot helium gas, a chemist used a diffraction grating, and found the first order maximum for each of three colours were at angles of 26.6°, 30.1° and 36.0°. If the grating he used had 1000 lines per mm, what were the wavelengths for these three colours?

4. 90° is the largest angle through which a bright spot maximum produced by a diffraction grating can be observed, but we can only observe a maximum for which *n* is an integer. For the experiment in question 3, calculate how many orders would be observed for each of the three colours.

SUBJECT VOCABULARY

diffraction when a wave passes close by an object or through a gap, the wave energy spreads out.
maximum (plural **maxima**) in a diffraction or interference pattern, the bright spots
minimum (plural **minima**) in a diffraction or interference pattern, the dark spots
monochromatic containing or using only one colour. Light of a single wavelength

4 WAVE INTERFERENCE

SPECIFICATION REFERENCE 2.3.39 2.3.40

LEARNING OBJECTIVES

- Understand what is meant by the terms coherence, path difference and interference.
- Interpret the relationship between phase difference and path difference.
- Explain examples of wave interference.

TWO-SOURCE INTERFERENCE

In **Section 3B.2** we saw that a wave meeting its own reflection would set up a standing wave pattern, also known as an **interference** pattern. Reflection is a convenient way to generate coherent waves that will produce a standing wave. However, any combination of waves that have the same frequency and a constant phase relationship will produce this result. For example, in **fig A**, the loudspeakers are connected to the same signal generator, so they simultaneously produce identical sound waves. Wherever the sound waves meet, they will superpose to produce a resultant displacement of air molecules. This means there will be points of maximum amplitude – antinodes – such as at position X, or points of zero amplitude – nodes – where there will be silence, such as position Y. The relative positions of nodes and antinodes will depend on the distance from the speakers, their separation and the wavelength of the waves.

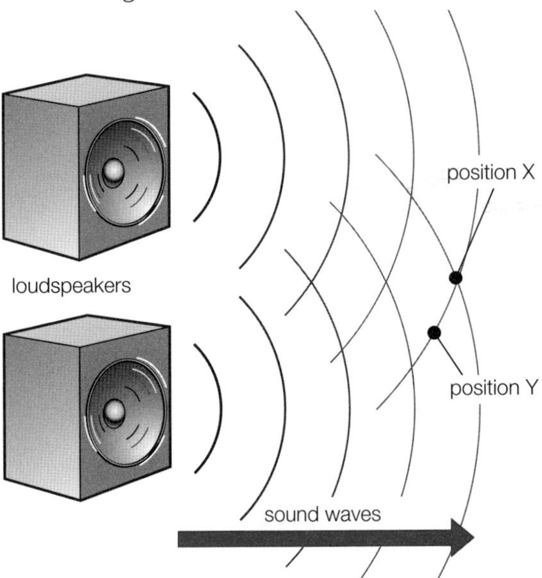

▲ **fig A** Two loudspeakers generating the same frequency sounds.

TWO-SLIT INTERFERENCE

Another interference pattern can be observed if light is shone into an experimental set-up that causes it to create two coherent wave sources that produce a standing wave pattern. This effect was first demonstrated by Thomas Young in 1803, using sunlight as the light source. A simple laboratory set-up using a torch, similar to that shown in **fig B**, would show a pattern of dark and light fringes that resembles the nodes and antinodes in a standing light wave.

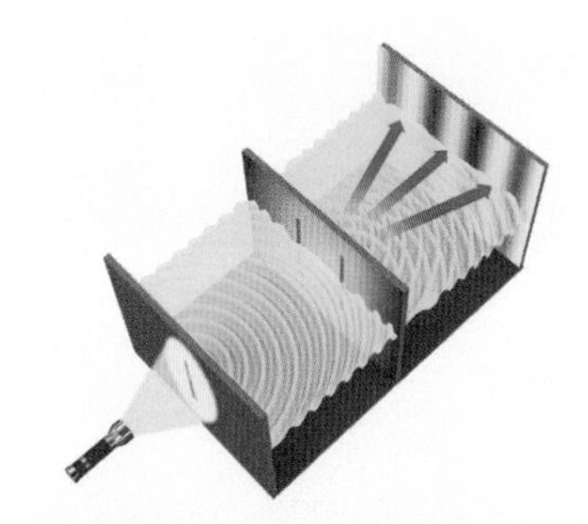

▲ **fig B** Two-slit interference of torch light. An interference pattern of dark and light fringes is seen on the screen.

Young explained his experiment using a theory that light behaves as a wave. In classical physics, interference is a phenomenon that can only occur with waves. Young's theory was highly controversial at the time, as he was contradicting Newton's theory that light behaved as a stream of particles. Despite the experimental evidence he demonstrated, most scientists at the time refused to believe Young's wave theory of light.

PRACTICAL SKILLS

Investigating two-source interference

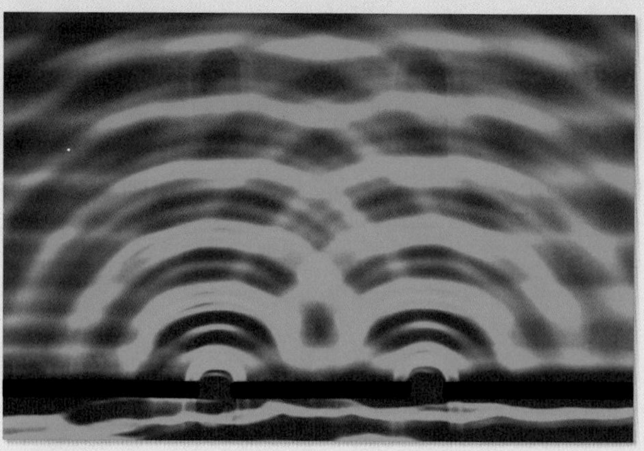

fig C Two-source interference demonstration in a ripple tank.

A ripple tank where waves are diffracted through two gaps will cause an interference pattern between the two water waves.

Another common experimental demonstration of two-slit interference is with laser light.

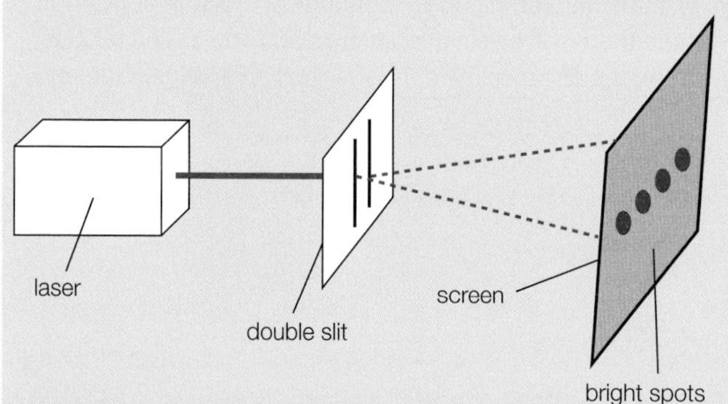

fig D Two-slit interference demonstration with a laser.

!

Safety Note: Ripple tank: do not handle electrical sockets, plugs and switches with wet hands or splash them with water. If a strobe light is used to freeze the wave action, persons with conditions adversely affected by flashing lights should be warned. Laser: do not view the laser beam directly or reflect it into eyes.

LEARNING TIP

The bright and dark spots in light interference experiments are often called 'fringes'.

LEARNING TIP

Two-source interference and two-slit interference are the same thing. In both cases, two waves are brought together to produce an interference pattern. Two-slit interference is a form of two-source interference, where the slits diffract existing waves in order to act as new sources of waves.

EXPLAINING TWO-SOURCE INTERFERENCE

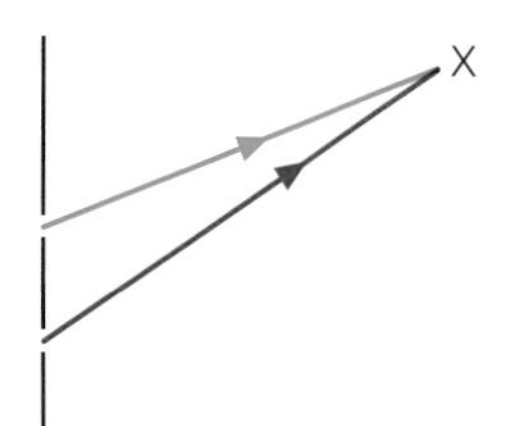

fig E Different distances from each slit to point X determine the resultant amplitude of the standing wave at X.

Each point in a two-source interference pattern will have a superposition result that depends on the phase difference between the waves coming from each source to that point. That phase difference will, in turn, depend on the relative distances to the point from each of the slits compared with the wavelength. In **fig E**, imagine the distance from the slit to X along the blue line is 100 wavelengths, and the distance along the red line is 115 wavelengths. As the path difference is exactly 15 wavelengths, the waves will be in phase at point X, and will constructively interfere, to produce a maximum resultant, as in an antinode. If we passed laser light through the two slits, then point X would show a bright spot on a screen placed there.

Alternatively, imagine the blue line distance to X is 97 wavelengths but the red line distance is 98.5 wavelengths. The path difference is 1.5 wavelengths. In this case, the waves will always be 3π radians out of phase – always in antiphase – and so will always cancel. This would look like a standing wave node, and if it were the laser light two-slit experiment, there would be a dark spot at point X.

Note that these wavelength numbers are examples only. As light waves are so short, the numbers should be millions of wavelengths for standard laboratory distances for X.

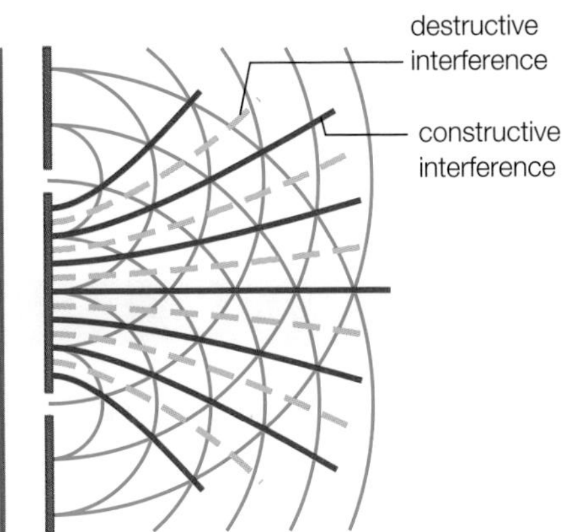

▲ **fig F** All points that are exactly in phase, where the path difference is a whole number of wavelengths, $n\lambda$, will produce constructive interference. Any points where the path difference is given by $\frac{(2n+1)\lambda}{2}$ will cause destructive interference.

The connection between the phase difference and the path difference comes if we remember that each complete cycle (or wavelength) corresponds to 2π radians in phase. Waves from one slit meeting waves from the other slit at a point will each have had to travel to that point, cycling through wavelengths as they go. By comparing the path difference, this can be converted into phase differences.

Points where the path difference is equal to $n\lambda$ exactly, will have a phase difference of $2n\pi$ exactly and will be in phase, producing constructive interference.

Points where the path difference is equal to $\frac{(2n+1)\lambda}{2}$ exactly, will have a phase difference of $(2n+1)\pi$ exactly and will be in antiphase, producing destructive interference.

CHECKPOINT

1. What are coherent waves?
2. In a laser light two-slit interference experiment, a bright fringe was observed at a point on the screen exactly halfway between the two slits. A dark fringe was observed on either side of this, and then another bright fringe on either side.
   (a) What is the phase difference between the waves arriving from each slit at the central maximum?
   (b) What is the path difference, and the phase difference, between waves arriving from each slit to the first minimum on the left of the central maximum?
   (c) What is the path difference, and the phase difference, between the waves arriving from each slit at the first maximum to the left of the central maximum?
   (d) Describe what difference would be observed in the fringe pattern if the screen were moved further away from the two slits.

SUBJECT VOCABULARY

interference the superposition outcome of a combination of waves. An interference pattern will only be observed under certain conditions, such as the waves being coherent

3B THINKING BIGGER

THE MARIMBA

CRITICAL THINKING, PROBLEM SOLVING, ANALYSIS, REASONING/ ARGUMENTATION, INTERPRETATION, ADAPTIVE LEARNING, ADAPTABILITY, PRODUCTIVITY, COMMUNICATION

The marimba is a percussion instrument like a xylophone. A standing wave is set up in the wood when it is struck with a hammer. The wooden bars vibrate with frequencies related to their sizes.

In this activity, we will look at some of the detail about tuning the marimba bars.

ONLINE ARTICLE

THE MARIMBA BAR

The marimba bar vibrates in complex patterns that produce a sound unique to the instrument. You can get an understanding of the sound quality (timbre) by studying the vibrations. Each type of vibration is called a mode of vibration. In a scientific study of a C3 (C below middle C) marimba bar, Bork et al. (1999) identified 25 modes of vibration in the range of 0 to 8000 Hz.

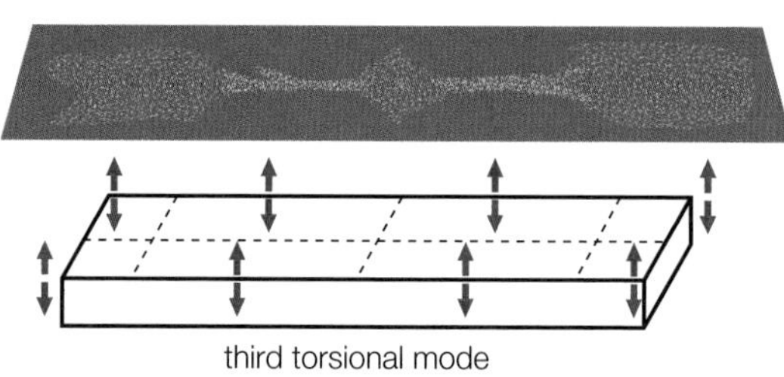

▲ **fig A** When the bar vibrates, the salt sprinkled all over it moves away from antinodes, leaving a pattern which identifies how the wood is vibrating. In this example, the bar is vibrating in a torsional mode, as well as the standard transverse vibration.

Drilling the Holes for the Bar – Finding the Nodes of the Fundamental Mode

Sprinkle some salt on the bar and then tap very lightly with a mallet at the centre of the bar. Keep tapping until the salt collects over each node as illustrated below.

▲ **fig B** Finding the nodes on a marimba bar, for the positions of mounting screws.

Human Hearing at Different Frequencies and the Graduation of Bar Width

You will note that marimba bars have different widths, growing wider with lower notes. One of the challenges in designing an instrument which sounds in the bass region is to provide tones that can be easily heard. The human ear has a large range of sensitivity to sound at different frequencies. Sensitivity in the bass region is much less than in the range around 1000 to 4000 Hz. In order to overcome this problem, the bass bars are wider so that they generate sound at a higher energy level.

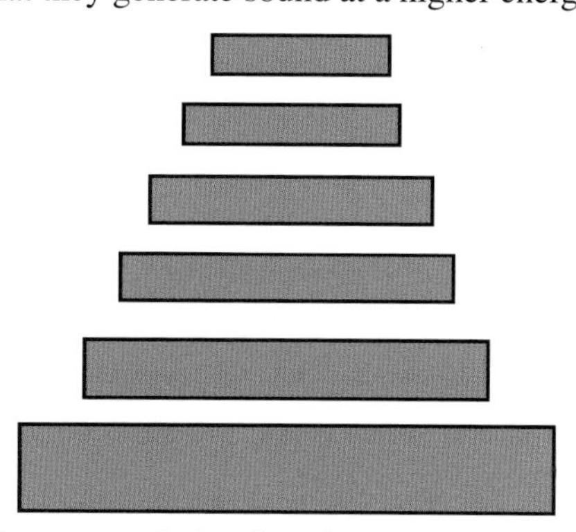

▲ **fig C** C-note marimba bars from five octaves.

The graph below shows the minimum levels of human hearing at different frequencies. Note that the human ear is most sensitive to frequencies in the range of 1000 to 4000 Hz. The energy level of the sound is graphed in decibels (dB), which is logarithmic. An increase of 10 dB represents a 10-fold increase in sound energy and an increase of 20 dB represents a 100-fold increase in sound energy.

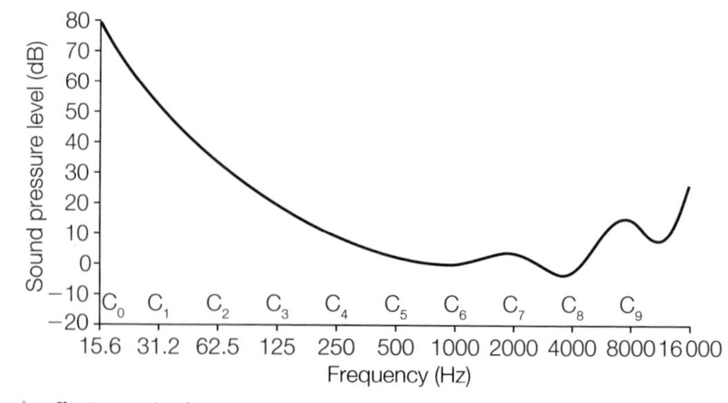

▲ **fig D** As the frequency of sound decreases, it must be played louder for us to hear it.

SCIENCE COMMUNICATION

1 The extract opposite is from a webpage explaining the complex physics of the vibrations and sounds produced by a marimba. This is a hobby of the author who is a scientist in a teaching university. Consider the extract and comment on the type of writing that is used. Then answer the following questions:

(a) Discuss the level of the science presented in the extract. Who might want to read this webpage?

(b) Comment on the fact that this webpage is several thousand words, and has many images and diagrams relating to the science of stationary waves. However, there is no picture of a complete marimba instrument on it.

(c) What does the phrase 'Bork et al' mean?

INTERPRETATION NOTE

This is an extract of the article only. You may find it helpful to visit the webpage and browse its different sections.

PHYSICS IN DETAIL

Now we will look at the physics in detail. Some of these questions link to topics earlier in this book, so you may need to combine concepts from different areas of physics to work out the answers.

2 Explain why the salt collects in the two places shown in **fig B**. Why would these be the best points to place the mounting screws for the bar?

3 What is the connection between frequency of the C-notes and the octave they are in, as shown on the *x*-axis in **fig D**?

4 Looking at **fig C**, estimate the ratios of the surface area of the bars. Compare these answers with the decibel values for each needed to hear each note on the graph in **fig D**. Why might these be linked?

ACTIVITY

Write a report to a school Music department which explains how they could make marimba bars in lessons. Include some of the physics behind what will alter their sound properties, and include outline plans for manufacture by school students. Also include instructions for experimentally testing the frequency of each bar produced. Your report should be written for the teacher in charge of the subject, whilst their lessons will be for IGCSE students.

THINKING BIGGER TIP

Think about how the vibrations of the bar will generate sound waves.

3B EXAM PRACTICE

1 What is the wavelength of the fundamental standing wave on a guitar string that is 74 cm long?

A $\lambda = 0.37$ m

B $\lambda = 0.74$ m

C $\lambda = 1.48$ m

D $\lambda = 2.96$ m [1]

(Total for Question 1 = 1 mark)

2 Which of the following is **not** caused by wave superposition?

A The fundamental note produced on a guitar string.

B Seismic P-waves cause surface damage before S-waves.

C Occasionally a theatre seat can be in a 'dead zone' where little can be heard from the stage.

D CDs can have a rainbow appearance. [1]

(Total for Question 2 = 1 mark)

3 Which of these correctly represents the same phase position in different units?

A 90° and π rad

B 180° and π rad

C 360° and π rad

D 180° and 2π rad [1]

(Total for Question 3 = 1 mark)

4 A student did experiments to show the diffraction of different types of waves.

Which row in the table correctly shows the experiment that would show the greatest diffraction effect?

	Type of wave	Wavelength / m	Gap size / m
A	water	0.02	0.10
B	light	5.5×10^{-7}	5.5×10^{-3}
C	water	0.05	0.08
D	sound	3.0	1.5

[1]

(Total for Question 4 = 1 mark)

5 In a two-slit interference of light experiment, dark and light fringes are observed on the screen in front of the two slits. What can you say about the light waves arriving from the two slits at any one of the dark fringes?

A The light waves are constructively interfering.

B The light waves are 90° out of phase.

C The light waves travel the same path length from each slit.

D The light waves are the same amplitude. [1]

(Total for Question 5 = 1 mark)

6 When oil floats on water, coloured interference patterns are often seen. The interference patterns are formed because of the thin film of oil. A thin film of oil can also produce interference patterns with monochromatic light. The diagram shows light from a monochromatic source, incident on a film of oil.

Explain why interference patterns may be seen. [5]

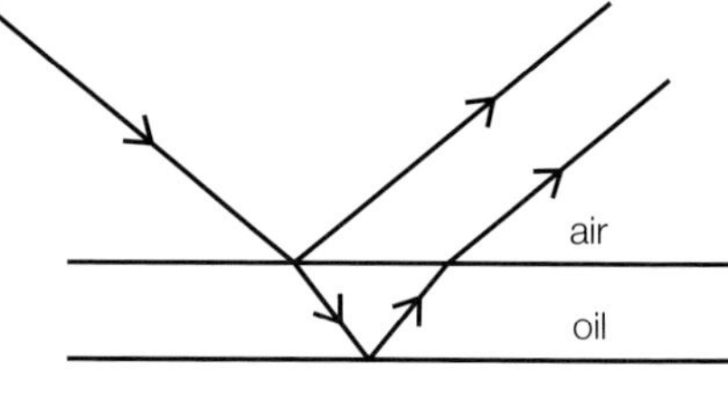

(Total for Question 6 = 5 marks)

7 In a school experiment, a laser is shone at right angles onto a diffraction grating to produce a pattern of light and dark spots on a screen several metres away. The grating is marked '500 lines/mm'.

(a) What is the distance between slits on the diffraction grating? [1]

(b) If the first bright spot from the centre is found to be at an angle of 15° from the original direction of the laser beam, what is the wavelength of the light? [2]

(c) The laser is changed for a different one which has a wavelength of 450 nm. How many orders of bright spot will be generated by the diffraction grating? [3]

(Total for Question 7 = 6 marks)

8 (a) A transverse wave travelling along a wire under tension has a speed v given by

$$v = \sqrt{\frac{T}{\mu}}$$

where T is the tension in the wire and μ is the mass per unit length of the wire.

Show that the units on both sides of the equation are the same. [3]

(b) The diagram shows a wire held under tension by hanging weights at one end and supported by a vibration generator at the other end. The frequency of the vibration generator is slowly increased from zero until a standing wave is formed.

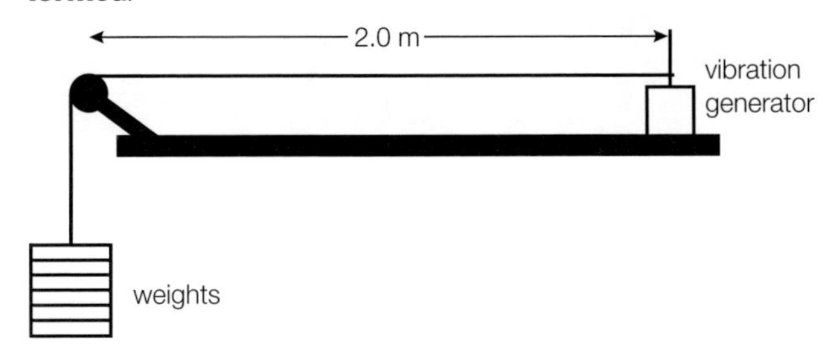

(i) Explain how the standing wave is produced. [3]

(ii) Calculate the wavelength of the standing wave. [1]

(iii) The weight is 150 N and the mass per unit length of the wire is 0.0050 kg m^{-1}.

Using the equation given in (a), calculate the speed of the transverse wave along the wire. [2]

(iv) The wire is observed as the frequency of the vibration generator is steadily increased to several times the frequency that produced the first standing wave. Describe and explain what is seen as the frequency is increased. [4]

(Total for Question 8 = 13 marks)

9 (a) State what is meant by diffraction. [2]

(b) State the principle of superposition of waves. [2]

(c) The photograph shows a beach in England. Waves can be seen passing rocks on their way to the beach. The uneven surface of the sand has formed as a result of diffraction and superposition of these waves.

(d) Use the ideas of diffraction and superposition to explain why the sand surface becomes uneven. [5]

(Total for Question 9 = 9 marks)

10 The diagram shows a metal wire held between fixed points S and T.

S ●————————————● T

(a) Draw the wire vibrating in its fundamental mode. [2]

(b) In part (a) the wire was vibrating at a frequency of 200 Hz. Draw the wire again, but now vibrating at a frequency of 600 Hz. [2]

(c) Explain why the wire sets up a stationary wave when it is plucked. [4]

(Total for Question 10 = 8 marks)

TOPIC 3 WAVES AND THE PARTICLE NATURE OF LIGHT

CHAPTER 3C MORE WAVE PROPERTIES OF LIGHT

Humans have long wondered at the appearance of rainbows and other ways light can trick our eyes. Developing our understanding of the physics behind such natural phenomena has allowed us to develop more and more complex imaging techniques. Deep drilling for oil exploration often uses optical fibres to gain an image of the rock at hundreds of metres depth. The same imaging physics allows doctors to view inside their patients with an endoscope.

This chapter will explain the details of refraction in different materials and how this leads to the properties of different types of lenses, as well as total internal reflection in optical fibres. Broadband internet communications are made possible by applying this idea.

MATHS SKILLS FOR THIS CHAPTER

- **Substituting numerical values into algebraic equations** (*e.g. finding the refractive index*)
- **Changing the subject of an equation** (*e.g. finding the critical angle*)
- **Plotting two variables from experimental data** (*e.g. plotting a graph of experimental results of refraction in glass*)
- **Determining the slope of a linear graph** (*e.g. using a graph of experimental results to find the refractive index of glass*)
- **Identifying uncertainties in measurements and using simple techniques to determine uncertainty** (*e.g. considering the uncertainty of experimental results in finding the refractive index of glass*)

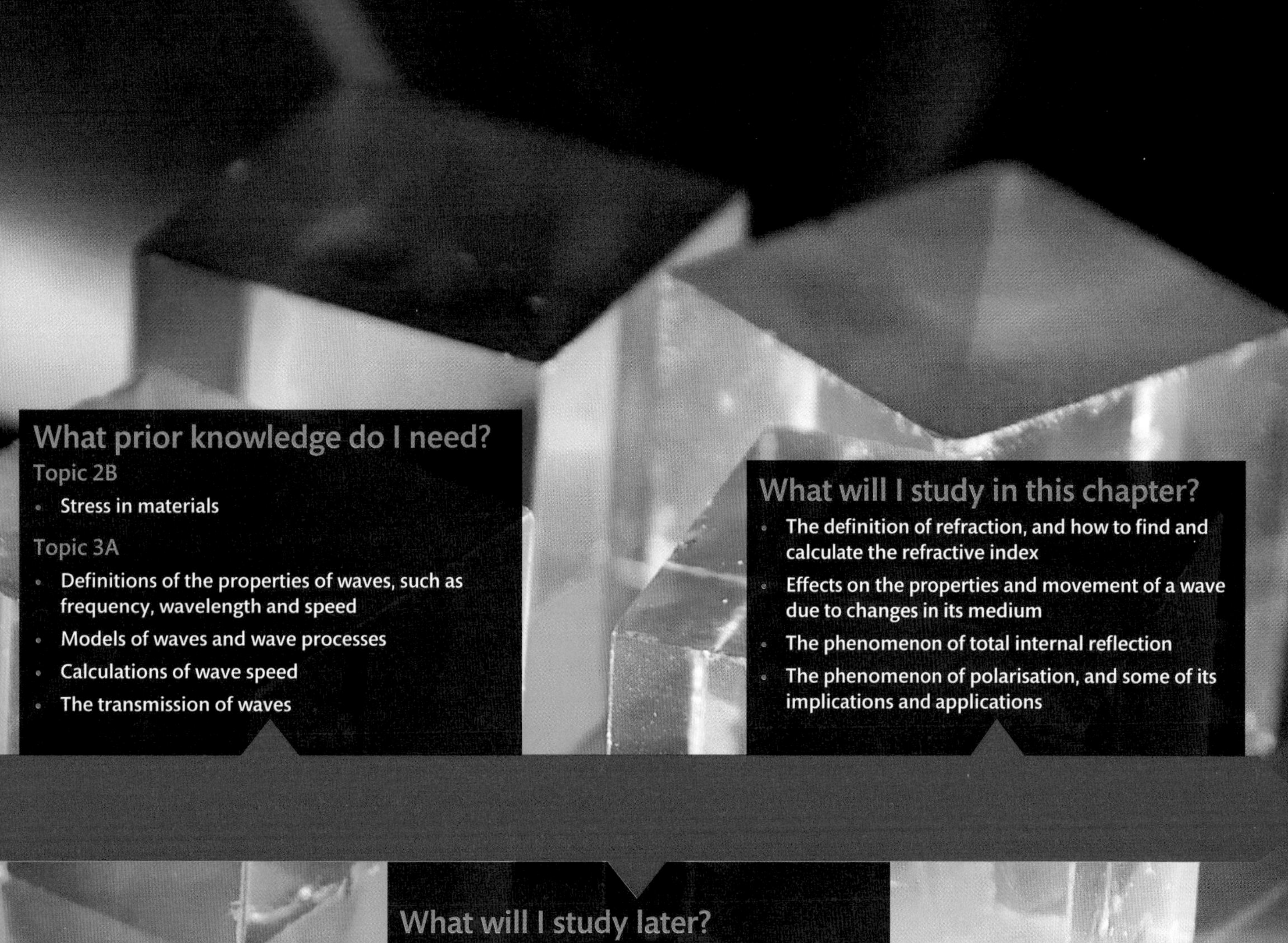

What prior knowledge do I need?

Topic 2B

- Stress in materials

Topic 3A

- Definitions of the properties of waves, such as frequency, wavelength and speed
- Models of waves and wave processes
- Calculations of wave speed
- The transmission of waves

What will I study in this chapter?

- The definition of refraction, and how to find and calculate the refractive index
- Effects on the properties and movement of a wave due to changes in its medium
- The phenomenon of total internal reflection
- The phenomenon of polarisation, and some of its implications and applications

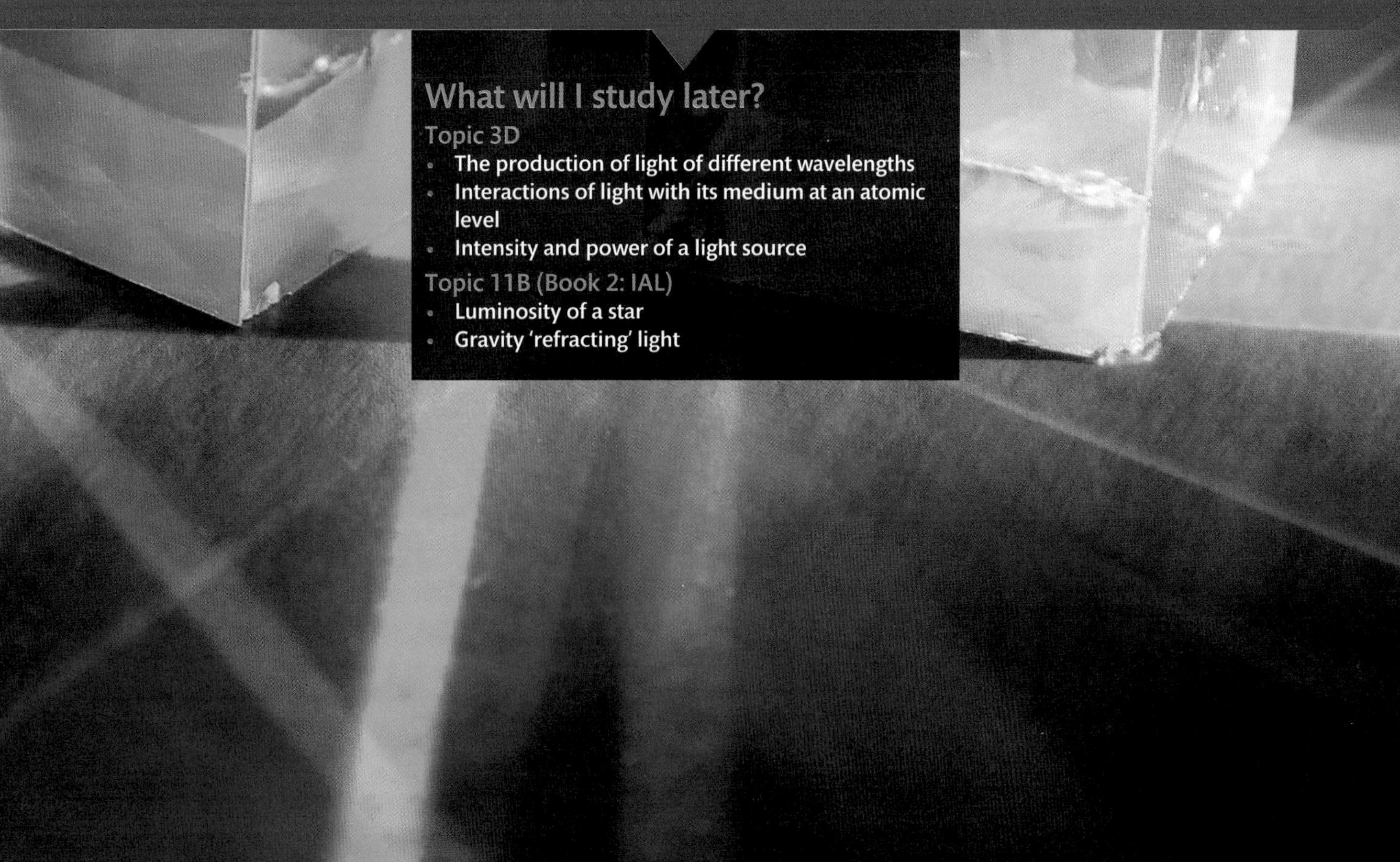

What will I study later?

Topic 3D

- The production of light of different wavelengths
- Interactions of light with its medium at an atomic level
- Intensity and power of a light source

Topic 11B (Book 2: IAL)

- Luminosity of a star
- Gravity 'refracting' light

3C 1 REFRACTION

SPECIFICATION REFERENCE 2.3.45 2.3.48 2.3.55

LEARNING OBJECTIVES

- Understand what is meant by refraction.
- Understand how to measure the refractive index of a solid material.
- Use two equations relating to refraction.

REFRACTION

When waves pass from one medium into another, there is a change in speed. The frequency remains constant, so the change in speed causes a change in wavelength. If the waves are approaching the interface between the two media at an angle then the change in speed causes a change in direction as well. This is called **refraction**.

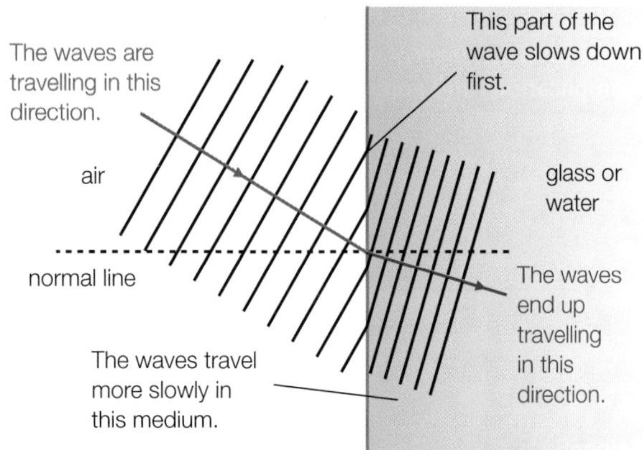

▲ **fig A** Wave speed depends on the medium, and refraction is caused by a change in medium changing the wave speed.

Any ray crossing an interface along the normal line does not change direction at all, as the wavefronts are parallel to the edge. Therefore their wavelength is equally changed along the length of the wavefront. If the wavefront is at an angle to the interface, then the part that hits first will change speed first. Then the wavefront becomes bent because different parts of it are travelling at different speeds.

The changes in direction caused by refraction are the basis for the functioning of lenses, and can lead to optical illusions.

A measure of the amount of refraction caused by different materials is called the **refractive index**, and its symbol is n. The refractive index, n, is equal to the ratio of the speed of light in a vacuum to the speed of light in the material:

$$n = \frac{c}{v}$$

Whilst it is difficult to measure the underlying change in the speed, at least for light waves, the effect on direction can be measured easily. The relationship between direction and refractive index is given by **Snell's law**:

$$n_1 \sin\theta_1 = n_2 \sin\theta_2$$

▲ **fig B** Refraction can alter perspective.

The values of n_1 and n_2 are the refractive indices in each medium. The values of θ_1 and θ_2 are the angles that the ray of light makes to the normal to the interface between the two media at the point the ray meets that interface, as shown in **fig C**.

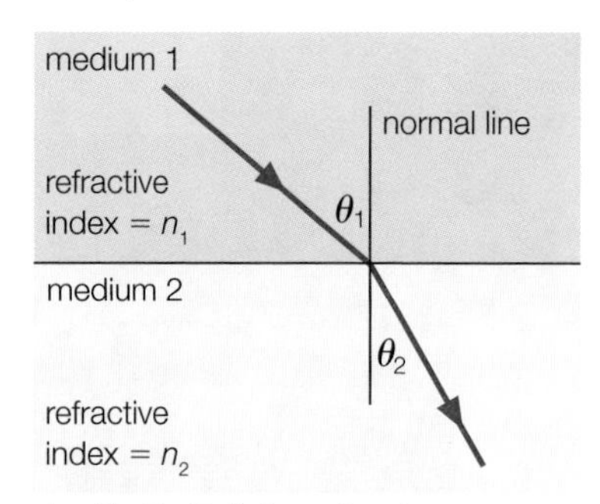

▲ **fig C** Defining refraction.

LEARNING TIP

In any optical system, light rays will follow exactly the same path if they are transmitted back from the end in the reverse direction. This means the angles involved in refraction are the same whether the light moves from the less dense medium into the more dense one and the opposite is also true. Remember though, apart from rays along the normal line, the angle to the normal will always be larger in the less dense medium.

WORKED EXAMPLE

What is the angle of refraction when a light ray passes from water (n_1 = 1.33) into glass (n_2 = 1.50), hitting the interface between the two at an angle of 48° to the normal?

$$n_1 \sin\theta_1 = n_2 \sin\theta_2$$

$$\sin\theta_2 = \frac{n_1 \sin\theta_1}{n_2}$$

$$\sin\theta_2 = \frac{1.33 \times \sin 48°}{1.50} = 0.659$$

$$\theta_2 = \sin^{-1} 0.659$$

$$\theta_2 = 41.2°$$

EXAM HINT

Note that the steps and layout of the solution in this worked example are suitable for refraction questions in the exam.

PRACTICAL SKILLS

Investigating refractive index

Using the equation for refractive index with experimental measurements will allow us to measure the refractive index of a material, as long as we know n for the other material. We usually do these experiments with air as the known material. As the speed of light in air is almost the same as in a vacuum, we take the refractive index of air to be $n_{air} = 1.00$.

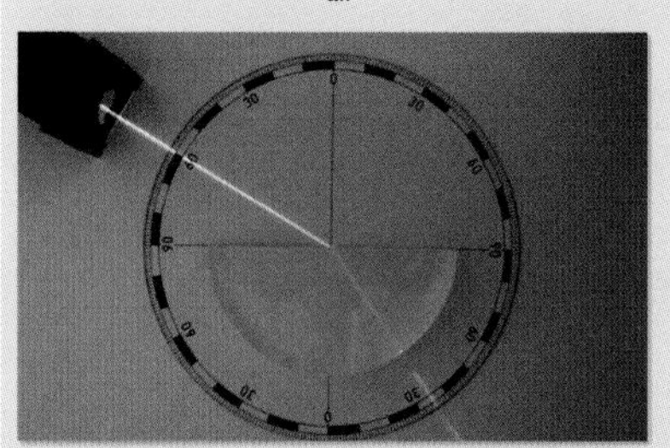

fig D Measuring angles of incidence and refraction in order to find the refractive index of a material.

Using a setup like that shown in **fig D**, we can take several different measurements of the angle of incidence (θ_1) and the corresponding angle of refraction (θ_2). The block used should be an exact semicircle, and we aim the ray to meet the glass at exactly the midpoint of the flat side. This means that the ray passing through the block will be travelling along a radius of the semicircle, and will leave the glass exactly along the normal to the circular edge. So the only change in direction occurs at the flat side of the block. Comparing the equation we have for refractive index with the equation for a straight line:

$$n_1 \sin\theta_1 = n_2 \sin\theta_2$$

$$n_1 = 1.00$$

$$\therefore \quad \sin\theta_1 = n_2 \sin\theta_2$$

$$\therefore \quad \sin\theta_2 = \frac{\sin\theta_1}{n_2}$$

$$y = mx + c$$

A plot of $\sin\theta_2$ on the y-axis against $\sin\theta_1$ on the x-axis should produce a straight best-fit line that passes through the origin. The gradient, m, of this line will be equal to the reciprocal of the refractive index for glass.

$$\therefore \quad n_{glass} = \frac{1}{m}$$

!

Safety Note: The raybox may get hot enough to burn skin. Do not knock glass blocks together as sharp chips may fly off and cause injury.

DISPERSION

One of the most well known phenomena in physics is the splitting of white light into a rainbow of colours by a prism, as shown in **fig E**.

$$n_1 \sin\theta_1 = n_2 \sin\theta_2 \qquad n = \frac{c}{v} = \frac{c}{f\lambda}$$

If the ray enters a prism from air ($n = 1.00$) our equations become:

$$\sin\theta_1 = \frac{c\sin\theta_2}{f\lambda} \qquad \sin\theta_2 = \frac{f\lambda\sin\theta_1}{c}$$

As the frequency stays constant throughout refraction, and the speed of light in a vacuum must be constant, if we keep the same angle of incidence then the sine of the angle of refraction will be proportional to the wavelength. Smaller wavelengths (violet) will be closer to the normal in the glass (compare this with **fig E**).

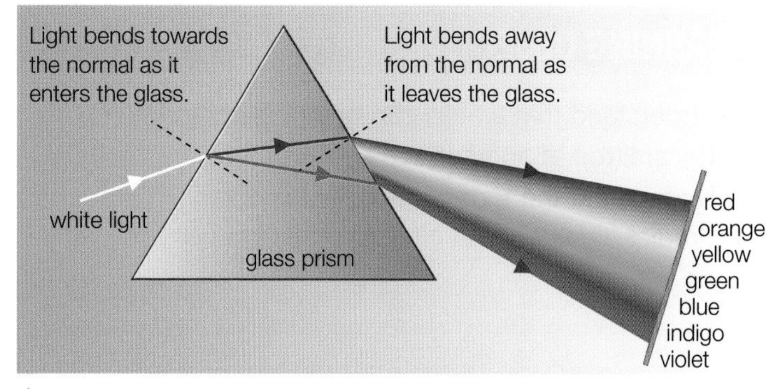

fig E Refraction affects different wavelengths (colours) differently.

On emergence from the glass prism, take care to continue with θ_1 as the angle in air:

$$\sin\theta_1 = \frac{c\sin\theta_2}{f\lambda}$$

The sine of the angle of refraction (θ_1 on emergence) will now be inversely proportional to the wavelength. Smaller wavelengths (violet) will be further from the normal in the air (compare this with **fig E**). As the angles of incidence on emergence are not all the same, due to the dispersion when they first entered the glass, the colour spreading effect is amplified, creating the familiar rainbow, or spectrum, of colours.

CHECKPOINT

SKILLS ADAPTIVE LEARNING, PROBLEM SOLVING

1. What happens to each of the following when a ray of light travels from air into clear plastic?
   (a) speed (b) wavelength (c) frequency
2. Why does the head of the giraffe in **fig B** appear to be disconnected from the body?
3. When light travels from air into glass ($n_{glass} = 1.50$) with an incident angle of 25°, what is the angle of refraction inside the glass?

SUBJECT VOCABULARY

refraction a change in wave speed when the wave moves from one medium to another. There is a corresponding change in wave direction, governed by Snell's law

refractive index, n, the amount that a material changes the speed of waves when they pass through the material from a different material:

$$\text{refractive index} = \frac{\text{speed of light in vacuum}}{\text{speed of light in the medium}} \qquad n = \frac{c}{v}$$

Snell's law the values of n_1 and n_2 are the refractive indices in each medium, and the values of θ_1 and θ_2 are the angles that the ray makes to the normal to the interface between the two media, at the point the ray meets that interface:

$$n_1 \sin\theta_1 = n_2 \sin\theta_2$$

3C 2 TOTAL INTERNAL REFLECTION

SPECIFICATION REFERENCE 2.3.46 2.3.47 2.3.55

LEARNING OBJECTIVES

- Understand that waves can be reflected and transmitted at an interface between media.
- Understand the term critical angle.
- Predict whether total internal reflection will occur at an interface.

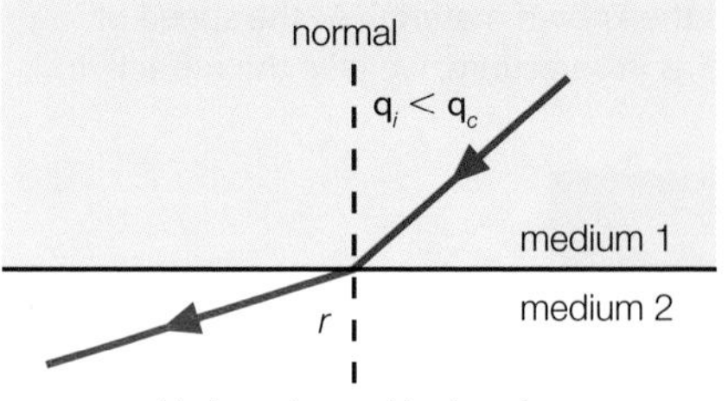

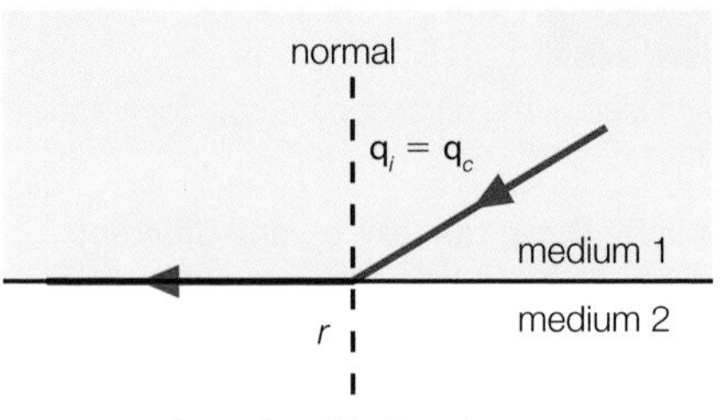

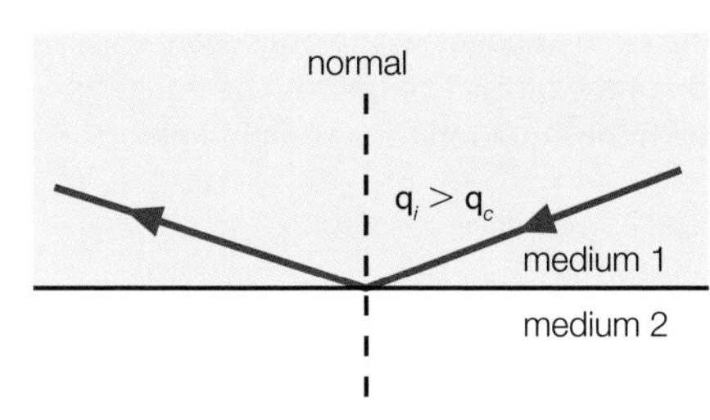

▲ **fig B** Moving from refraction to total internal reflection.

PARTIAL REFRACTION

When waves pass from one medium into another, some wave energy will pass through and some will be reflected. The proportions will depend on the amount of refraction and the angle of incidence.

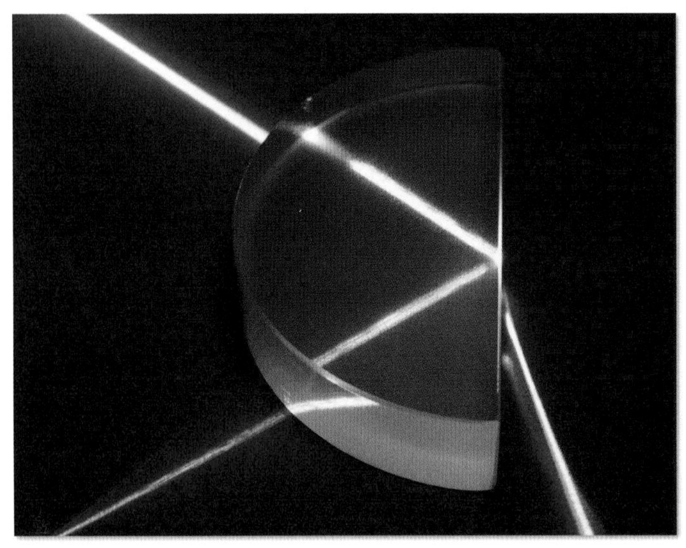

▲ **fig A** Partial transmission and partial reflection of light at a media interface.

In the diagrams of **fig B** we can see a ray of light trying to leave an optically more dense medium. In these examples, the partial reflection is not drawn. The first diagram shows simple refraction, for which we could use Snell's law to make some calculations. The angle in the less dense medium is greater than the incident angle inside the more dense medium. In the second diagram, the incident angle has been increased. From Snell's law, we find that at this **critical angle**, the ray would emerge in the less dense medium at an angle of 90° – it would emerge exactly along the interface. If the angle within the more dense medium is increased further, the emergent angle should increase also, but this would then be greater than 90°. This means the ray emerges within the more dense medium. This is not refraction, as that requires a change in medium. This means that Snell's law cannot apply. In this case, what happens is that all the wave energy is reflected inside the more dense medium, and the angles follow the law of reflection. This is **total internal reflection (TIR)**.

CRITICAL ANGLE CALCULATIONS

From Snell's law, we could find the critical angle for a material:

$$n_1 \sin\theta_1 = n_2 \sin\theta_2$$

If we take medium 1 to be the optically more dense material, then θ_2 must be 90° when the light is at the critical angle, θ_c, in medium 1.

$$n_1 \sin\theta_1 = n_2 \sin\theta_2$$

$$n_1 \sin\theta_c = n_2 \sin 90°$$

$$\sin 90° = 1$$

$$\therefore \quad n_1 \sin\theta_c = n_2 \qquad \sin\theta_c = \frac{n_2}{n_1}$$

If the situation involves a light ray emerging into air, then the equation becomes:

$$\sin\theta_c = \frac{1}{n_1}$$

If we know the critical angle, then that will give us the refractive index for the material:

$$n_1 = \frac{1}{\sin\theta_c}$$

APPLICATIONS OF TIR

Fig C shows the movement of light through 45° prisms. The left-hand example is commonly used in periscopes, and the right-hand diagram is the fundamental basis for reflective signs.

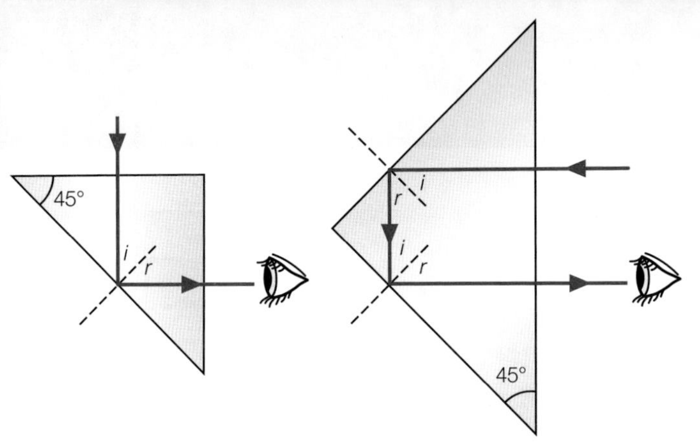

▲ **fig C** Total internal reflection helps us direct light usefully.

A more complex use for TIR is in fibre optics. A thin glass fibre can guide light along its length by the repeated TIR at the internal edges. This may just be used for decorative lighting, Fibre optics can be used to guide sunlight to the interior of large buildings. Alternatively, optical fibres can be used to carry information as light pulses (as in fibre broadband) or as actual images (in uses such as a medical endoscope – see **fig E**).

PRACTICAL SKILLS

Investigating total internal reflection

Using a setup like that shown in **fig D**, we can steadily increase the angle of incidence (θ_1) within the more dense glass and observe the emerging angle of refraction (θ_2) in the air. As the angle inside the glass increases, we will see a partial reflection of the ray within the glass growing stronger in intensity, until the critical angle is reached inside the glass.

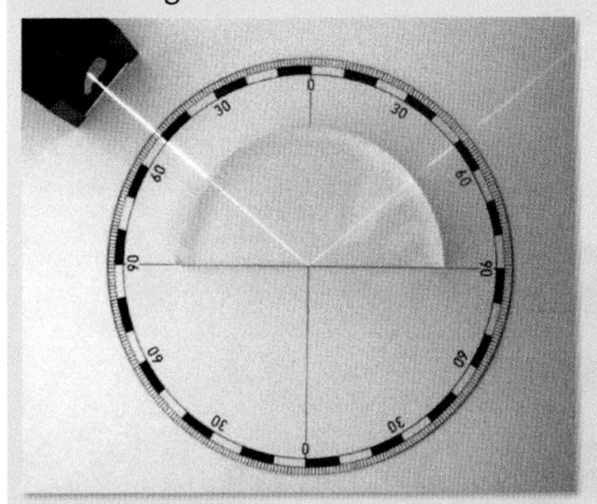

▲ **fig D** Measuring angles of incidence and refraction in order to find the critical angle and refractive index for glass.

The block used should be an exact semicircle, and we aim the ray to travel through the glass and hit exactly the midpoint of the flat side. This means that the ray passing through the glass will be travelling along a radius of the semicircle and will have entered the glass exactly along the normal to the circular edge. So the only change in direction occurs at the flat side of the block. By carefully recording the critical angle, when the light emerges, we can calculate the refractive index for the glass.

$$n_{glass} = \frac{1}{\sin\theta_c}$$

Safety Note: The raybox may get hot enough to burn skin. Do not knock glass blocks together as sharp chips may fly off and cause injury.

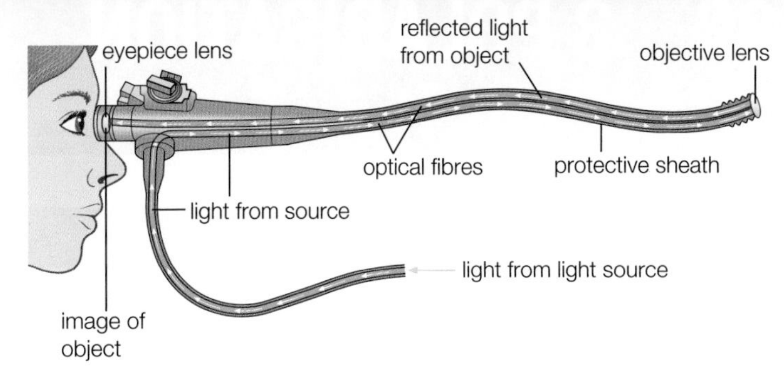

▲ **fig E** A medical endoscope is used to view the body's internal organs without cutting the patient open. Light is sent in along one optic fibre, and the reflection is carried away along the other for viewing by medical staff.

CHECKPOINT

SKILLS INTERPRETATION, ADAPTIVE LEARNING

1. What is the meaning of the term *critical angle*?
2. Explain how the reflections shown in **fig D** are possible.
3. Water has a refractive index of 1.33. What is the critical angle for water?
4. Explain why fish can observe predator birds in any part of the sky by only viewing a cone above them, as shown in **fig F**.

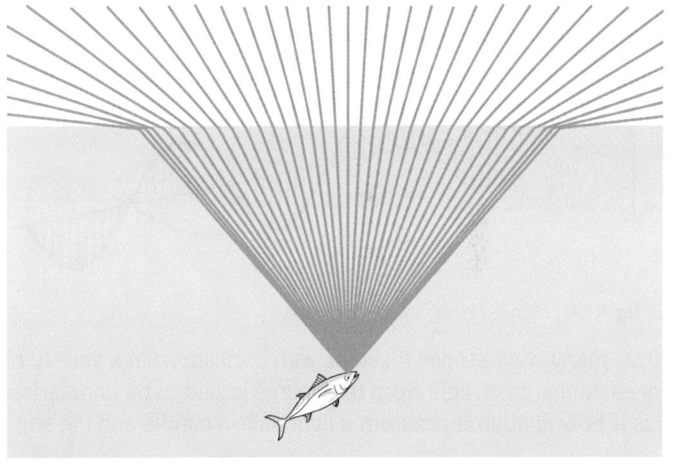

▲ **fig F** Fish can easily see all of the world above the water.

SUBJECT VOCABULARY

critical angle the largest angle of incidence that a ray in a more optically dense medium can have and still emerge into a less dense medium. Beyond this angle, the ray will be totally internally reflected

total internal reflection (TIR) waves reflect back into the same medium at a boundary between two media. This requires two conditions to be met:

- the ray is attempting to emerge from the more dense medium
- the angle between the ray and the normal to the interface is greater than the critical angle

3C 3 POLARISATION

SPECIFICATION REFERENCE 2.3.49

LEARNING OBJECTIVES

- Understand what is meant by plane polarisation.
- Describe how polarisation can be used with models to investigate stresses in structures.

PLANE POLARISATION

Transverse waves have oscillations at right angles to the direction of motion. In many cases, the plane of these oscillations might be in one fixed orientation. **Fig A** shows the electric (red) and magnetic (blue) fields in an electromagnetic wave. In this example of a light wave, the electric fields only oscillate in the vertical plane. The wave is said to be plane polarised or, more precisely, vertically plane polarised. For electromagnetic waves, the plane of the electric field's oscillations is the one that defines its plane of **polarisation**.

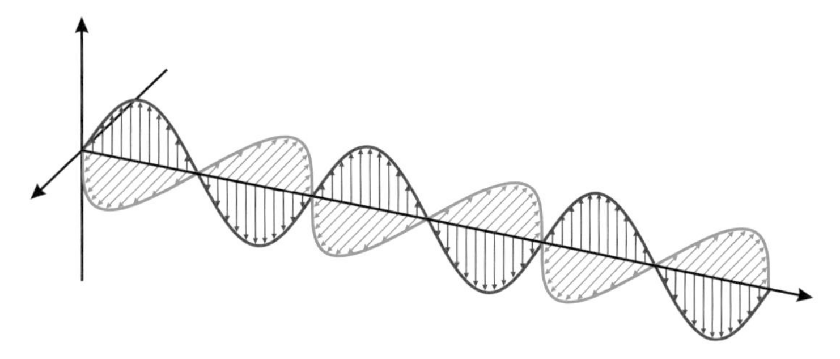

▲ **fig A** A polarised electromagnetic wave.

Often, many waves travel together, with oscillations in a variety of planes. In this case, light from this source is said to be unpolarised. This is how light emerges from a light bulb, a candle and the sun.

LEARNING TIP

Polarisation is only possible with transverse waves. If a wave is polarised, it must be a transverse wave.

POLARISING FILTERS

Unpolarised radiation can be passed through a filter that will transmit only those waves that are polarised in a particular plane. Waves on a string could be polarised simply by passing the string through a card with a slit in it, which will then only allow oscillations to pass through if they are in line with the slit. For light waves, the polariser is a piece of plastic soaked with chemicals with long chain molecules, called a Polaroid sheet. The Polaroid filter will only allow light waves to pass if their electric field oscillations are orientated in one direction.

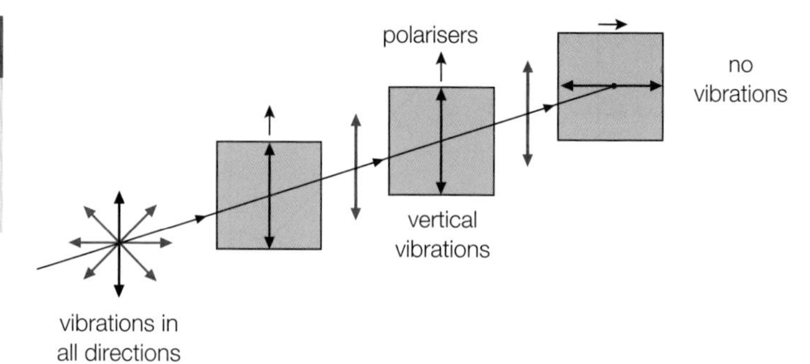

▲ **fig B** Polaroid filters transmit light waves if their plane of polarisation matches with the orientation of the filter.

Fig B illustrates the effects of Polaroid filters on light from a bulb. It starts unpolarised, with vibrations in all directions. The first filter only permits vertical vibrations, so they are selected, and the light is transmitted vertically plane polarised. The second filter is oriented in the same vertical direction, and the light will pass through this without change. The third Polaroid filter is oriented horizontally. This blocks vertically polarised light, and so no waves are transmitted beyond it. The second and third filters are referred to as 'crossed Polaroids', as their orientations are at right angles and, together, they will always block all light.

PRACTICAL SKILLS

Investigating structural stresses

We can use crossed Polaroids to observe stress concentrations in clear plastic samples.

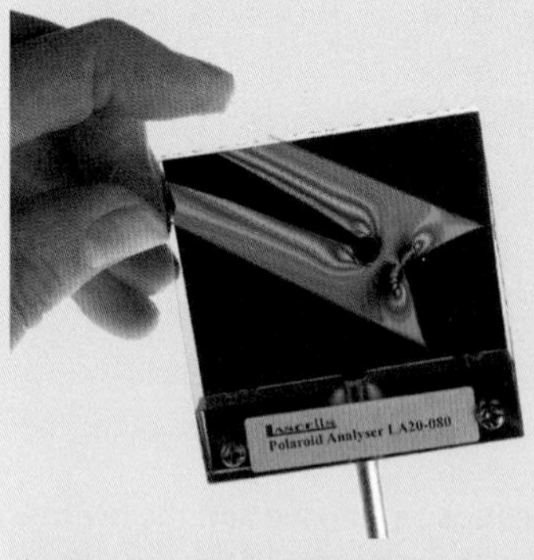

▲ **fig C** Stress analysis using crossed Polaroids.

The first Polaroid produces polarised light, which passes into the plastic sample. Stressed areas have their molecules in slightly different orientations, and this will affect the passage of the light through the plastic. This effect varies with the colour of the light. When the second Polaroid acts on the emerging light, some of the light will have travelled slightly more slowly through the plastic and will destructively interfere with other light waves of the same colour. Thus, depending on the degree of stress in the plastic, the colours that emerge will vary. Differently stressed areas appear as different colours through the second Polaroid. Changing the stresses on the plastic will alter the internal stresses, and the interference pattern will change. Engineers use this to see stress concentrations in models of structures, and to observe how the stress concentrations change when the amount of stress changes. This allows them to alter the design to strengthen a structure in regions of highest stress.

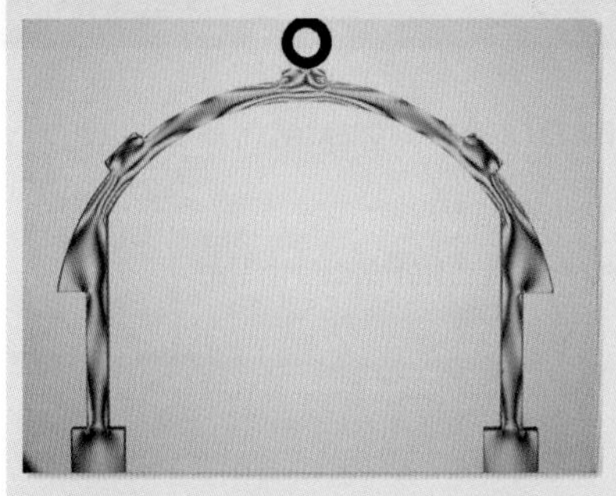

▲ **fig D** Engineering stress analysis using crossed Polaroids.

POLARISATION BY REFLECTION AND REFRACTION

When unpolarised light reflects from a surface, such as a road, the waves will become polarised. The degree of polarisation depends on the angle of incidence, but it is always tending towards horizontal plane polarisation, as shown in **fig E**. Sunglasses often have lenses with Polaroid filters. The Polaroid filters will block light that is horizontally polarised. If a driver has these, then they will block polarised light reflected from the road. This makes it easier for a driver to see clearly.

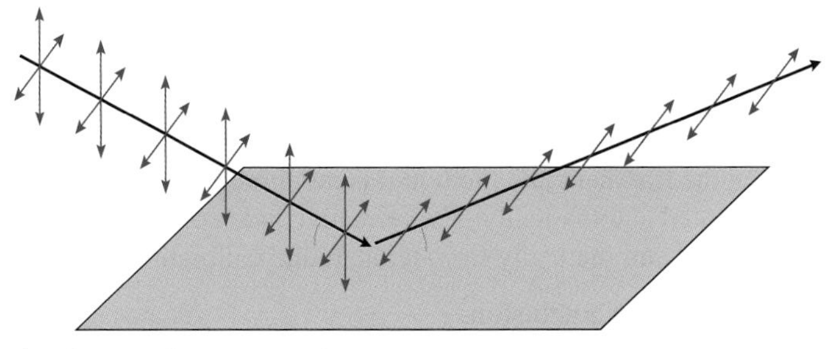

▲ **fig E** Reflection can polarise waves.

Light waves incident on a surface into which they can refract (see **Section 3C.1**), such as a pond, will reflect partially horizontally polarised light as in **fig E**, but will also transmit partially vertically polarised light into the new medium, as in **fig F**.

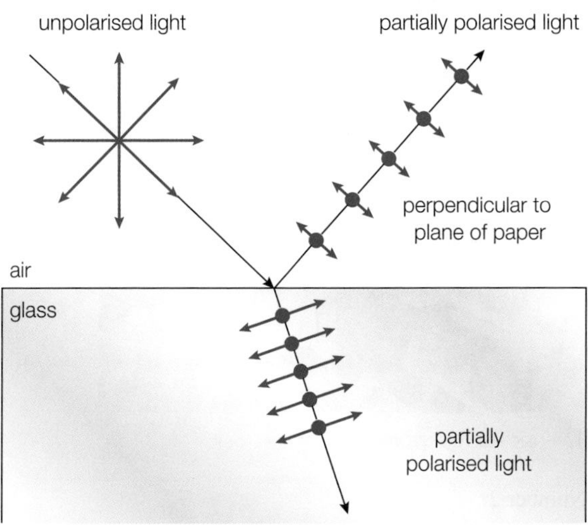

▲ **fig F** Refraction can polarise waves.

POLARISATION BY CHEMICAL SOLUTIONS

The analysis of stress concentrations investigated above works because different parts of the plastic model have different effects on polarised light. This is also the case with some chemicals, such as sugar solution. The amount of the concentration of the sugar solution varies the angle to which it rotates the polarisation of the light. We can use Polaroid filters to analyse the strength of the sugar solution, by measuring the angle at which the light polarisation emerges after passing through the solution.

▲ **fig G** Sugar solutions can rotate the plane of polarisation. The degree of polarisation on a particular wavelength (colour) depends on the concentration of the solution, and how far the light has had to pass through it. This gives rise to the changing colours seen along the length of this tube of sugar solution.

CHECKPOINT

SKILLS CRITICAL THINKING

1. Silbab says that he has a 'sound wave polariser' on his new music system. Explain why his claim is incorrect.
2. What does it mean to say that a wave is unpolarised?
3. Why do ski goggles often have Polaroid filters with a vertical orientation?
4. Explain the benefits of being able to use polarisation to analyse stress concentrations in engineering models.

EXAM HINT

When asked about the orientation of polarising filters, think carefully about whether you want to block the light, or for the filter to transmit it.

SUBJECT VOCABULARY

polarisation the orientation of the plane of oscillation of a transverse wave. If the wave is (plane) polarised, all its oscillations occur in one single plane

3C THINKING BIGGER

GLASS FORENSICS

SKILLS CRITICAL THINKING, PROBLEM SOLVING, CREATIVITY, PRODUCTIVITY

Refractive indices can be determined to several decimal places, and tiny variations can highlight very small differences in the compositions of materials like glass.

In this activity, we will look at some ideas about identifying glass in crime scene investigations.

SCIENTIFIC JOURNAL PAPER

FORENSIC GLASS COMPARISON

Introduction

Glass can be found in most places. It is produced in a wide variety of forms and compositions, and these affect the properties of this material. It can occur as evidence when it is broken during a crime. Broken glass fragments ranging in size from large pieces to tiny fragments may be transferred to and retained by nearby persons or objects. The presence of fragments of glass on the clothing of an alleged burglar in a case involving entry through a broken window may be important evidence if fragments are found. The importance of such evidence will be enhanced if the fragments are found to have the same properties as the broken window. However, if the recovered fragments are different in their measured properties from the glass from the broken window, then that window can be eliminated as a possible source of the glass on the subject's clothing (Koons et al. 2002).

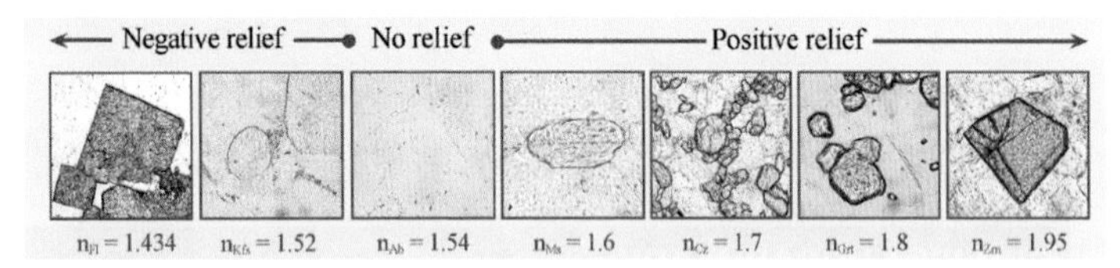

fig A Crystals with differing refractive indices viewed within an oil bath with refractive index, n = 1.54. Those with higher or lower refractive index are easily visible, whereas the quartz with the same refractive index is virtually invisible.

Optical Properties

Refractive index is a function of chemical composition and atomic arrangement (Stoiber and Morse 1981). Refractive index is the most commonly measured property in the forensic examination of glass fragments (Koons et al. 2002), because:

- It can be measured quickly from small pieces.
- It can aid in the characterization of glass.
- It provides a good chance to identify the glass exactly. (Koons et al. 2002)

Immersion methods are used to measure refractive index in some laboratories. These methods take advantage of the fact that when using monochromatic light, a particle immersed in a liquid of identical refractive index will become invisible (Bloss 1961).

In this method, a video camera captures the image of the particle edge, and a computer calculates the point of minimum contrast – the match point – across the particle edge while automatically varying the temperature. The temperature is adjusted so that the refractive index of the liquid is higher than that of the glass sample. The instrument lowers the temperature of the preparation through the match point for the glass. The contrast between the fragment and the liquid is monitored, and the match point is noted automatically. The refractive index of the sample is calculated automatically from temperature calibration data.

Interpretations/Conclusions

The variations of the properties of glass within a single source are usually small (Bottrell et al. 2007), typically below the resolving power of the techniques used to measure them. The variation among glass objects from different sources can be observed and measured (Hickman 1983; Koons et al. 1988; Koons and Buscaglia 1999; Koons and Buscaglia 2002) and are usually much greater than the variation within a single object.

fig B A forensic lab refractometer.

By Maureen Bottrell, from *Forensic Science Communications*, April 2009 – Volume 11 – Number 2, published by the Federal Bureau of Investigation, Quantico, Virginia, USA

SCIENCE COMMUNICATION

1 The extract opposite is from a paper in a scientific journal on the forensic lab investigations that can be done on glass.

(a) The paper quoted is a type of scientific paper known as a 'review paper', in which the author does not present new research, but summarises other scientists' primary research in order to present a large amount of current knowledge. How does the way the paper is written tell you that this is a review paper?

(b) Is there any bias present in the report? Explain how you decided on your answer to this.

PHYSICS IN DETAIL

Now we will look at the physics in detail. Some of these questions link to topics earlier in this book, so you may need to combine concepts from different areas of physics to work out the answers.

2 Explain why a crystal in oil will become invisible if the two have identical refractive indices.

3 Why will a change in the temperature of oil change its refractive index?

4 Why does a refractometer need a powerful microscope system as part of its set up?

5 Glass or plastic from different sunglasses lenses can often have identical refractive indices. Suggest another method by which two large (>1 cm^2) pieces of glass from different sunglasses could be identified to match up with their original owners.

WRITING SCIENTIFICALLY

As you read these articles, identify everything that contributes to writing a scientific article. For example, the vocabulary, sources, the way information is presented. Make sure your own answers are written scientifically.

ACTIVITY

The Young modulus can be affected by the atomic impurities in a material. Prepare a set of experimental instructions for use by a forensic lab technician. These are to help the technician to identify the Young modulus of a sample of glass in order to compare a sample found on a suspect with a sample from a broken window. Explain what quantities should be measured, how to analyse the measurements to determine the Young modulus, and what level of accuracy should be expected in the comparisons. Include instructions for the preparation of the sample, including cutting to the best size and shape.

THINKING BIGGER TIP

Remember that the Young modulus can be determined as the stiffness constant for a material in compression as well as for one in tension.

3C EXAM PRACTICE

1 Which of these is the correct unit for refractive index?

A °　　**B** $m\,s^{-1}$

C m　　**D** no units [1]

(Total for Question 1 = 1 mark)

2 What is the critical angle for Perspex which has a refractive index of 1.48?

A 0.026°　　**B** 0.74°

C 42.5°　　**D** 47.5° [1]

(Total for Question 2 = 1 mark)

3 A student did an experiment to measure the critical angle in a glass block.
Which row in the table correctly shows the readings for which he could find the critical angle.

	Angle of incident ray / °	Angle of resulting ray / °
A	10	15
B	24	45
C	42	90
D	57	57

[1]

(Total for Question 3 = 1 mark)

4 Which of the following cannot be experimentally demonstrated with sound waves?

A Diffraction

B Polarisation

C Two source interference

D Ultrasound [1]

(Total for Question 4 = 1 mark)

5 Two parallel rays of light, one blue, one red, are travelling in air and are incident on one side of a glass prism. The blue light passes into the prism and meets the second face at the critical angle as shown in the diagram.

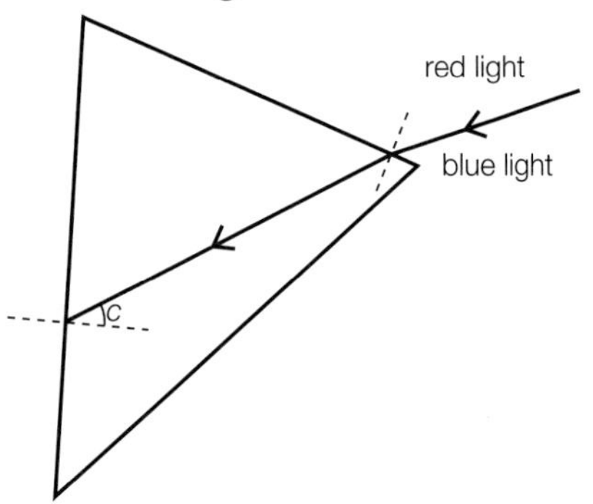

(a) Add to the diagram the path of the blue light after it meets the second face. Label this path X. [1]

(b) (i) The speed of blue light in the glass prism is $1.96 \times 10^8\,m\,s^{-1}$.
Calculate the refractive index of this glass for blue light. [2]

(ii) Calculate the critical angle for blue light in this glass prism. [2]

(c) The refractive index of this glass for red light is less than for blue light. Add to the diagram to complete the path of the red light through the prism. Label this path Y. [2]

(Total for Question 5 = 7 marks)

6 If you look into a fish pond on a bright sunny day, you sometimes cannot see the fish because of the glare of light reflected off the surface. When the sunlight is reflected off the surface of the water it is plane-polarised.

(a) State the difference between plane-polarised and unpolarised light. [1]

(b) Explain how Polaroid sunglasses can enable the fish to be seen. [3]

(c) State why sound waves cannot be polarised. [1]

(Total for Question 6 = 5 marks)

7 A student carries out an experiment to measure the refractive index of glass. He does this by shining a ray of light through a semicircular glass block and into the air as shown.

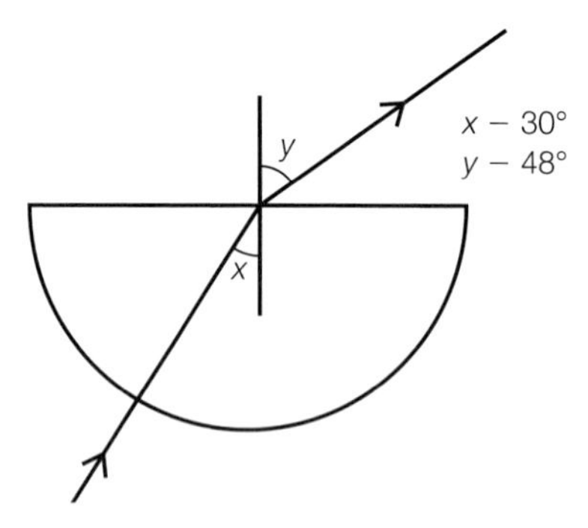

(a) Calculate the refractive index from air to glass, ${}_an_g$. [2]

(b) (i) The student steadily increases the angle x in glass and finds that eventually the light does not pass into the air. Explain this observation. [3]

(ii) Calculate the largest value of angle x that allows the light to pass out of the block into the air. [2]

(Total for Question 7 = 7 marks)

8 Optical fibres have many uses in medicine and communications. They can also be incorporated into materials such as fabrics.

Some optical fibres are made from a central core of transparent material surrounded by a material of a different refractive index as a cladding.

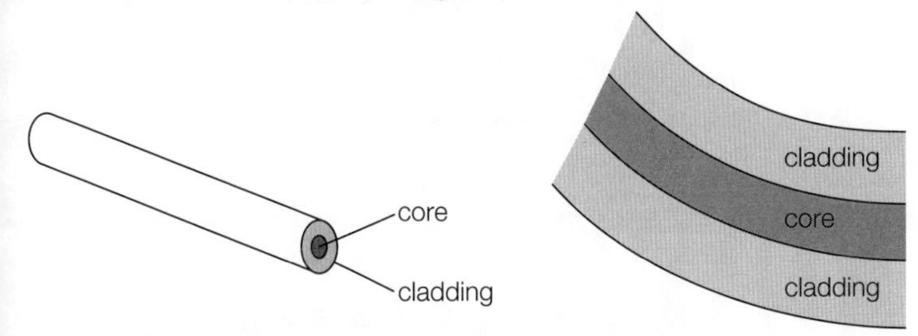

speed of light in the core = $1.96 \times 10^8\,\mathrm{m\,s^{-1}}$
speed of light in the cladding = $2.03 \times 10^8\,\mathrm{m\,s^{-1}}$

(a) Calculate the critical angle for the core-cladding boundary. [3]

(b) The diagram below shows a ray of light inside the core of a fibre. The ray is incident on the core-cladding boundary at an angle of 80°.

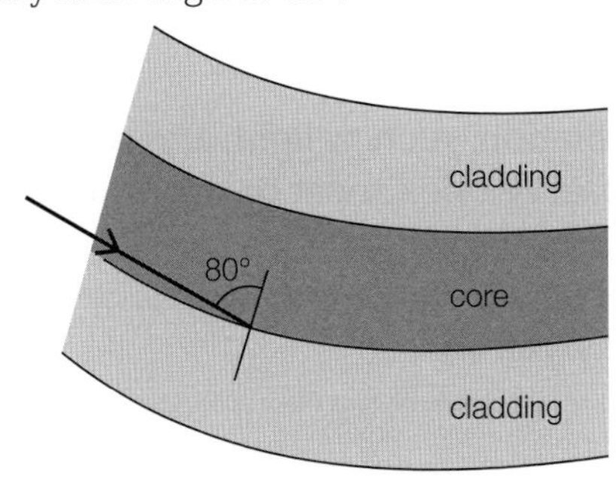

State what happens to this ray of light when it is incident on the core-cladding boundary as shown. [1]

(c) The light source for these curtains is at the top. Suggest why the bottom of the curtain is much brighter than the rest of the curtain. [2]

(Total for Question 8 = 6 marks)

9 Rainbows are seen when sunlight is dispersed by raindrops. The light is separated into different colours because they each take different paths through raindrops.
A ray of white light is incident on a raindrop. The diagram shows the subsequent path of the red light.

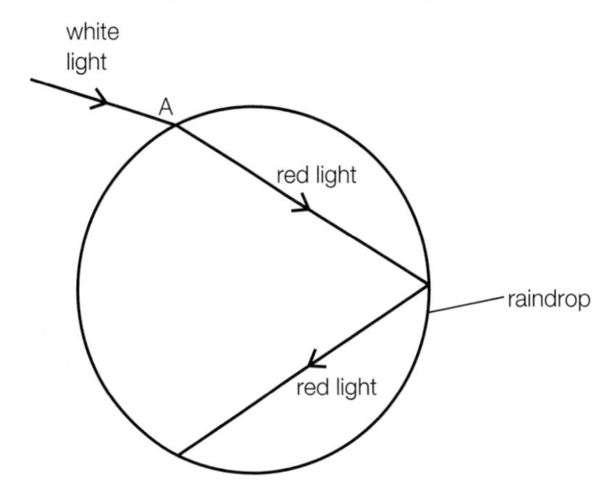

(a) Name the effect that is experienced by the red light at A. [1]

(b) (i) On the diagram, label an angle of incidence with an i and an angle of refraction with an r. [2]

(ii) On the diagram, draw the path that a violet ray of light would take, through the raindrop and into the air. [2]

(c) (i) State what is meant by the critical angle. [1]

(ii) Calculate the critical angle for red light in the raindrop.
refractive index for red light in water = 1.3 [2]

(d) Red light has a frequency of $4.2 \times 10^{14}\,\mathrm{Hz}$ and travels at a speed of $2.2 \times 10^8\,\mathrm{m\,s^{-1}}$ in the raindrop.
Calculate the wavelength of the red light in the raindrop. [2]

(Total for Question 9 = 10 marks)

10 Explain how infrared radiation can travel the length of an optical fibre, and describe and explain applications of this. [6]

(Total for Question 10 = 6 marks)

TOPIC 3 WAVES AND THE PARTICLE NATURE OF LIGHT

CHAPTER 3D QUANTUM PHYSICS

Albert Einstein is one of the world's most famous physicists. So what did a mainly self-educated office worker in the Swiss patent office produce that could make him so important? In 1905, Einstein had what is often referred to as his Annus Mirabilis – year of miracles – in which he published scientific papers on four separate topics. All four papers made significant breakthroughs in the topics they were concerned with, and were diverse, not directly related to each other. His paper on the photoelectric effect won the 1921 Nobel Prize for Physics.

In this chapter, we will be dealing with the ideas that electromagnetic radiation, such as light, and particles, such as electrons, can behave as both waves and particles. Usually we only observe one aspect or the other, such as the photoelectric effect demonstrating that light can be a particle. Scientists are still not clear on an explanation for both light and electrons that can universally describe their behaviour under all circumstances. Einstein helped us along the way, but he did not solve the puzzle. Maybe someday you will.

MATHS SKILLS FOR THIS CHAPTER

- **Units of measurement** (*e.g. the electronvolt, eV*)
- **Use of standard form and ordinary form** (*e.g. using a wavelength of light*)
- **Changing the subject of an equation** (*e.g. using the photon energy equation*)
- **Solving algebraic equations** (*e.g. finding the colour of light emitted by an excited atom*)
- **Substituting numerical values into algebraic equations** (*e.g. a photoelectric effect calculation*)
- **Determining the slope and intercept of a linear graph** (*e.g. analysing photoelectric cell data*)

What prior knowledge do I need?

Topic 3A

- Models of waves and wave processes
- The existence and properties of the electron
- The electronic structure of atoms
- Macroscopic emission and absorption of light

Topic 3B

- Diffraction and interference of light
- Detailed knowledge of the parts of the electromagnetic spectrum, and the frequencies of electromagnetic radiations

What will I study in this chapter?

- The idea of wave–particle duality, and the nature of light
- Photon energy
- The photoelectric effect
- Electron diffraction and interference
- How light is created

What will I study later?

Topic 7 (Book 2: iAL)

- Experimental investigations of the structure of subatomic particles

Topic 7A (Book 2: IAL)

- How to calculate the wavelength of an electron

Topic 7B (Book 2: IAL)

- The process of thermionic emission, and its similarities to and differences from the photoelectric effect

Topic 7C (Book 2: IAL)

- Details about the smallest fundamental particles yet found

Topic 11B (Book 2: IAL)

- Blackbody radiation of all wavelengths of electromagnetic radiation, and the colours of stars

3D 1 WAVE-PARTICLE DUALITY

SPECIFICATION REFERENCE 2.3.50 2.3.57 2.3.58

LEARNING OBJECTIVES

- Understand how the behaviour of electromagnetic radiation can be described in terms of both waves and photons.
- Use the equation $E = hf$ for the energy of a photon of electromagnetic radiation.

HUYGENS' PRINCIPLE

The Dutch scientist Christiaan Huygens came up with a principle for predicting the future movement of waves if we know the current position of a wavefront. The basic idea is to consider that any and every point on the wavefront is a new source of circular waves travelling forwards from that point. When the movement of these numerous circular waves is plotted, and then their superposition considered, the resultant wave will be the new position of the original wavefront.

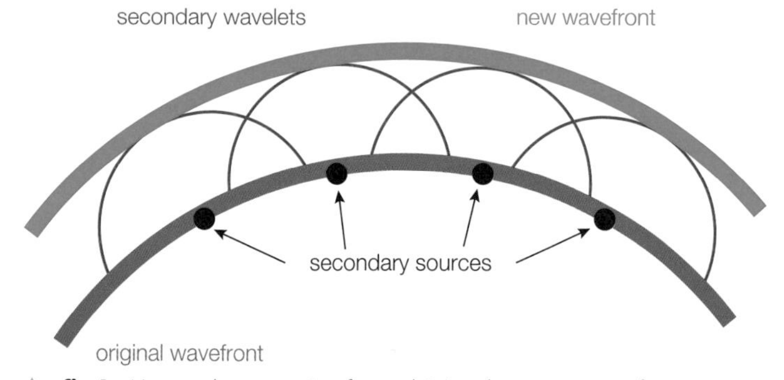

▲ **fig A** Huygens' construction for explaining the movement of waves.

Huygens' geometrical system exactly explains all the basic phenomena that we see with light waves. We can make drawings like **fig A** to correctly predict the movement of wavefronts in reflection, refraction, diffraction (**fig B**), interference and straight-line propagation of light.

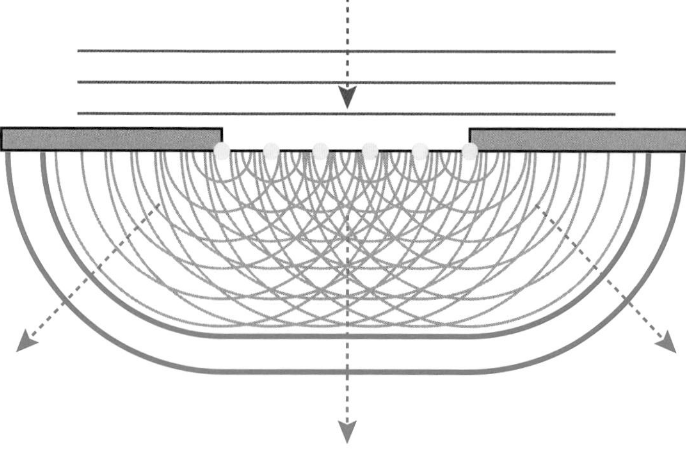

▲ **fig B** Huygen's construction correctly predicts diffraction through a gap.

LIGHT IS A WAVE

We have seen numerous instances of light undergoing wave-like activities.

EVIDENCE THAT LIGHT IS A WAVE

The interference pattern produced by diffraction and Young's two-slit experiment both require the superposition of wave displacements to generate the standing wave pattern seen. This is only possible if light is behaving as a wave. It must have repeating cycles of displacement that cause this superposition. The constructive and destructive interference gives a pattern of antinodes and nodes (maxima and minima). Particles cannot superpose in this way.

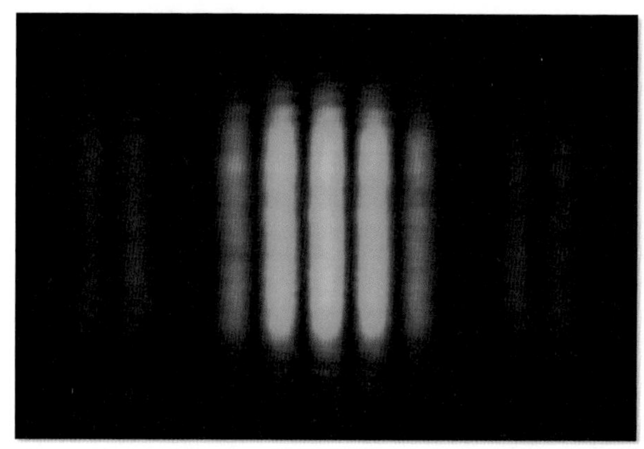

▲ **fig C** Superposition effects, such as standing wave interference patterns, are only possible if light behaves as a wave.

Polarisation is another phenomenon exhibited by electromagnetic waves, including light, that can only be explained in classical physics by using the ideas of waves.

LIGHT IS A PARTICLE

In some cases, the examples of light acting as a wave are phenomena that can only occur for waves. In **Section 3D.2** we will look in detail at the photoelectric effect. This was first observed in 1887 by Heinrich Hertz (after whom the unit for frequency is named). However, it was not until the beginning of the twentieth century that an adequate explanation of the phenomenon was made. Working on the basis of Max Planck's 1901 suggestion that light could exist as quantised packets of energy called **photons**, Einstein produced a paper called, in English, *Concerning an Heuristic Point of View Toward the Emission and Transformation of Light.* Einstein wrote four important papers in 1905, including one on relativity and one regarding $E = mc^2$, but it was the explanation of the photoelectric effect for which he was awarded the 1921 Nobel Prize in Physics. Crucially, the photoelectric effect cannot be explained using a wave theory for light. However, the idea that light travels as particles, or photons, whose energy is proportional to the frequency it would have when considered as a wave, fits all the observations exactly.

PHOTON ENERGY

For electromagnetic radiation, the energy of the photon can be calculated by multiplying the frequency by Planck's constant, h. This constant has an extremely small value because it represents the fundamental minimum possible step in energy. In SI units, $h = 6.63 \times 10^{-34}$ J s. If we work on very small scales, photons cannot have energy values that differ by less than the Planck constant. This means that there are some energy values that are impossible in our universe. Such a system of minimum sized steps is called **quantisation**.

The energy of a photon is given by the equation:

photon energy (J) = Planck's constant (J s) × frequency (Hz)

$$E = hf$$

WORKED EXAMPLE

Calculate the energy of a photon of yellow light that has a wavelength of 550 nm.

$$E = hf$$

$$c = f\lambda$$

$$\therefore f = \frac{c}{\lambda} = \frac{3 \times 10^{8}}{550 \times 10^{-9}} = 5.45 \times 10^{14}\text{ Hz}$$

$$E = 6.63 \times 10^{-34} \times 5.45 \times 10^{14}$$

$$E = 3.6 \times 10^{-19}\text{ J}$$

Calculate the frequency of a gamma ray photon that has 2.24×10^{-14} joules of energy.

$$E = hf$$

$$\therefore f = \frac{E}{h} = \frac{2.24 \times 10^{-14}}{6.63 \times 10^{-34}}$$

$$f = 3.38 \times 10^{19}\text{ Hz}$$

EXAM HINT

Note that the steps and layout of the solution in this worked example are suitable for questions on photon energy and frequency in the exam.

ELECTRONS ARE PARTICLES

Experiments that produce ions can demonstrate electrons behaving as particles because a fixed lump of mass and charge is removed from the atom in order to change the atom into an ion. The charge to mass ratio is a unique identifying property of particles, and was first demonstrated for the electron by J.J. Thomson in 1897. Robert Millikan, in an experiment finally published in 1913, took this one step further to find the electron charge itself. The fact that electrons hold a fixed amount of charge and a fixed mass indicates they are localised particles.

▲ **fig D** Robert Millikan's oil drop apparatus showed that electrons have a fixed charge of 1.5924×10^{-19} coulombs. The difference with the currently accepted value is caused by a mistake in Millikan's calculations, because he used a slightly incorrect value for the viscosity of air.

ELECTRONS ARE WAVES

If electrons are made to travel at very high speeds, they will pass through gaps and produce a diffraction pattern. They will also interact with a double-slit apparatus to produce the interference pattern seen when waves pass through two slits. Diffraction and interference are not expected by classical particles, as they should simply travel straight through the slits. Observation of these experimental results proves that electrons can behave as waves.

WAVES OR PARTICLES?

The experimental observations highlighted in this section have suggested that both electrons and electromagnetic radiation seem to contradict our explanations of them. Physicists have good theoretical descriptions of the wave and particle natures for electrons and also for electromagnetic radiation. However, we do not have a complete and perfect single theory that explains both correctly for either electrons or EM radiation.

PHENOMENON	EVIDENCE FOR WAVES	EVIDENCE FOR PARTICLES
light	diffraction, interference, polarisation	photoelectric effect
electron	diffraction, interference	ionisation

table A Summary of the evidence for light and electrons behaving as waves or particles.

The idea that these things behave as waves under certain circumstances and as particles under other circumstances is known as **wave–particle duality**.

CHECKPOINT

SKILLS DECISION MAKING

1. Suggest an experiment that demonstrates:
   (a) light behaving as a wave
   (b) electrons behaving as waves.
2. What is the energy of a photon of UV light that has a wavelength of 2×10^{-7} m?

SKILLS INNOVATION

3. Demonstrating the polarisation of a beam of electrons requires a complex experiment. If electrons were shown to be travelling like polarised light, explain whether this would confirm that the electrons were particles or waves.
4. Is light a wave or a particle? Explain your answer.
5. Draw a picture illustrating how Huygens' construction could explain diffraction of sea waves around a large rock.

SUBJECT VOCABULARY

photons 'packets' of electromagnetic radiation energy where the amount of energy $E = hf$, which is Planck's constant multiplied by the frequency of the radiation: the quantum unit that is being considered when electromagnetic radiation is understood using a particle model

quantisation the concept that there is a minimum smallest amount by which a quantity can change: infinitesimal changes are not permitted in a quantum universe. The quantisation of a quantity is like the idea of the precision of an instrument measuring it

wave–particle duality the principle that the behaviour of electromagnetic radiation can be described in terms of both waves and photons

3D 2 THE PHOTOELECTRIC EFFECT

SPECIFICATION REFERENCE 2.3.59 2.3.60 2.3.61 2.3.62

LEARNING OBJECTIVES

- Explain the photoelectric effect and experimental observations of it.
- Understand how the photoelectric effect provides evidence for the photon model of light.
- Use the photoelectric effect equation.

PHOTOELECTRONS

If ultraviolet light is shone onto a negatively charged zinc plate, the plate loses its charge. The explanation for this is that the light causes electrons to leave the metal, removing the negative charge. The electrons that are released are called **photoelectrons**. However, variations on this experiment produce some surprising observations. There is a certain minimum amount of energy that an electron needs in order to escape the surface of the metal. This energy is called the **work function**, and has the symbol, ϕ.

The wave theory of light would allow some of the wave energy to be passed to the electrons to enable them to gain the work function and escape. In the wave theory, the energy carried by a wave depends on its amplitude, which for light means its brightness. So the wave theory would predict that any colour of light, if made sufficiently bright, should enable the release of photoelectrons. Alternatively, if the wave were of smaller amplitude, it could shine for longer and slowly pass energy to the electrons, until they had gained as much as the work function and escape. However, this is not what is observed. The same negatively charged zinc plate cannot be discharged by red light, no matter how bright or how long the illumination. No photoelectrons are emitted under red light. There is a maximum wavelength for the light, above which no photoelectrons are ever emitted. This can be thought of equivalently as a minimum frequency for the light, known as the **threshold frequency**.

PRACTICAL SKILLS

Investigating photoelectrons

If you charge a gold leaf electroscope then it will show a deflection on the gold leaf. If you place a zinc plate on top of the electroscope, the photoelectric effect will allow the electroscope to be discharged if it held a negative charge. If you shine UV light onto the zinc plate, the gold leaf will fall immediately. This observation shows that electrons on the zinc plate have escaped. Those electrons remaining on the gold leaf and electroscope stalk spread out more, so that there is no longer enough mutual repulsion to hold up the gold leaf. You may need to clean the surface of the zinc, as it oxidises quite easily in air. As such, there may be a layer of oxide on top of the pure metal, stopping the light absorption.

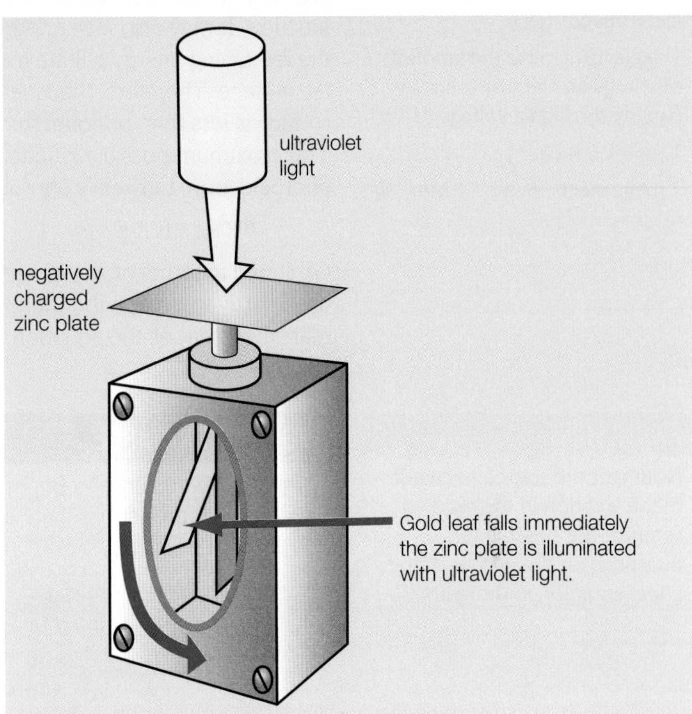

▲ **fig A** Demonstrating the photoelectric effect.

If you attempt to discharge the electroscope with an ordinary desk lamp, or sunlight, there will be very little effect. This is because these sources provide little UV light intensity, and longer wavelengths will not cause the photoelectric effect with zinc. Even if you left the electroscope in front of the desk lamp for several hours, it would not discharge. However, the UV light need not be very bright, and it will still immediately cause discharge.

Safety Note: Do not view ultraviolet light directly or by reflection to avoid retinal damage. If electroscope is charged with a high voltage connection, use a safe connection method to avoid electric shock.

The explanation for these experimental observations can be summarised as:

- Light travels as photons, with a photon's energy proportional to the frequency.
- When a photon encounters an electron, it transfers all its energy to the electron (the photon ceases to exist).
- If an electron gains sufficient energy – more than the work function – it can escape the surface of the metal as a photoelectron.
- Brighter illumination means more photons per second, which will mean a greater number of photoelectrons emitted per second.
- If an electron does not gain sufficient energy from an encounter with a photon to escape the metal surface, it will transfer the energy gained from the photon to the metal as a whole before it can interact with another photon. Thus, if the photon energy is too low, no photoelectrons are observed.

LEARNING TIP

The SI units for the terms in the photoelectric effect equation are all joules (J), but it is often used in units of electronvolts (eV).

The electronvolt is the amount of energy an electron gains by passing through a voltage of 1 V.

1 eV = 1.6×10^{-19} J

1 mega electronvolt = 1 MeV = 1.6×10^{-13} J

THE PHOTOELECTRIC EFFECT EQUATION

Einstein developed an equation to explain the photoelectric effect, which is based on the conservation of energy. We have previously seen that the energy of a photon can be found from $E = hf$. During the photoelectric effect, the energy of a photon is transferred to an electron. To escape a metal surface, an electron requires at least an amount of energy equal to the work function. It may also lose energy before escaping the metal due to other physical processes. Only the remaining energy will be available for the electron emission kinetic energy to be emitted from the surface. Therefore, the kinetic energy that the electron can have on emission from the metal surface is less than or equal to the difference between the photon energy and the work function. This maximum possible kinetic energy $\frac{1}{2}mv^2_{max}$ determines the maximum initial velocity v_{max} of the photoelectron. Einstein's photoelectric effect equation then is:

$$\frac{1}{2}mv^2_{max} = hf - \phi$$

Or, stated in terms of the conservation of energy, the photon energy transferred to the electron is used partly to escape the metal surface and partly for its kinetic energy on leaving the metal. So, a rearrangement of the equation is:

$$hf = \phi + \frac{1}{2}mv^2_{max}$$

EXAM HINT

Note that the steps and layout of the solution in this worked example are suitable for questions on the photoelectric effect equation in the exam.

WORKED EXAMPLE

What is the work function of potassium if green light with a wavelength of 510 nm shining on it produces photoelectrons that have a maximum kinetic energy of 0.14 eV?

$$\frac{1}{2}mv^2_{max} = hf - \phi$$

$$\therefore \quad \phi = hf - \frac{1}{2}mv^2_{max}$$

$$\frac{1}{2}mv^2_{max} = 0.14\text{ eV} = 0.14 \times 1.6 \times 10^{-19} = 2.24 \times 10^{-20}\text{ J}$$

$$hf = \frac{hc}{\lambda} = \frac{6.63 \times 10^{-34} \times 3 \times 10^8}{510 \times 10^{-9}} = 3.9 \times 10^{-19}\text{ J}$$

$$\phi = hf - \frac{1}{2}mv^2_{max} = 3.9 \times 10^{-19} - 2.24 \times 10^{-20}$$

$$\phi = 3.68 \times 10^{-19}\text{ J}$$

THE PHOTOELECTRIC CELL EXPERIMENT

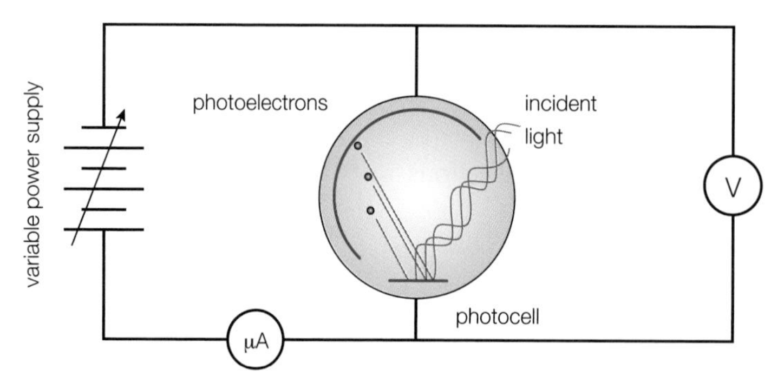

▲ **fig B** A photoelectric effect cell experiment circuit.

We could use the photoelectric effect equation to measure Planck's constant and the work function for a metal. In a vacuum, we place the metal as the anode in a cell that has a gap to the cathode – see **fig B**. When we shine light of a known frequency onto the anode, photoelectrons will be emitted and the current registered on the ammeter. If we slowly increase the pd across the photoelectric cell, eventually the anode will become sufficiently positive that all photoelectrons will be stopped and attracted back to it, so the photoelectric current will be zero. This **stopping voltage**, V_s, will give us the maximum kinetic energy of the photoelectrons, from the definition of voltage:

$$\frac{1}{2}mv^2_{max} = e \times V_s$$

If we use a range of light frequencies and find the stopping voltage for each, we can plot a graph of the photoelectron maximum kinetic energy, on the y-axis, against frequency, on the x-axis. Comparison with the equation for a straight line shows us that the graph should produce a straight best-fit line and the gradient will be equal to Planck's constant.

$$\frac{1}{2}mv^2_{max} = hf - \phi$$

$$y = mx + c$$

The y-intercept will represent the value of the work function, ϕ. Also, when the value of y is zero (the x-intercept) then the photon energy must equal the work function.

$$\frac{1}{2}mv^2_{max} = hf - \phi$$
$$0 = hf - \phi$$
$$\therefore \quad hf = \phi$$

This means that the value of the x-intercept will give the threshold frequency for the anode metal.

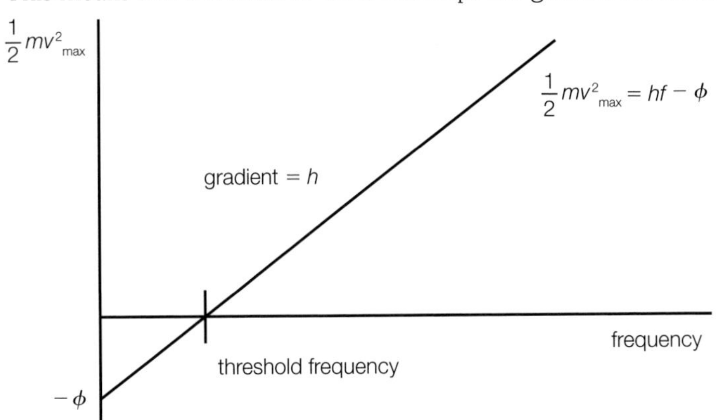

▲ **fig C** Graphical analysis of results from a photoelectric cell experiment can determine Planck's constant and the work function for the anode metal.

EXAM HINT

When you have a graph with plotted data points, start by drawing a best-fit line. The question may not ask you to do this, but it will be expected.

METAL	WORK FUNCTION / eV
cadmium	4.07
caesium	2.10
iron	4.50
nickel	5.01
zinc	4.30

table A Examples of photoelectric work function values.

CHECKPOINT

SKILLS CRITICAL THINKING, INNOVATION

1. What is the work function of potassium in electronvolts?
2. (a) Why would iron not release photoelectrons if red light were shone on it?
   (b) How would this result change if the red light were turned up to double brightness?
3. How does the photoelectric cell experiment described above get over the problem that the electrons released will have a variety of kinetic energies, depending on how deep in the metal they start?
4. What is the threshold frequency of caesium?

SUBJECT VOCABULARY

photoelectrons electrons released from a metal surface as a result of its exposure to electromagnetic radiation
work function the minimum energy needed by an electron at the surface of a metal to escape from the metal
threshold frequency the minimum frequency of electromagnetic radiation that can cause the emission of photoelectrons from the metal
stopping voltage the minimum voltage needed to reduce the photoelectric current to zero, when illuminated with a particular frequency of light

3D 3 ELECTRON DIFFRACTION AND INTERFERENCE

SPECIFICATION REFERENCE 2.3.53 2.3.54

LEARNING OBJECTIVES

- Explain how diffraction experiments provide evidence for the wave nature of electrons.
- Describe other evidence for the wave nature of electrons.
- Be able to use the de Broglie equation $\lambda = \frac{h}{p}$.

We have seen that waves passing through a gap will produce a diffraction pattern with bright and dark fringes as the diffracted waves from each side of the gap superpose to produce the interference pattern.

▲ **fig A** Circular diffraction pattern caused by a laser passing through a pinhole.

ELECTRON DIFFRACTION

As a diffraction pattern is a wave phenomenon, to observe this from a beam of electrons means that they must be behaving as waves. This is true whether we are passing waves through a gap, as in **fig A**, or reflecting waves from a grating, as we saw in **fig H** of **Section 3B.3**.

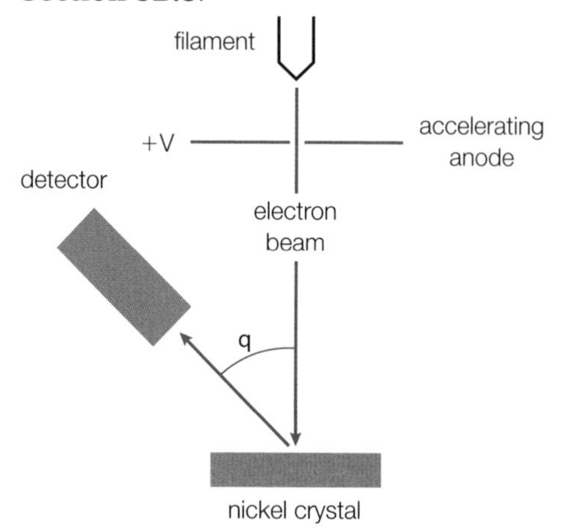

▲ **fig B** Davisson and Germer reflected a beam of electrons from a nickel crystal and measured the intensity of the reflection at different angles.

In 1927, Davisson and Germer tried to detect diffraction of electron 'waves' when they reflected from a crystal of nickel. **Fig B** illustrates the experimental setup. They measured the intensity of the beam at different angles for various accelerating voltages, and plotted a graph of their results. **Fig C** shows the graph for a beam accelerated through 54 volts.

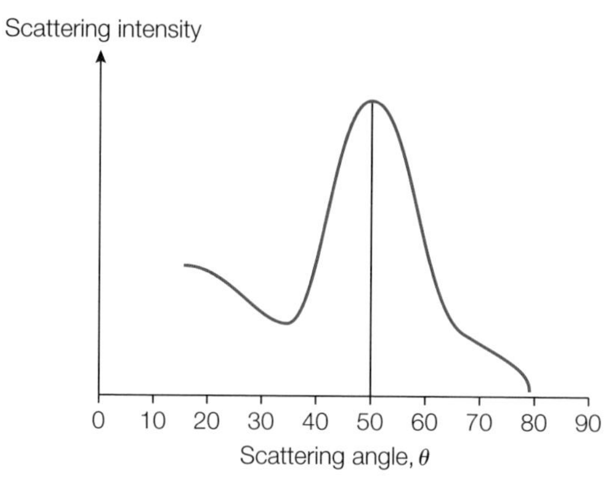

▲ **fig C** Davisson and Germer's electron beam reflection from a surface of atoms, acting as a reflection grating, showed variable intensity at different angles, exactly as is observed with waves.

Not only did Davisson and Germer prove experimentally that electrons can behave as waves, but their results also allow calculation of the distance between atoms in the nickel crystal. This has given rise to advances in the study of atomic structures using electron beam crystallography.

DE BROGLIE EQUATION

In 1924, a French prince called Louis de Broglie suggested electrons could behave as waves and proposed an equation to calculate their wavelength. Their wavelength is inversely proportional to the momentum they have when considered as particles.

$$\text{electron wavelength (m)} = \frac{\text{Planck's constant (J s)}}{\text{momentum (kg m s}^{-1}\text{)}}$$

$$\lambda = \frac{h}{p}$$

WORKED EXAMPLE

Calculate the wavelength of electrons travelling at 10% of the speed of light.

$$\lambda = \frac{h}{p} = \frac{6.63 \times 10^{-34}}{9.11 \times 10^{-31} \times 0.1 \times 3 \times 10^{8}}$$

$$= 2.43 \times 10^{-11}\ \text{m}$$

EXAM HINT

Note that the steps and layout of the solution in this worked example are suitable for questions requiring the de Broglie equation in the exam.

PRACTICAL SKILLS

Investigating electron diffraction

Although they are quite expensive, some school laboratories have an electron beam diffraction tube. This accelerates a beam of electrons through a high voltage, and then passes the beam through a thin piece of graphite. The carbon atoms in the graphite acts as a diffraction grating in two dimensions, which produces a circular diffraction pattern. The front end of the tube has a phosphorescent screen that will show up the diffraction pattern.

▲ **fig D** An electron beam diffraction tube demonstrates the wave nature of electrons.

If we make careful measurements of the dimensions of the tube and the diffraction pattern produced with different accelerating voltages, we can carry out an approximate calculation of the atom spacing in graphite.

Safety Note: Electron beam tubes operate at a hazardous voltage. Operate with shrouded connections to minimise the risk of electric shock.

TWO-SLIT ELECTRON INTERFERENCE

In 1965, Richard Feynman suggested that electrons should also be able to produce the two-slit interference pattern seen with light, as they can behave as waves. Recently, this has been shown to be the case, giving further evidence for the wave nature of electrons.

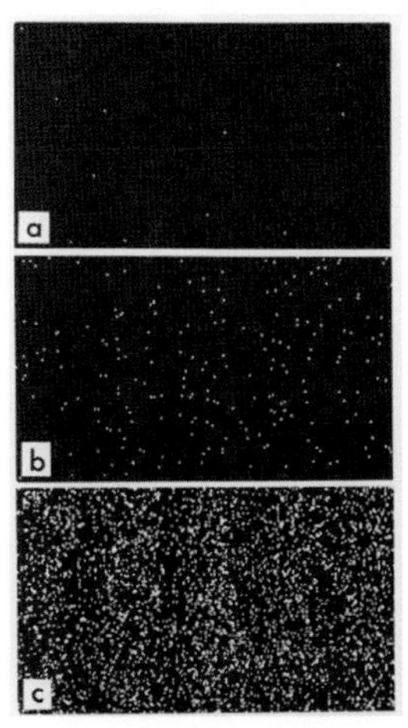

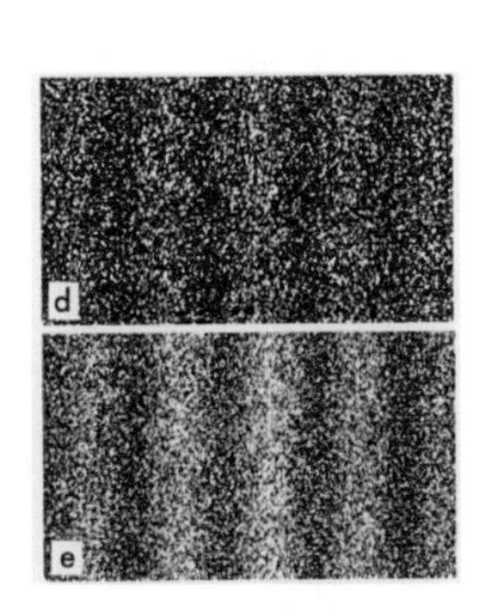

▲ **fig E** Recent experiments have been able to demonstrate electrons producing an interference pattern when passed through a two-slit-type experiment. The five images here show the same image after different amounts of time. The particle electrons hit the screen in places that slowly build up the wave interference pattern.

However, the latest research has allowed scientists to observe that the interference pattern is built up by the movement through the apparatus of the individual electrons. They are behaving as both individual particles and waves at the same time. They are showing wave–particle duality.

ELECTRON MICROSCOPY

One of the most important applications of the wave nature of electrons is its use to study objects at very small scales. An electron beam can work like a beam of light, but with some different properties that can make the electron beam much more useful for microscopy. The main advantage is that the wavelength of an electron beam can be controlled by altering the voltage applied to accelerate the electrons. In some instances, this can mean the wavelength of electrons in the beam can be 10^{-10} m – the same size as atoms. Approximately, the minimum-sized object that can be imaged by any wave is about the same size as the wavelength. In this example, the electron waves are 1000 times shorter than light waves, and so they can produce images of objects 1000 times smaller than visible light.

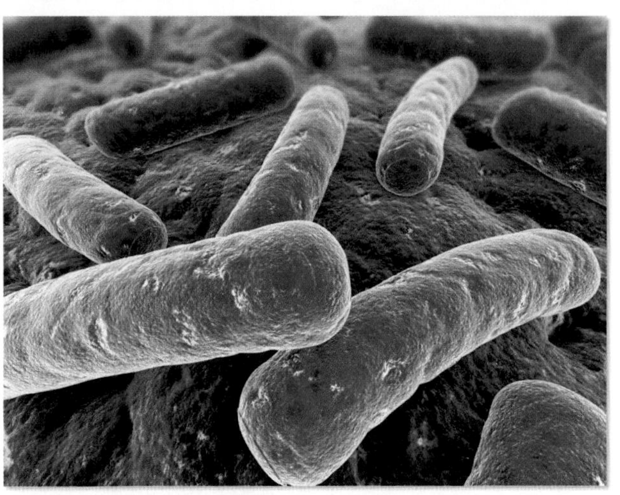

▲ **fig F** Electron microscopes can have a magnification up to 10 million times. These bacteria are about 1 micrometre long.

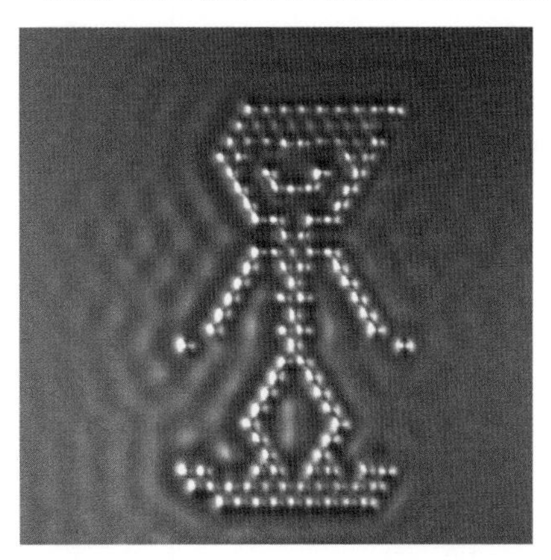

▲ **fig G** Electron microscopes have sufficient magnification to see the results of manipulation of individual atoms. This Atom Boy image was made by researchers at IBM who moved individual carbon monoxide molecules on a surface of copper metal.

CHECKPOINT

SKILLS PROBLEM SOLVING

1. Give an example of electrons behaving as a wave, and explain why this can only occur for waves.
2. Why would an electron microscope beam need to be accelerated through a higher voltage in order to view a smaller object?
3. What would be the wavelength of electrons that had been accelerated through a 400 V pd?
4. Estimate the de Broglie wavelength of a football moving towards the goal.

LEARNING OBJECTIVES

- Understand atomic line spectra in terms of energy level transitions.
- Calculate the frequency of radiation emitted or absorbed in an electron energy transition.
- Calculate the intensity of a source of light.

ELECTRON ENERGY LEVELS

In **Section 4A.6** we will see that electrons in semiconductors can have varying amounts of energy, and that the energy they have can put them into the valence band or conduction band. These energy bands are wide in solids – there is a large range of values of energy that the electron could have and still be in that band. In free atoms, such as those in a gas, the energy values that the electrons could have are limited to a small number of exact values, often called energy levels.

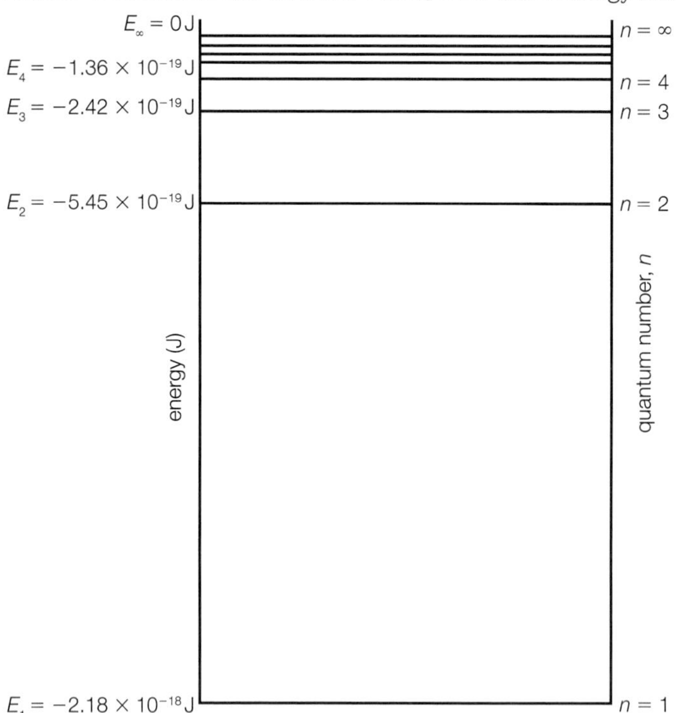

▲ **fig A** An energy level diagram for hydrogen. The energy values are negative, as we have to put energy in to lift the electron up from the ground state.

EXCITATION AND DE-EXCITATION

Fig A illustrates the energy values that electrons of hydrogen could have. Under normal circumstances, an electron in an atom of hydrogen would be in its **ground state**. This is the lowest energy level, with a quantum number (or level) of $n = 1$. In order to move up energy levels, the electron must take in some energy. This is called **excitation**. Electrons can become excited if the atom collides with another particle. Alternatively, if the electron absorbs a photon that has exactly the correct amount of energy, the electron can jump to a higher energy level. For example, the difference in energy between the ground state (-2.18×10^{-18} J) and the $n = 2$ state (-5.45×10^{-19} J) is 1.635×10^{-18} J.

A photon with exactly this energy could be absorbed by an electron in the ground state of hydrogen, and this would lift the electron up to the energy level above the ground state, $n = 2$, and the photon would no longer exist. This is illustrated in **fig B**.

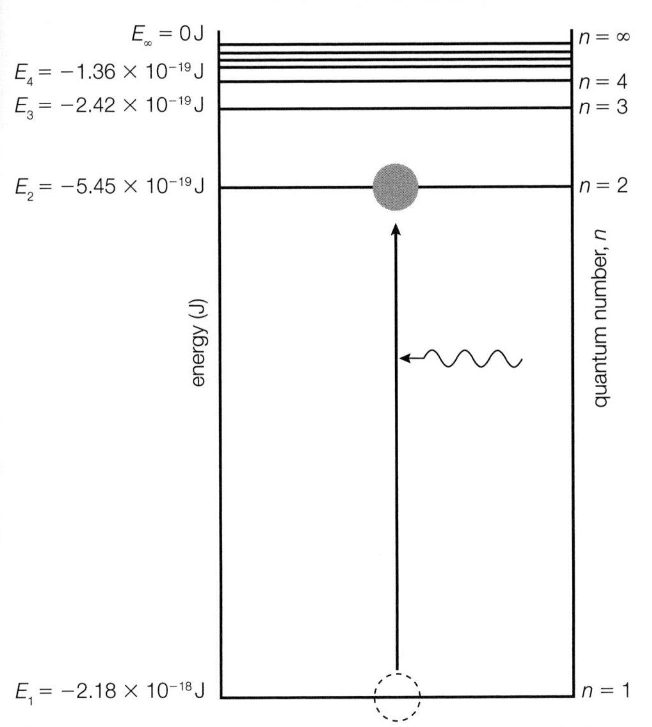

▲ **fig B** Electron excitation in a hydrogen atom. For ease of reference, the energy levels are numbered with integers from $n = 1$ for the ground state, upwards.

An incident photon that does not have the energy exactly equivalent to a jump between the current position of the electron and one of the higher levels will not be absorbed – the photon and electron will not interact at all. If gas atoms are illuminated by a range of frequencies (colours), those with the correct frequency values will be absorbed, so there will be some colours missing from the light after it passes through the gas.

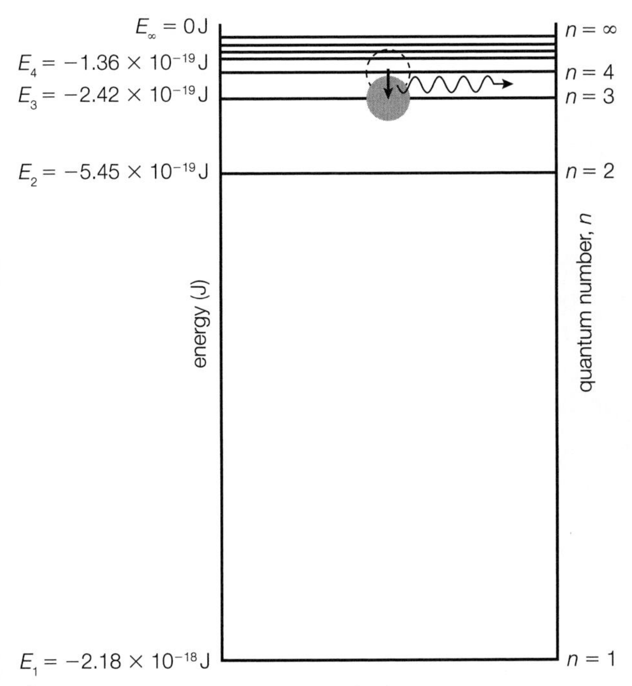

▲ **fig C** Electron de-excitation in a hydrogen atom.

If an electron is already excited, after a random amount of time it will de-excite. This may involve dropping straight down to the ground state, or it may drop to an intermediate level if there is one. **Fig C** shows an excited hydrogen electron dropping from energy level $n = 4$ down to $n = 3$. As the electron ends up with less energy than it had, the conservation of energy requires that this energy is emitted as a photon with exactly the energy difference between the levels. The frequency of the emitted photon can be calculated from the equation for photon energy, $E = hf$. So, a collection of gas atoms that are excited will emit light with a particular collection of frequencies, dependent on which element the gas atoms are.

WORKED EXAMPLE

What are the frequencies of the photon absorbed in **fig B** and the photon emitted in **fig C**?

In **fig B**, the electron is raised from level $n = 1$ up to level $n = 2$. The energy difference between the levels – the energy gained by the electron – must equal the energy of the absorbed photon.
From the diagram, the energy level difference is 1.635×10^{-18} J.

$$\Delta E = E_2 - E_1 = 1.635 \times 10^{-18}\text{ J}$$

$$\Delta E = hf$$

$$\therefore f = \frac{\Delta E}{h}$$

$$f = \frac{1.635 \times 10^{-18}}{6.63 \times 10^{-34}}$$

$$f = 2.47 \times 10^{15}\text{ Hz}$$

In **fig C**, the electron drops from level $n = 4$ down to level $n = 3$. The energy difference between the levels – the energy lost by the electron – must equal the energy of the emitted photon.
From the diagram, the energy level difference is from -1.36×10^{-19} J down to -2.42×10^{-19} J.

$$\Delta E = E_3 - E_4 = -2.42 \times 10^{-19} - -1.36 \times 10^{-19} = -1.06 \times 10^{-19}\text{ J}$$

The fact that this energy value is negative indicates that energy is given out here.

$$f = \frac{\Delta E}{h}$$

$$f = \frac{1.06 \times 10^{-19}}{6.63 \times 10^{-34}}$$

$$f = 1.60 \times 10^{14}\text{ Hz}$$

EXAM HINT

Note that the steps and layout of the solution in this worked example are suitable for exam questions requiring the calculation of frequencies of photons absorbed or emitted by a change in electron energy level.

IONISATION

Fig A shows the ground state energy of a hydrogen atom is 2.18 × 10^{-18} J, which is equivalent to 13.6 electronvolts. The $n = \infty$ level at the top of the diagram has an energy value of zero. At this level, the electron has left the atom – the hydrogen is ionised. This means that the energy required to ionise an atom of hydrogen in its ground state, its **ionisation energy**, is 13.6 eV.

LEARNING TIP

Hydrogen has the simplest energy level system as it only has one electron, but everything explained here is applicable to an electron in any atom. All we need is the energy level diagram and we can calculate the frequencies of photons emitted and absorbed for a variety of energy level transitions.

LINE SPECTRA

Light made up of multiple wavelengths (colours) can be split up to show which colours are present. This could be done using a diffraction grating in which the amount of diffraction is dependent on the wavelength, and so the various colours will spread different amounts. The resulting spectrum will often be a series of individual lines, if the original light contained only a few wavelengths. Such a **line spectrum** is the typical result of exciting the atoms of a gas, perhaps by heating the gas.

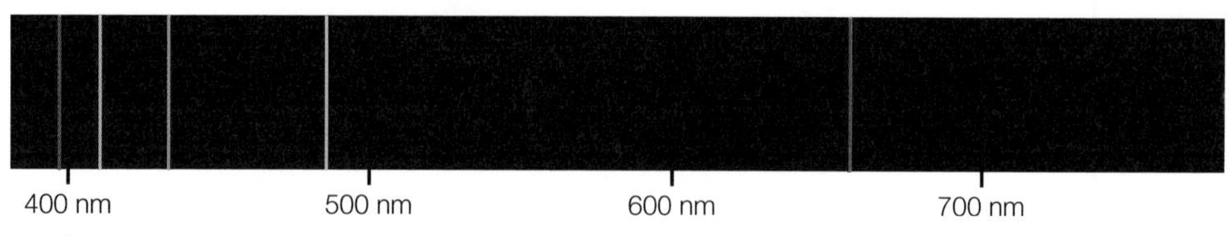

▲ **fig D** Hydrogen emission line spectrum.

LEARNING TIP

Remember that $c = f\lambda$, so photons that all have the same frequency will also all have the same wavelength (colour).

For example, **fig D** shows the line spectrum from hydrogen gas. The gas has electricity passed through it, and this will excite the hydrogen atoms. Each coloured line is a wavelength of light given off as a result of an electron dropping between two energy levels. The different energy gaps cause the difference in wavelengths emitted. When viewed all together, the colours merge, and the gas glows a purple colour.

PRACTICAL SKILLS

Investigating gas discharge spectra

A high voltage will cause an electric current to pass through a gas in a discharge tube. The electrical energy excites the electrons in each atom of the gas, and they then drop energy levels at random times, giving off a photon for each energy level transition. As there are so many atoms, it appears as if the gas is continuously emitting light, and of all the possible colours – all the possible transitions within its energy level ladder. Using a diffraction grating, we can analyse the light emitted from the tube to detect the separate colours, or wavelengths, being emitted. We observe a line spectrum.

▲ **fig E** A gas discharge tube demonstrates the emission of photons by excited electrons of a gas. A diffraction grating shows the separate lines in the spectrum.

If we change the tube for one containing a different gas, we will observe a different line spectrum.

Safety Note: Gas discharge tubes operate at a hazardous voltage. Operate with shrouded connections to minimise the risk of electric shock.

INTENSITY OF RADIATION

When a lamp emits light, we could measure its intensity. This is the amount of energy it carries, per unit area, and per unit time. Power is the rate of transfer of energy, so this becomes:

$$\text{intensity (W m}^{-2}) = \frac{\text{power (W)}}{\text{area (m}^2)}$$

$$I = \frac{P}{A}$$

WORKED EXAMPLE

On a particular day, sunlight lands on Earth with an intensity of 960 W m^{-2}. What is the power received by the solar cell on a calculator placed in this sunlight, if the cell is a rectangle 1 cm by 4 cm?

$$I = \frac{P}{A}$$

$$\therefore\ P = I \times A$$

$$= 960 \times 0.01 \times 0.04$$

$$\therefore\ P = 0.38\ \text{W}$$

EXAM HINT

Note that the steps and layout of the solution in this worked example are suitable for intensity of radiation questions in the exam.

EXAM HINT

In questions about energy level diagrams, make sure you emphasise that electrons can only have the energy values of the levels shown, so the *only possible* energy transfers match *exactly* with the differences between the levels shown.

CHECKPOINT

SKILLS INTERPRETATION

1. What is the wavelength of light emitted when an excited hydrogen electron falls from energy level $n = 2$ down to its ground state?
2. Why couldn't a photon of light with an energy of 1.43×10^{-18} joules excite hydrogen from its ground state?
3. **Fig F** shows the two important energy levels (out of the large number of available levels) of a neon atom, inside a He-Ne laser. Calculate the energy of the photon emitted when a neon atom electron excited to the n = A level drops to level B. Give your answer in electronvolts.
4. Other than by absorbing photons, explain how a ground state atom of mercury could be excited if it were in the tube of a compact fluorescent light bulb.
5. Explain why the light from a compact fluorescent light bulb containing mercury gas will produce a spectrum of various colours when electricity passes through the gas.
6. Estimate the number of photons hitting your face each second when you stand in the sunshine.

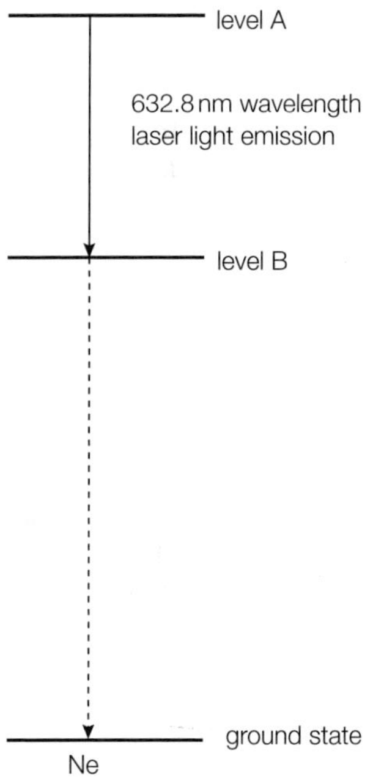

level A

632.8 nm wavelength laser light emission

level B

ground state

Ne

▲ **fig F** Electron de-excitation in a neon atom.

SUBJECT VOCABULARY

ground state the lowest energy level for a system, for example, when all the electrons in an atom are in the lowest energy levels they can occupy, the atom is said to be in its ground state

excitation an energy state for a system that is higher energy than the ground state, for example, in an atom, if an electron is in a higher energy level than the ground state, the atom is said to be 'excited'

ionisation energy the minimum energy required by an electron in an atom's ground state in order to remove the electron completely from the atom

line spectrum a series of individual lines of colour showing the frequencies present in a light source

EXAM HINT

In questions about photons and energy level diagrams, make sure you emphasise that the *only possible* photons have frequencies that match *exactly* with the differences between the energy levels shown, using $E = hf$

SOLAR CELLS TO POWER THE USA?

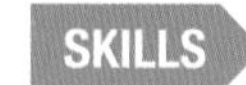

CRITICAL THINKING, PROBLEM SOLVING, ANALYSIS, INTERPRETATION, ADAPTIVE LEARNING, PRODUCTIVITY, ETHICS, COMMUNICATION, ASSERTIVE COMMUNICATION

In this activity, you will consider the quantity of solar cells that might be needed to generate enough electricity for the USA, analyse the limitations on making such a calculation, and discuss the practical difficulties in implementing such a plan.

POPULAR BLOG

AMAZING MAP: TOTAL SOLAR PANELS TO POWER THE UNITED STATES OF AMERICA

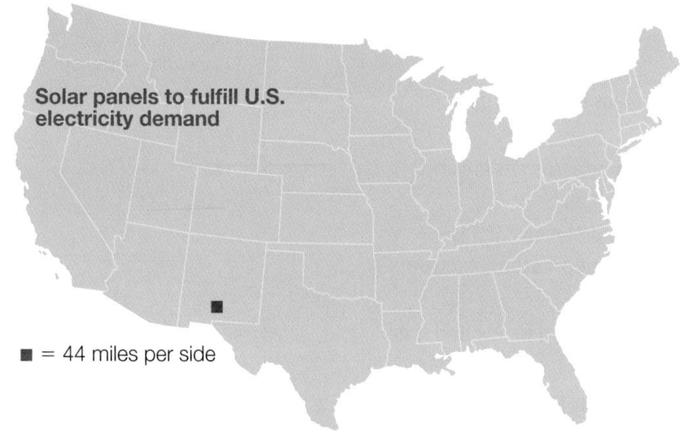

This map illustrates the total area of solar panels that would be needed to fulfill the electricity demands of the United States. The United States Energy Information Association (eia.gov) reveals in their December 2013 'Electrical Power Annual' report, in the table below, that the peak load for all interconnections of electricity during the summer of 2012 was 767 762 Megawatts within the mainland United States. It is also reported that total electricity consumption for the United States during 2012 was 3 694 650 million kW h (million kilowatt-hours).

3 694 650 000 000	kW h	Total 2012
10 122 328 767 123	Wh	Daily Average
421 763 698 630	W	Hourly Average
767 762 000 000	W	Peak Load

I imagined using a 250-watt Sharp ND–250QCS solar panel (made in the USA) 65″ × 39″ each.

767 762 000 000 watts / 250 watts per panel

3 071 048 000 total panels

PER PANEL	**TOTAL NEEDED**
2535 sq.″	54 063 240 833 sq.′
17.6 sq.′	1939 sq. miles
	44 miles per side of square

PROBLEMS

That is a lot of solar panels! 3 billion of them. However it is interesting to consider the total approximate cost of these panels compared with what the United States government spends every year on everything else… If one approximates each panel to cost $250, the total cost would be $767 billion. To put it in perspective, the government spent $3600 billion in 2012.

The raw materials and manufacturing capability would be enormous to build that many panels.

The infrastructure and additional equipment needed would be tremendous.

It's dark at night – no solar power generation… so traditional power would be required during those times. A battery storage bank would be huge for night storage at these power levels…

Solar panels and inverters are not 100% efficient. There will be cloudy days.

Less sun during the winter months unless this was built on the equator.

A SOLUTION

While it may not be possible to build a central location for our nation's electricity needs, it is possible to add to one's own electricity needs with solar photovoltaic panels. Even building an off-grid system is a possibility for many people.

In any event, it was an interesting exercise to determine how much total area of solar panels would fulfill the nation's electricity demands. It's a smaller footprint than I initially thought it would be…

By Ken Jorgustin, from Modern Survival Blog, https://modernsurvivalblog.com/alternative-energy/amazing-total-area-of-solar-panels-to-power-the-united-states/

SCIENCE COMMUNICATION

1 The article opposite was written for an American website blog. Consider the article and comment on the type of writing being used. Think about the following, for example. Is this a scientist reporting the results of their experiments, a scientific review of data, a newspaper or magazine style article for a specific audience? Then answer the following questions:

(a) Is there any bias present in the report? What parts of the article suggest any bias?

(b) How has the author adapted their use of language to suit their audience? Would the wording be different, for example, if they were trying to explain the same ideas to primary school children? Or to a US government committee for energy infrastructure?

(c) Discuss the extent to which the 'problems' highlighted by the author undermine the basic idea in the article.

PHYSICS IN DETAIL

Now we will look at the physics in, or connected to, this article. Some of these questions link to topics in much earlier sections of this book, so you may need to combine concepts from different areas of physics to work out the answers.

2 Convert all the dimensions quoted in the article into SI units: (1 inch, 1″ = 2.54 cm; 1 foot, 1′ = 12 inches; 1 mile = 1.6 km)

3 The radiation from the sun lands on Earth with a power of approximately 1000 W m^{-2}. How much sunlight power would land on one of the solar panels considered in the article?

4 If the average wavelength of light from the sun landing at the surface of the Earth is 550 nm, how many photons will be landing on this one solar panel per second? Discuss whether it is reasonable to make this wavelength average for this calculation.

5 If the panel produces 250 W, what is the efficiency of one solar panel?

6 (a) If the density of air is 1 kg m^{-3}, and the average wind speed is 6.0 ms^{-1}, what area of wind turbines would be needed to transfer the equivalent kinetic energy from the wind as the power needed in the solar cells calculation in the blog article?

(b) A wind turbine manufacturer claims that their new 'super-turbines' are 80% efficient, have 100 m diameter rotors and need 1 square kilometre of land so as not to interfere with each other's wind. How much land would be needed for a giant wind farm to power the electricity needs of the USA?

ACTIVITY

Many working environments require people to be able to communicate scientific ideas in persuasive ways. Imagine you work for a US electricity generating company that specialises in nuclear power stations. Prepare a critique of this blog article which could be used to persuade the public that the author's idea, for solar power to be used to generate all of America's electricity needs, is impractical and too expensive. You should not explain any benefits of using other means, such as nuclear power: your brief is simply to put the public off the idea of large solar power stations.

THINKING BIGGER TIP

We have not yet considered the practicalities of connecting the giant solar power station to the national grid to transport the electricity all over the USA. Some calculations about quantity and cost of cabling and pylons could add some weight to your critique.

3D EXAM PRACTICE

1 What is the energy of a photon of yellow light, with a wavelength of 550 nm?

A 5.79×10^{-38} J **B** 3.62×10^{-21} J

C 3.62×10^{-19} J **D** 2.26 J [1]

(Total for Question 1 = 1 mark)

2 Use information in the diagram, showing energy levels for electrons in a hydrogen atom, to calculate the frequency of a photon of light emitted by an electron moving from the level $n = 4$ to the ground state.

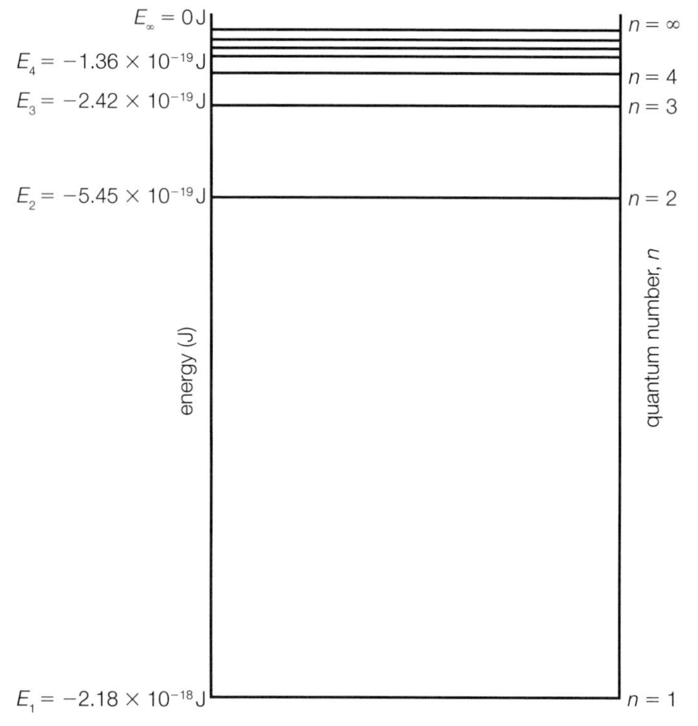

A 8.20×10^{-20} Hz **B** 2.42×10^{-6} Hz

C 3.08×10^{15} Hz **D** 2.05×10^{14} Hz [1]

(Total for Question 2 = 1 mark)

3 A zinc plate illuminated by ultraviolet light will emit photoelectrons. Which of the following would **not** be observed using the same zinc plate?

A Photoelectrons would still be emitted using shorter wavelength UV light.

B More photoelectrons would be observed each second using brighter UV of the original wavelength.

C The stopping voltage for the photoelectrons would be decreased using UV light with a longer wavelength.

D The maximum kinetic energy of the photoelectrons would be increased using brighter UV of the original wavelength. [1]

(Total for Question 3 = 1 mark)

4 Which row in the table correctly shows observed evidence for both light and electrons behaving as waves?

	Light experimental evidence	Electron experimental evidence
A	gas discharge spectra	electron diffraction patterns through graphite
B	Millikan's oil drop experiment	ionisation
C	solar eclipses demonstrate shadows	gas discharge spectra
D	two slit interference patterns	electron diffraction patterns through graphite

[1]

(Total for Question 4 = 1 mark)

5 An electronics student is using light emitting diodes (LEDs) to make a traffic light model. He uses red, orange and green LEDs. The table gives information about these LEDs. They are identified as 1, 2 and 3.

LED	Frequency / 10^{14} Hz	Wavelength / 10^{-9} m	Colour
1	5.66	530	
2	5.00	600	
3	4.41	680	

(a) Complete the table by filling in the colour of light emitted by each LED. [1]

(b) Calculate the energy of the lowest energy photon emitted by this traffic light model. [3]

(Total for Question 5 = 4 marks)

6 In 1921, Albert Einstein won the Nobel Prize for his work on the photoelectric effect.

The results of experiments on the photoelectric effect show that:

- photoelectrons are not released when the incident radiation is below a certain threshold frequency
- the kinetic energy of the photoelectrons released depends on the frequency of the incident light and not its intensity.

Explain how these results support a particle theory, but not a wave theory of light. [6]

(Total for Question 6 = 6 marks)

7 The graph shows how the maximum kinetic energy E of photoelectrons emitted from the surface of aluminium varies with the frequency f of the incident radiation.

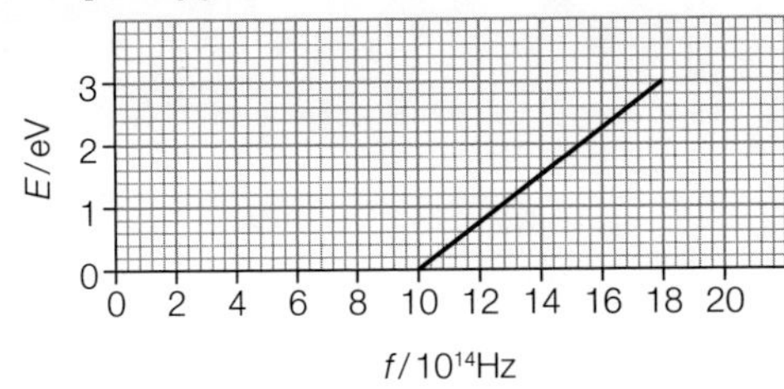

(a) Explain why no photoelectrons are emitted below a frequency of 10×10^{14} Hz. [1]

(b) Calculate the work function of aluminium in electron volts. [3]

(c) State the quantity represented by the gradient of the graph. [1]

(d) Add a second line to the graph to show how E varies with f for a metal which has a work function less than aluminium. [2]

(Total for Question 7 = 7 marks)

8 The diagram shows the lowest three energy levels of a hydrogen atom.

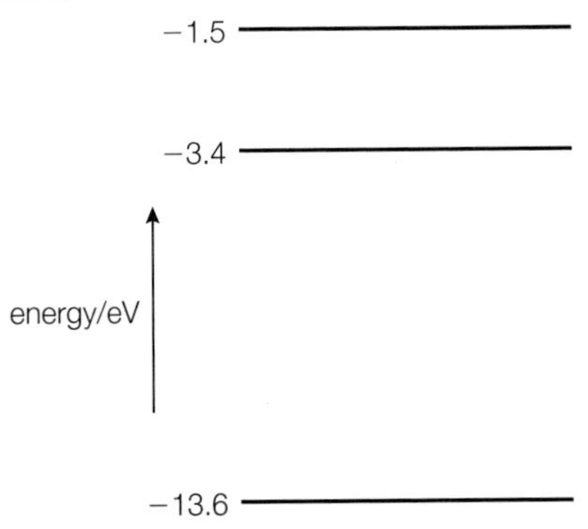

(a) Excited hydrogen atoms can emit light of wavelength 6.56×10^{-7} m.

(i) Calculate the frequency of this light. [2]

(ii) The energy of a photon of this frequency is 3.03×10^{-19} J.
By means of a calculation determine which electron transition emits this photon. [2]

(b) The spectrum of light from the sun has a dark line at a wavelength of 656 nm. Explain why this wavelength is missing from the spectrum. [2]

(Total for Question 8 = 6 marks)

9 A gold leaf electroscope is used to detect very small amounts of charge. When the electroscope cap is negatively charged, electrons spread along the metal rod and the gold leaf so they both become negatively charged. The rod and leaf repel each other, so the gold leaf rises up.

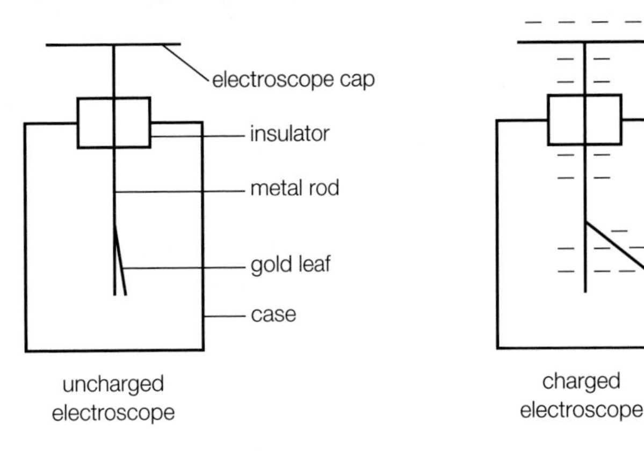

A gold leaf electroscope can be used to demonstrate the photoelectric effect. A clean zinc plate is placed onto the cap of the electroscope and the plate and electroscope are charged negatively. Ultraviolet radiation is shone onto the zinc plate.

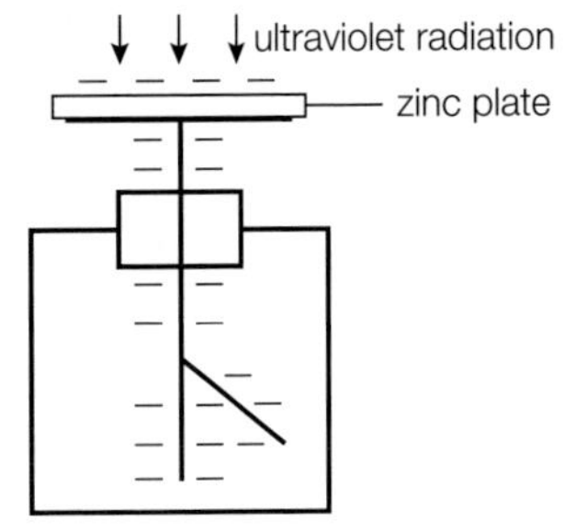

(a) The gold leaf slowly falls.
Explain, with reference to the work function of zinc, why this happens. [4]

(b) Why is the effect not observed if the ultraviolet radiation is replaced by visible light? [1]

(c) Ultraviolet radiation of wavelength 2.00×10^{-7} m is shone onto the zinc plate.
Calculate the maximum speed of the electrons emitted from the plate.
work function of zinc = 6.88×10^{-19} J [4]

(d) The source of ultraviolet radiation is moved further away from the zinc plate.
State what will happen to the maximum speed of the electrons emitted from the plate. Justify your answer. [2]

(Total for Question 9 = 11 marks)

TOPIC 4 ELECTRIC CIRCUITS

CHAPTER 4A ELECTRICAL QUANTITIES

In 1881, the streets of Godalming, a town in England, were the first in the world to be lit by a public electricity service. The power was generated from a water wheel in a nearby river. Electricity has revolutionised the way people live, but we have only been using it so widely for about 100 years. Today, it is difficult to imagine what life would have been like when the night was not artificially lit and music could only be heard if the instruments were played live.

It is extraordinary how influential the use of electricity has become in such a short time, and just how difficult it might be for us today to adapt to life without electricity. Arguably, the first electrical battery was built by Alessandro Volta in 1800. A little over 200 years later, almost everything around you will either use electricity directly or would have been impossible to manufacture without electricity.

In this chapter, you will learn about the fundamental electrical quantities that can be measured and how they are related to each other, along with the essential material structures that allow, or restrict, the flow of electricity.

MATHS SKILLS FOR THIS CHAPTER

- **Units of measurement** (*e.g. the coulomb, C*)
- **Calculating areas of circles** (*e.g. finding the cross-sectional area of a wire in order to find resistivity*)
- **Use of standard form and ordinary form** (*e.g. representing the charge on an electron*)
- **Changing the subject of an equation** (*e.g. finding the resistivity of a material from an object's resistance*)
- **Substituting numerical values into algebraic equations** (*e.g. using the transport equation*)
- **Determining the slope of a linear graph** (*e.g. finding the resistivity from experimental results*)

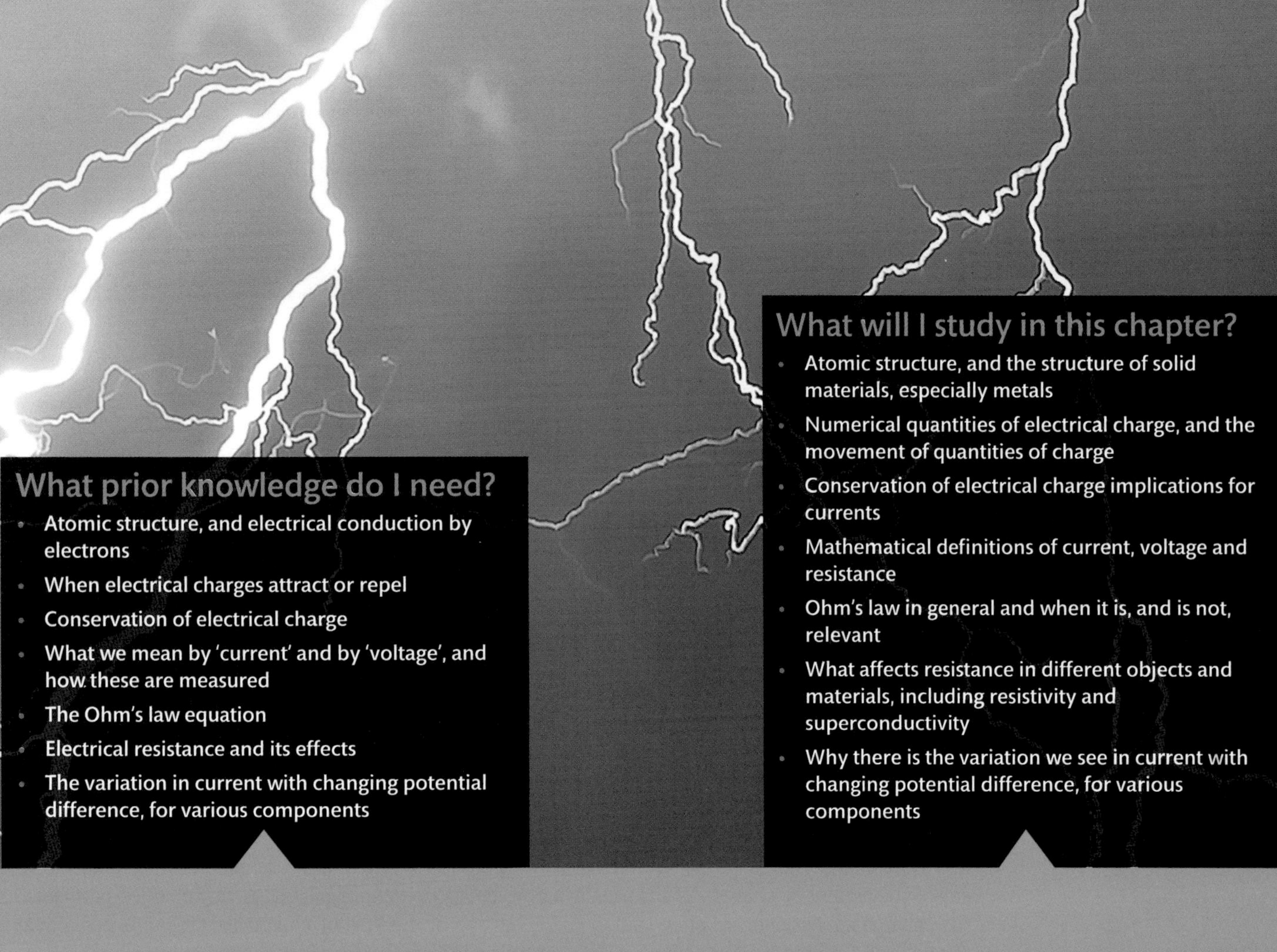

What prior knowledge do I need?

- Atomic structure, and electrical conduction by electrons
- When electrical charges attract or repel
- Conservation of electrical charge
- What we mean by 'current' and by 'voltage', and how these are measured
- The Ohm's law equation
- Electrical resistance and its effects
- The variation in current with changing potential difference, for various components

What will I study in this chapter?

- Atomic structure, and the structure of solid materials, especially metals
- Numerical quantities of electrical charge, and the movement of quantities of charge
- Conservation of electrical charge implications for currents
- Mathematical definitions of current, voltage and resistance
- Ohm's law in general and when it is, and is not, relevant
- What affects resistance in different objects and materials, including resistivity and superconductivity
- Why there is the variation we see in current with changing potential difference, for various components

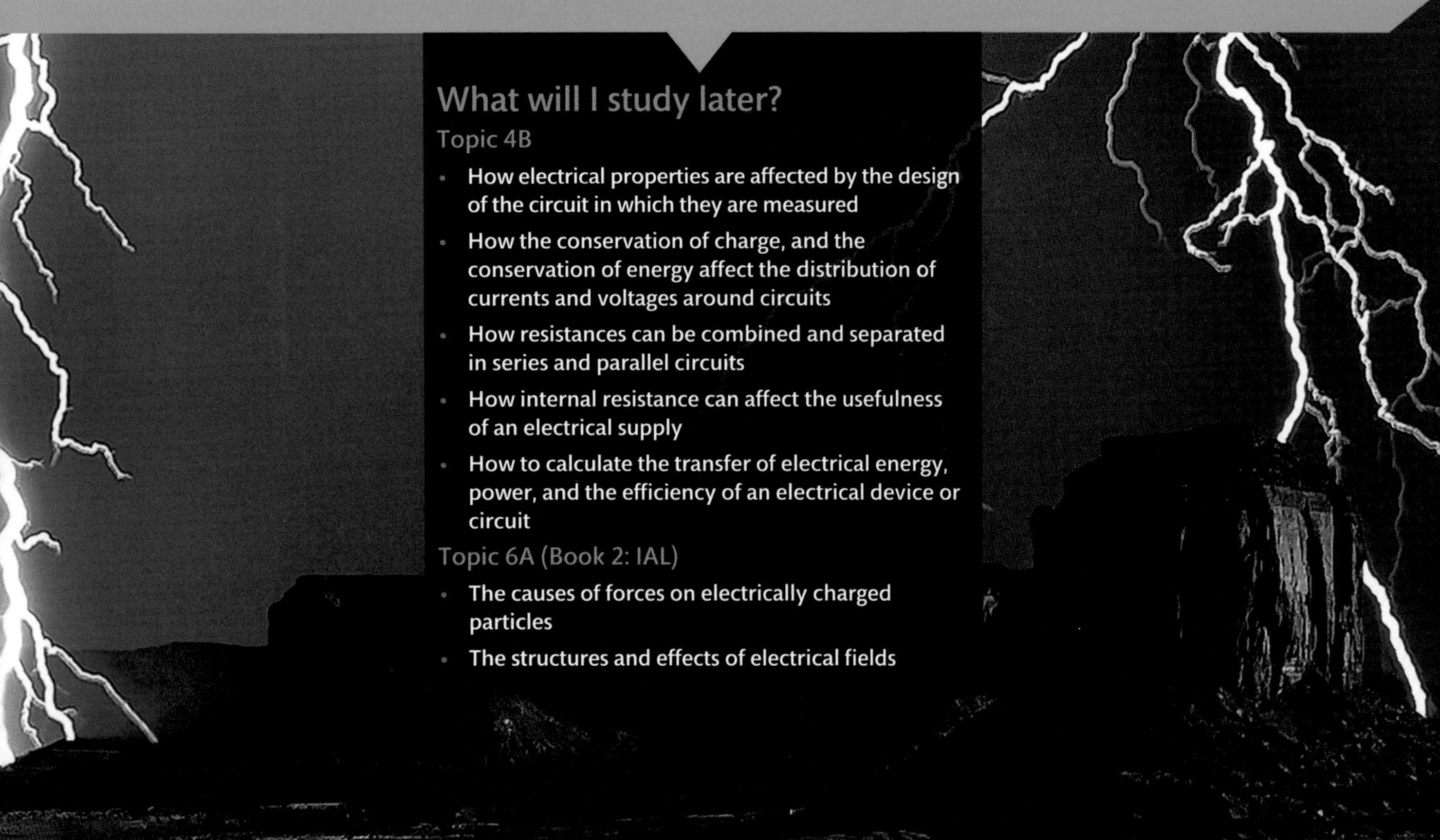

What will I study later?

Topic 4B

- How electrical properties are affected by the design of the circuit in which they are measured
- How the conservation of charge, and the conservation of energy affect the distribution of currents and voltages around circuits
- How resistances can be combined and separated in series and parallel circuits
- How internal resistance can affect the usefulness of an electrical supply
- How to calculate the transfer of electrical energy, power, and the efficiency of an electrical device or circuit

Topic 6A (Book 2: IAL)

- The causes of forces on electrically charged particles
- The structures and effects of electrical fields

1 ELECTRIC CURRENT

LEARNING OBJECTIVES

- Describe electric current as the rate of flow of charged particles.
- Make calculations of electric current.

▲ **fig A** Electric current is a measure of the rate of movement of electric charge. Lightning strikes can have currents of 10 000 amperes.

ELECTRIC CHARGE

Some particles have an electric **charge**. For example, the electron has a negative charge. In SI units, electric charge is measured in **coulombs** (C) and the amount of charge on a single electron in these units is -1.6×10^{-19} C.

$$e = -1.6 \times 10^{-19}\,\text{C}$$

This means that you would have one coulomb of negative charge if you collected together 6.25×10^{18} electrons, as shown in this calculation:

$$\text{total charge, } Q = ne = 6.25 \times 10^{18} \times -1.6 \times 10^{-19} = -1\,\text{C}$$

The charges on fundamental particles such as electrons are fixed properties of these particles. It is impossible to create or destroy charge – the total charge must always be conserved.

ELECTRIC CURRENT

If electric charge moves, this is referred to as an electric **current**, and the definition of current is the *rate* of movement of charge. As it is usually a physical movement of billions of tiny charged particles, such charge movements are often said to *flow*. Thinking of the flow of charge like the current in a river can be useful, and we will see later how current splits and recombines at circuit junctions in a manner that is like water flow. Also, as the total amount of water in a river at a given time does not change, even if the river splits, this again shows the conservation of charge.

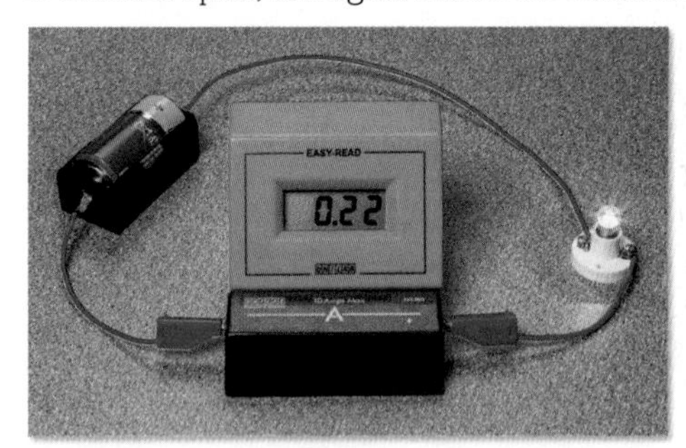

▲ **fig B** We can measure electric current through a component using an ammeter connected in series.

Electric current occurs when a charged particle, which is free to move, experiences an electric force. If it can move it will be accelerated by the force. This movement of charge forms the electric current. Most electric circuits are made from metal wiring in which there are electrons that are free to move. These *conduction electrons* then form the current.

Any source of electrical energy can create an electric force in order to produce a current. In **fig C**, the cell causes the electric force experienced by the negative conduction electrons so they move through the metal – they are attracted to the positive anode of the cell.

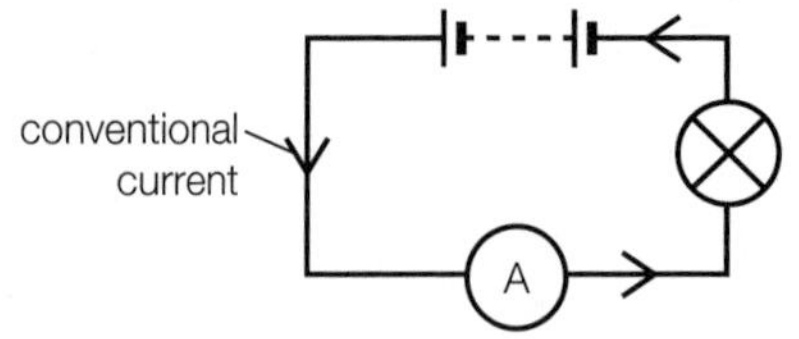

▲ **fig C** Conventional current flows from positive to negative. Negative electrons would move in the opposite direction.

CALCULATING CURRENT

The SI unit for electric current is the **ampere**, A. Current can be calculated from the equation:

$$\text{current (A)} = \frac{\text{charge passing a point (C)}}{\text{time for that charge to pass (s)}}$$

$$I = \frac{\Delta Q}{\Delta t}$$

Thus one ampere (1 A) is the movement of one coulomb (1 C) of charge per second (1 s).
For example, if the lightning in **fig A** takes 0.1 seconds to transfer 1150 coulombs of charge, we could calculate the current in the lightning:

$$I = \frac{\Delta Q}{\Delta t} = \frac{1150}{0.1} = 11\,500\text{ A}$$

The equation that defines electric current is often used in a rearranged form to find ΔQ, the amount of charge that has moved through a component in a given time, Δt.

$$\Delta Q = I\Delta t$$

PRACTICAL SKILLS

Observing charge flow

We can monitor small movements of charge, to see how they form a current, using a hanging ball that will conduct electricity.

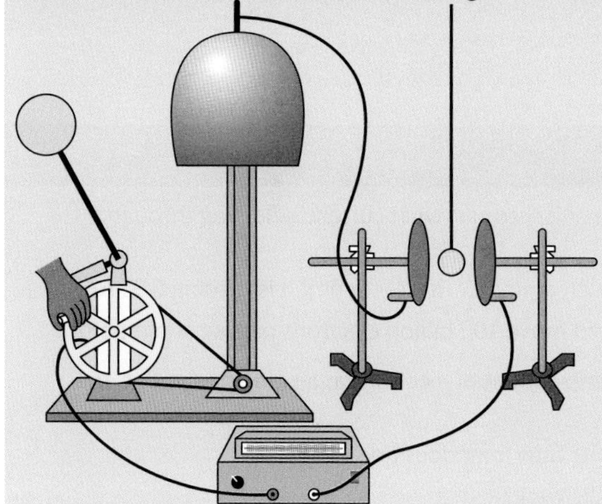

fig D A ball on a string can carry small numbers of electrons across a high voltage gap, and this current is measured using a spot galvanometer.

The high voltage set up across the air gap between the metal plates encourages negative electrons to want to move towards the positive side. The hanging ball is painted with conducting paint and swings backwards and forwards across the gap. It carries a small quantity of electrons from one metal plate to the other each time. We can measure this small movement of charge on a very sensitive ammeter. If we time the period of oscillation of the moving ball and measure the tiny current, we can calculate how many electrons would pass across on each journey of the ball. If the ball is moving too fast to be timed by eye, then we can use a stroboscope to measure the frequency of oscillations.

Example calculation:

If the spot galvanometer registers 6.5 microamperes and the ball takes 2.0 milliseconds for one swing, how many electrons does it carry on each pass from the negative plate to the positive one?

$$\Delta Q = I \times \Delta t = 6.5 \times 10^{-6} \times 2 \times 10^{-3} = 13 \times 10^{-9}\text{ C} = 13\text{ nC}$$

$\therefore$ no of electrons per swing:

$$N = \frac{\Delta Q}{e} = \frac{-13 \times 10^{-9}}{-1.6 \times 10^{-19}}$$

$$N = 8.1 \times 10^{10}\text{ electrons}$$

Note that on the ball's return journey it will be positively charged, having lost an excess of electrons when contacting the positively charged plate. However, positive charge moving in the opposite direction still is a current, and it will be at the same rate – the same current – because the ball speed is constant.

Safety Note: The Van de Graaff generator will give unpleasant electric shocks but these should not harm healthy individuals. Anyone with a cardiac condition must not receive electrical discharges as they may suffer ventricular fibrillation which could prove fatal. Persons with epilepsy may also be at risk of a seizure and should be advised accordingly.

EXAM HINT

It is useful if you can learn the value for the charge on an electron - you will need to use it so often that you will save a lot of time if you don't have to look it up.

$$e = -1.6 \times 10^{-19}\,C$$

Note that whilst this is a negative value as the electron is negative, we usually ignore the negative sign in calculations and give positive current values.

IONIC CHARGE CARRIERS

If the circuit is more unusual, there may be other charged particles, *charge carriers*, which can move to form the electric current. For example, in the electrolytic processing of bauxite ore to produce aluminium metal, the bauxite is dissolved in cryolite – another aluminium compound – and this solution then has free aluminium ions (charge carriers) that can move through the liquid as an electric current. These ions are positively charged, and will move because of the electric force towards the negative cathode. This is still an electric circuit that must obey the rule of conservation of charge, and in which we can measure the current as the rate of flow of charge.

As the charge on an electron is a fixed negative amount, we can easily calculate the charge on any ion. This would be important in a situation where ions were moving as the charge carriers in an electric current, for example, in electrolysis.

LEARNING TIP

The charge on a proton is the same magnitude as that on an electron, but is positive: $q_p = +1.6 \times 10^{-19}\,C$.

EXAM HINT

Note that the steps and layout of the solution in this worked example are suitable for questions on calculating the net charge in the exam.

WORKED EXAMPLE

What would be the charge on an iron (III) ion, Fe^{3+}?

Three electrons have been lost, so the net charge is that of the ion's three excess protons:

$$q_{Fe^{3+}} = 3 \times 1.6 \times 10^{-19} = +4.8 \times 10^{-19}\,C$$

CHECKPOINT

1. (a) If 12.5×10^{18} electrons move through a lamp in 3.2 seconds, what current is flowing through the lamp?

(b) How many coulombs of charge move through an ammeter in 8 seconds if it is reading 0.95 A?

(c) How long would it take for a current of 0.68 A to move 100 billion electrons past a certain point?

2. In a shuttling ball experiment to demonstrate the movement of electrons as an electric current, the galvanometer showed a current of 0.12 μA.

(a) If the ball carried 20 nC of charge on each swing, how long does one swing take?

(b) How many electrons does the ball carry in each swing?

3. In an electrolysis experiment, 0.01 moles of Cu^{2+} ions moves through the copper sulfate solution in 23 minutes. How much current flows?

(One mole contains Avogadro's number of any particle, and this is $N_A = 6.02 \times 10^{23}$.)

SUBJECT VOCABULARY

charge a fundamental property of some particles. It is the cause of the electromagnetic force, and it is a basic aspect of describing electrical effects

coulomb, C the unit of measurement for charge: one coulomb is the quantity of charge that passes a point in a conductor per second when one ampere of current is flowing in the conductor. The amount of charge on a single electron in these units is $-1.6 \times 10^{-19}\,C$

electric **current** the rate of flow of charge. Current can be calculated from the equation:

$$\text{current (A)} = \frac{\text{charge passing a point (C)}}{\text{time for that charge to pass (s)}}$$

$$I = \frac{\Delta Q}{\Delta t}$$

ampere the unit of measurement for electric current: one ampere (1 A) is the movement of one coulomb (1 C) of charge per second (1 s)

LEARNING OBJECTIVES

- Define electromotive force and potential difference.
- Make calculations of voltage and the energy transfer in components.
- Evaluate a model of electrical circuits.

For any electrical circuit to be of use, it has to act as a means to transfer energy usefully. The circuit must have at least one component that can supply electrical energy. It will also have components that transfer this electrical energy to other stores, and at least one of these stores or transfers of energy will be useful to its purpose.

For example, in a bicycle light there will be a cell that transfers electrical energy from a chemical reaction, and there will be a lamp that transfers some of the electrical energy as light radiation.

LEARNING TIP

Almost every aspect of the study of physics has connections with transfers of energy.

VOLTAGES

The electrical quantity **voltage** is a measure of the amount of energy a component transfers per unit of charge passing through it. It can be calculated from the equation:

$$\text{voltage (V)} = \frac{\text{energy transferred (J)}}{\text{charge passing (C)}}$$

$$V = \frac{E}{Q}$$

▲ **fig A** Electric circuits transfer energy.

ELECTROMOTIVE FORCE

For a supply voltage – a component which is putting electrical energy into a circuit – the correct term for the voltage is **electromotive force**, or **emf**. If a cell supplies one joule (1 J) of energy per coulomb of charge (1 C) that passes through it, it has an emf of 1 volt (1 V).

$$\text{emf (V)} = \frac{\text{energy transferred (J)}}{\text{charge passing (C)}}$$

$$\varepsilon = \frac{E}{Q}$$

For example, if the bicycle lamp in **fig A** was powered by a cell with an emf of 1.5 V, then this cell would transfer 1.5 J of electrical energy to each coulomb of charge that passed through it.

POTENTIAL DIFFERENCE

For a component which is using electrical energy in a circuit and transferring this energy, the correct term for the voltage is **potential difference**, or **pd**. If a component uses one joule (1 J) of energy per coulomb of charge (1 C) that passes through it, it has a pd of 1 volt (1 V). The energy being used by the component could be referred to as work done, W.

$$\text{pd (V)} = \frac{\text{energy transferred (J)}}{\text{charge passing (C)}}$$

$$V = \frac{W}{Q}$$

For example, if the bicycle light in **fig A** produced light from an LED with a pd of 1.5 V, then this LED would transfer 1.5 J of electrical energy from each coulomb of charge that passed through it to light energy and a little thermal energy.

WORKED EXAMPLE

A cell transfers 40 J of chemical energy for every 15 C of electrons that pass through it, in order to power an electrical circuit. What is the cell's emf?

$$\varepsilon = \frac{E}{Q}$$

$$\varepsilon = \frac{40}{15}$$

$$\varepsilon = 2.7\,\text{V}$$

EXAM HINT

Note that the steps and layout of the solution in this worked example are suitable for questions on calculating the emf and pd in the exam.

For every 2000 C of charge passing through it, a loudspeaker transfers 18 500 J of electrical energy to sound and thermal energy. What is the loudspeaker's pd?

$$V = \frac{W}{Q}$$

$$V = \frac{18\,500}{2000}$$

$$V = 9.25\,\text{V}$$

THE ELECTRONVOLT

The **electronvolt**, eV, is a unit of energy that is generally used with sub-atomic particles. Its definition comes from the equation defining voltage:

$$V = \frac{E}{Q}$$

If an electron is accelerated by a potential difference of 1 V, the energy transferred to it will be:

$$E = V \times e$$

$$\therefore \quad E = 1 \times 1.6 \times 10^{-19} = 1.6 \times 10^{-19}\,\text{J}$$

The amount of energy transferred to an electron by passing through a voltage of 1 V is an electronvolt. So 1 eV = 1.6×10^{-19} J.

▲ **fig B** We can measure voltage across a component with a voltmeter connected in parallel.

ELECTRICAL MODELS

A model is a way of thinking about an idea or phenomenon in order to help us understand it better. For example, as the structure of an atom is too small to see, it is common for people to imagine it to be like that of planets orbiting the sun in the solar system. There are many aspects of this model that do not correctly represent an atom. However, there are some aspects that help people to think about atoms more clearly.

Electricity has many aspects that are not visible to us in everyday life, and physicists often use models to explain some of these. All models will have limitations, so it is important to evaluate the strengths and weaknesses of any model to make sure that you do not rely too heavily on it.

MODELLING VOLTAGE

One model that could be used to try and understand the transfers of energy in an electric circuit could be to think of an electric circuit as a a ski area. When people go skiing, they take a lift which carries them up to the top of the hill. From there, they then slide back down to the bottom. Some skiers may slide down different routes, but they all finish back at the same lower level they started at.

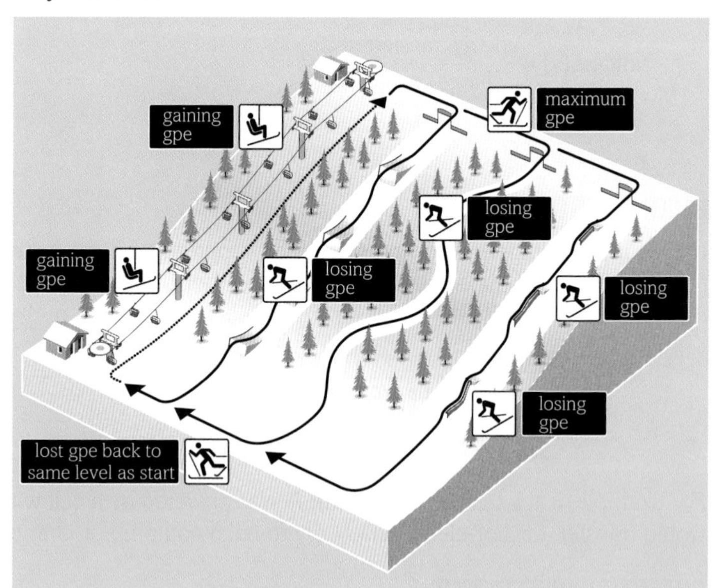

▲ **fig C** Modelling an electric circuit as a ski area.

The skiers on the lift gain gravitational potential energy, which is representing electrical energy in this analogy. By the time they have all skied back down through the runs and over the obstacles, they have lost all that gpe again and are back to the same level as before they started. This illustrates the principle of conservation of energy in an electric circuit. If moving charge is given energy by a source of emf, it will transfer all that energy on its journey around the circuit, through the various pds. Around the circuit, the total of all the emfs will be the same as the total of all the pds – total energy supplied will equal total energy used.

The ski lift takes skiers to the top of the slope – it gives them potential energy like a battery transfers electrons electrical energy. From there, they can slide down a number of possible routes – like separate, parallel, loops in a circuit. If the skier does a trick off an obstacle, then we could imagine that he was passing through a component and transferring energy to the component. We could think about measuring the emf of the chairlift, and the pd of each obstacle on the run.

Notice that although the picture in **fig D** has skiers and snowboarders, for simplicity we only refer to 'skiers'. This is a weakness of the model, as both would represent charge carriers flowing in our circuit, but usually electric circuits have only one type of charge carrier that flows – the electrons.

▲ **fig D** Modelling charge carriers as skiers.

CHECKPOINT

SKILLS INTERPRETATION, ADAPTIVE LEARNING

1. Explain how potential difference is different from electromotive force.
2. (a) A cell provides 76 C of charge with 120 J of energy. What is the cell's emf?
   (b) When a lamp carries a current of 2.4 A for 5.4 s, it transfers 120 J of energy. What is the pd across the lamp?
3. Convert the following:
   (a) 9.6×10^{-19} J into electronvolts
   (b) 4.8 MeV into joules.
4. Write an evaluation of the strengths and weaknesses of the ski area model of an electric circuit. Include discussion of at least the following ideas:
   - the representation of charge carriers
   - a complete circuit is needed
   - gravitational potential as the analogy for electrical energy
   - the ski lift as a cell
   - the snowpark obstacles as components using electrical energy
   - differing speeds by different skiers.
5. Write a suggestion for a different model of an electric circuit, explaining what electrical aspect is represented by the various sections of your model.

SUBJECT VOCABULARY

voltage a measure of the amount of energy a component transfers per unit of charge passing through it. It can be calculated by the equation:

$$\text{voltage (V)} = \frac{\text{energy transferred (J)}}{\text{charge passing (C)}}$$

$$V = \frac{E}{Q}$$

electromotive force, or **emf** a voltage as defined above, with the energy coming into the circuit
potential difference or **pd** the correct term for the voltage of a component that is using electrical energy in a circuit and transferring this energy to other stores
electronvolt the amount of energy an electron gains by passing through a voltage of 1 V

$1 \text{ eV} = 1.6 \times 10^{-19}$ J

1 mega electronvolt = 1 MeV = 1.6×10^{-13} J

4A 3 CURRENT AND VOLTAGE RELATIONSHIPS

SPECIFICATION REFERENCE 2.4.66 2.4.70

LEARNING OBJECTIVES

- State Ohm's law.
- Calculate resistances.
- Explain the I–V characteristics of components.

The movement of electrons through a circuit is caused by the electric force that an emf generates. Thus, an emf could be said to *drive* the current around a circuit. A higher voltage emf would be expected to do this more as it creates a stronger force on the charges so more of them are driven round the circuit faster – resulting in an increased current. The electric force is directly proportional to the voltage, so we would expect current to increase in direct proportion to voltage. In many cases this is true – $I \propto V$ – but in some situations other factors become important and alter the relationship.

RESISTANCE CALCULATIONS

Electrical **resistance** is considered to be the opposition to the flow of current within a conductor. It can be calculated from the equation:

$$\text{resistance } (\Omega) = \frac{\text{potential difference } (V)}{\text{current } (I)}$$

$$R = \frac{V}{I}$$

OHM'S LAW

fig A Measuring the resistance of an ohmic conductor.

If the current is proportional to the voltage driving it through a component, this component is called an *ohmic conductor*, as this means it follows **Ohm's law**. The proportional relationship can be used to find the resistance of the component.

For example, what is the resistance of the resistor shown in **fig A**?

$$R = \frac{V}{I} = \frac{4.0}{0.18}$$

$$R = 22\,\Omega$$

For an ohmic conductor, the answer to the calculation of resistance would be the same for all voltages and their corresponding current values (providing the temperature remains constant – see **Section 4A.5**).

PRACTICAL SKILLS

Investigating I–V relationships

We can do an experiment to investigate whether or not a component follows Ohm's law using the equipment in **fig A**. The component under test is the resistor, and we could replace this resistor to allow testing of other components.

Use various values of supply emf and measure the potential difference across and current through the resistor for each one. This should include reversing the terminals on the power supply in order to measure the effects of negative pds across the resistor. In the case of a resistor, negative pds correspond to negative currents - a voltage creating an electric force in the opposite direction in the wires will attract electrons in the opposite direction around the circuit.

In this example, the results would produce a graph like the one shown in **fig B**. The straight line illustrates that I is proportional to V for all tested values, telling us that the resistor is an ohmic conductor.

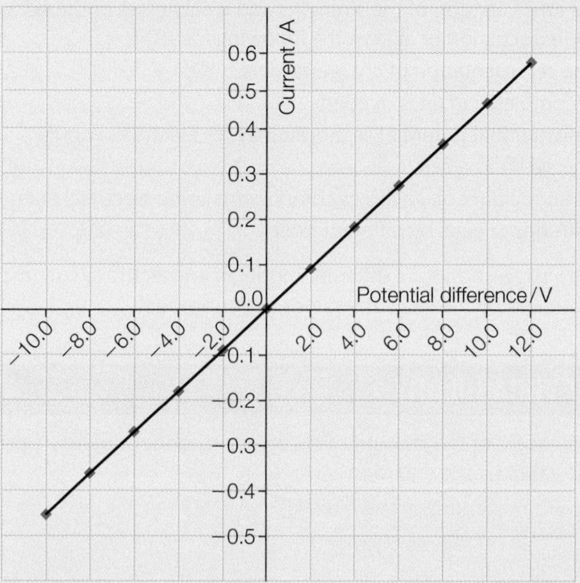

fig B I–V graph for a resistor.

You can calculate the resistance of the resistor from the graph in **fig B**. First calculate the gradient of the straight line from:

$$m = \frac{\Delta I}{\Delta V}$$

This calculation is the reciprocal of the one for resistance given above. So, you can find the resistance here by taking the reciprocal of the gradient:

$$R = \frac{1}{m}$$

Note that using the gradient is only effective for an ohmic conductor. If the line is not straight, the resistance changes and can be calculated using Ohm's law at each specific V/I value on the line.

Safety Note: Some components will get hot enough to cause burns especially if their normal 'working range' is exceeded.

CURRENT–VOLTAGE CHARACTERISTICS

For an electric circuit design, we need to know how components will react when the pd across them changes, in order to ensure that the circuit performs its intended function under all circumstances. Part of the specification of any component is a graph of its $I-V$ characteristics. We have already seen the graph for a simple resistor, and a metal at constant temperature would produce the same straight-line result. The only difference would be that the gradient will be different in each case, as it corresponds to the specific resistance of the resistor or wire.

$I-V$ GRAPH FOR A FILAMENT BULB

For the filament lamp $I-V$ graph in **fig C**, we can see that for a small voltage, the current is proportional to it, as shown by the straight-line portion of the graph through the origin. At higher voltages, a larger current is driven through the lamp filament wire, and this heats it up. At hotter temperatures, metals have higher resistance. The gradient of the graph in **fig B** was given as the reciprocal of the resistance, and so the gradient here becomes less towards higher voltages: higher resistance means a lower gradient. In **Section 4A.5**, we will see why the resistance increases with temperature and why this reduces current.

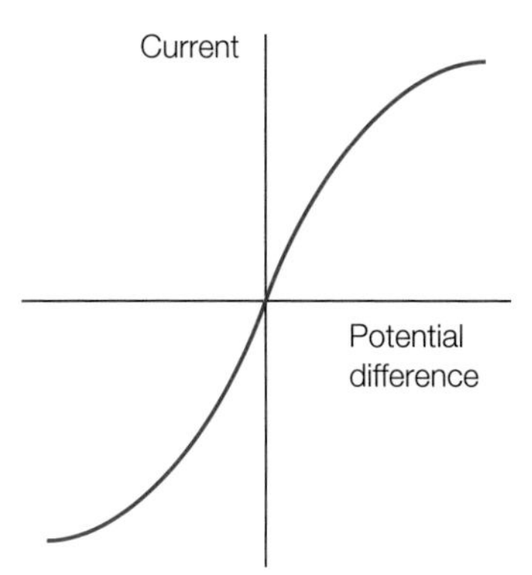

▲ **fig C** A filament lamp is an example of a non-ohmic conductor. This is because the current through it affects its own temperature – higher current means a higher temperature – and controlling temperature is part of the definition of Ohm's law.

$I-V$ GRAPH FOR A DIODE

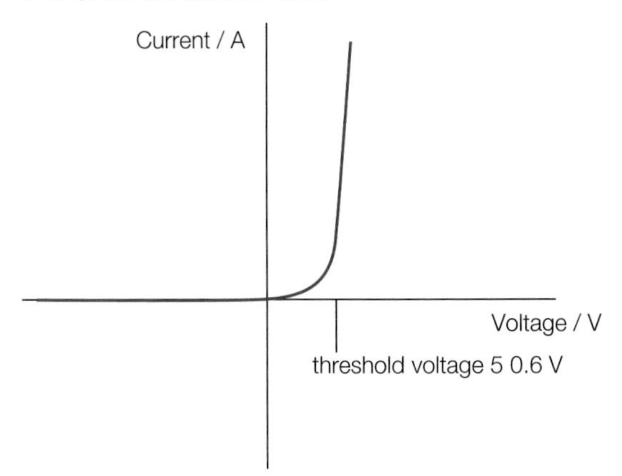

▲ **fig D** A semiconductor diode is designed to only pass current in one direction, so the line follows the x-axis, $I = 0$, for negative voltages.

For the diode in **fig D**, we can see the basic idea that a diode only conducts in the forwards direction, and so there is zero current for negative voltages. It also requires a minimum voltage in the forwards direction. This *threshold voltage* is typically around 0.6 V. The curve for a diode will be looked at in detail in **Section 4A.6**, which covers the ideas of electrical conduction in different materials.

$I-V$ GRAPH FOR A THERMISTOR

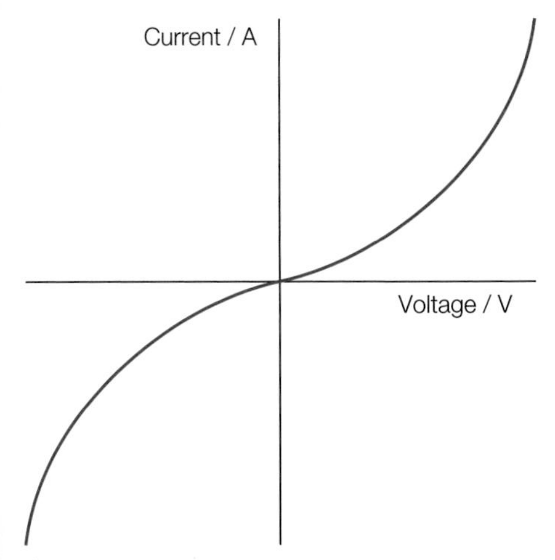

▲ **fig E** A thermistor is designed to alter its resistance with temperature; in the reverse manner to a filament bulb.

EXAM HINT

To explain why resistance in a metal increases with temperature, you must include both points:

- The metal atoms vibrate faster and further
- so, there are more collisions between electrons and atoms.

For the thermistor in **fig E**, its resistance reduces with the temperature. The gradient of the line increases with the heating effect of the increasing current. The gradient represents the reciprocal of the resistance: larger gradient value means lower resistance. This is a result of its manufacture from semiconductor materials, whose atoms release more conduction electrons as the temperature rises. This will be explained in detail in **Section 4A.6**.

CHECKPOINT

1. State Ohm's law.
2. A loudspeaker in a circuit allows a current of 0.05 A to pass through it when the pd across it is 6 V. What is the loudspeaker's resistance?
3. What is the difference between an ohmic and a non-ohmic conductor?

SKILLS PROBLEM SOLVING

4. Graphs of current against voltage are usually drawn with V on the x-axis, as the experimental data would be collected by having pd as the independent variable. This is just standard procedure. Considering the calculation of resistance, why might it be more useful to plot these data with pd on the y-axis?

SKILLS ANALYSIS

5. What is the difference between component A and component B shown in the I–V plot of **fig F**? Explain your answer.

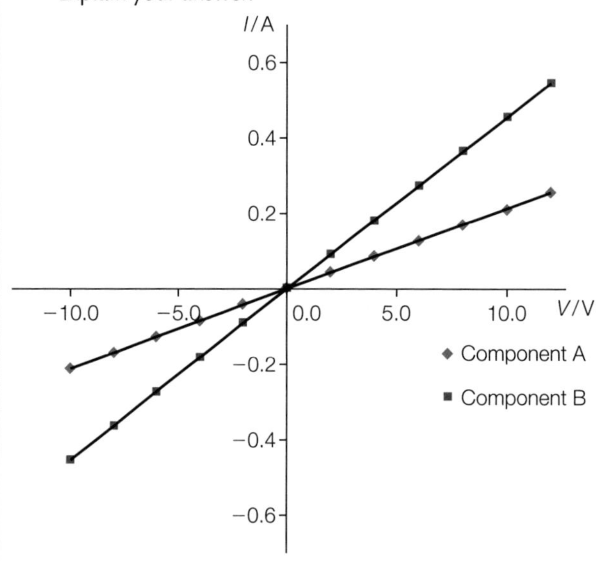

fig F I–V graphs for two components.

SUBJECT VOCABULARY

resistance the opposition to the flow of electrical current. It can be calculated from the equation:

$$\text{resistance } (\Omega) = \frac{\text{potential difference } (V)}{\text{current } (A)}$$

$$R = \frac{V}{I}$$

Ohm's law the current through a component is directly proportional to the voltage across it, providing the temperature remains the same. The equation for this is often expressed as:

$$\text{voltage } (V) = \text{current } (A) \times \text{resistance } (\Omega)$$

$$V = I \times R$$

4 RESISTIVITY

SPECIFICATION REFERENCE

2.4.71 2.4.72 CP7

CP7 LAB BOOK PAGE 28

LEARNING OBJECTIVES

- Define resistivity.
- Explain how to measure resistivity experimentally.
- Make calculations of resistance using resistivity.

Resistance is the result of collisions between charge carriers and atoms in the current's path. This effect will vary depending on the density of charge carriers and the density of fixed atoms, as well as the strength of the forces between them. So, pieces of different materials with identical dimensions will have differing resistances. The general property of a material to resist the flow of electric current is called **resistivity**, which has the symbol rho, ρ, and SI units ohm metres, Ω m.

LEARNING TIP

Resistivity is a property of a material. All samples of the same material, regardless of their shape and size, will have the same resistivity, whilst their resistances may be very different.

The resistance of an object is dependent on its dimensions and the material from which it is made. This gives rise to an equation for calculating the resistance of a uniform sample of material if we know the resistivity of the material:

$$\text{resistance } (\Omega) = \frac{\text{resistivity } (\Omega\text{ m}) \times \text{sample length (m)}}{\text{cross-sectional area (m}^2)}$$

$$R = \frac{\rho l}{A}$$

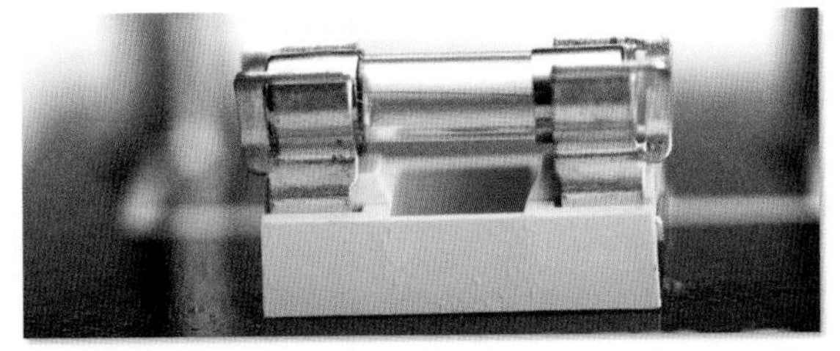

fig A With knowledge of the resistivity, ρ, of a material, we can calculate the resistance of objects made from it.

RESISTIVITY EQUATION EXAMPLE

See **fig A** and **table A**. What is the resistance of a piece of copper fuse wire if it is 0.40 mm in diameter and 2 cm long?

wire radius = 0.20 mm = 2.0×10^{-4} m

cross-sectional area, $A = \pi r^2 = \pi \times (2.0 \times 10^{-4})^2 = 1.3 \times 10^{-7}\text{ m}^2$

$$R = \frac{\rho l}{A}$$

$$R = \frac{1.7 \times 10^{-8} \times 0.02}{1.3 \times 10^{-7}}$$

$$R = 2.6 \times 10^{-3}\ \Omega$$

MATERIAL	RESISTIVITY, ρ / Ω m AT 20 °C	$\frac{\Delta\rho}{\Delta t}$ / % °C^{-1}
silver	1.6×10^{-8}	+0.38
copper	1.7×10^{-8}	+0.40
aluminium	2.8×10^{-8}	+0.38
constantan	4.9×10^{-7}	+0.003
germanium	4.2×10^{-1}	−5.0
silicon	2.6×10^{3}	−7.0
polyethene	2×10^{11}	
glass	~10^{12}	
epoxy resin	~10^{15}	

table A Resistivity varies greatly between materials, and is also dependent on temperature. Note the small change in resistivity with temperature for constantan, an alloy of copper and nickel; this information is used where accurately known resistance is important.

PRACTICAL SKILLS CP7

Investigating resistivity

You can investigate the resistivity for a metal in the school laboratory using a simple circuit.

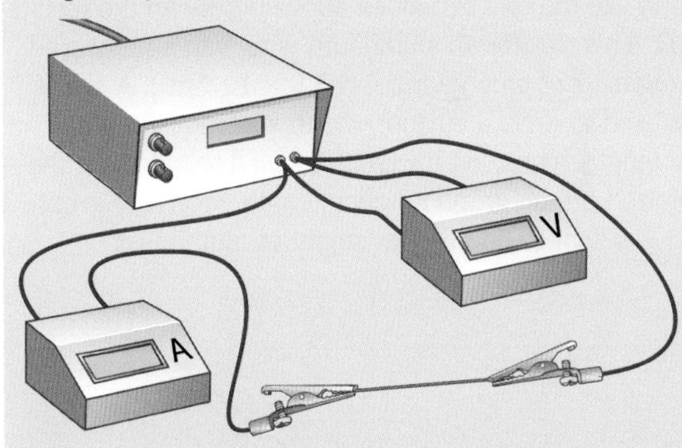

fig B Measuring the resistance for various lengths of a wire will allow us to plot a graph to find its resistivity.

We will need to use a micrometer screw gauge to measure the wire's diameter. For improved accuracy, this is done in right-angled pairs at several places along the length of the wire, and then we take the mean diameter measurement.

For several different lengths of the wire, the wire's resistance should be measured using the voltmeter–ammeter method ($R = \frac{V}{I}$). The resistance will be small, so care must be taken to ensure currents are safely low.

$$R = \frac{\rho l}{A}$$

The equation involving resistivity means that we could calculate a value for it by re-arranging the equation and taking one of the results and making the calculation. However, it is always more reliable to produce a straight-line graph of experimental results and calculate our answer from the gradient. In this case, plotting resistance on the y-axis against length on the x-axis will give the gradient as $\frac{\rho}{A}$, and from this we could calculate the resistivity with more confidence in our conclusion.

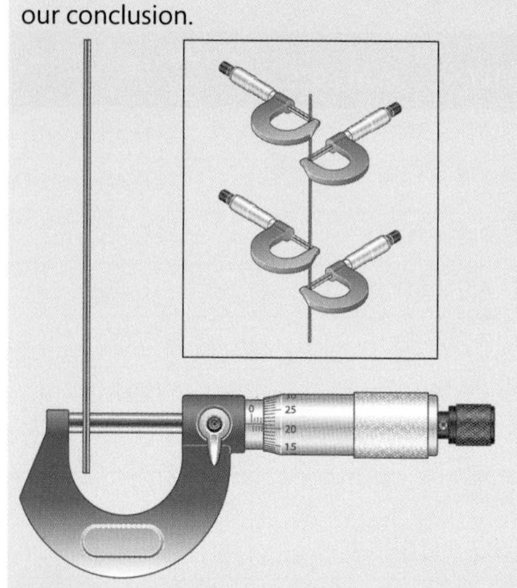

▲ **fig C** Accurate measurement of the wire's diameter is vital.

Safety Note: The wire will get hot enough to cause burns especially if too large a voltage/current is applied.

RESISTIVITY INVESTIGATION EXAMPLE

An investigation like the one explained above produced the results shown in **fig D**. This was for an aluminium wire with a measured diameter averaging 0.22 mm (giving $A = 3.8 \times 10^{-8}\,\text{m}^2$). A single 1.5 V cell was used to drive a current, which was measured at each different length. These data were then used to calculate the resistances plotted on the graph for each length. From the graph, we can use its gradient to find the resistivity of aluminium.

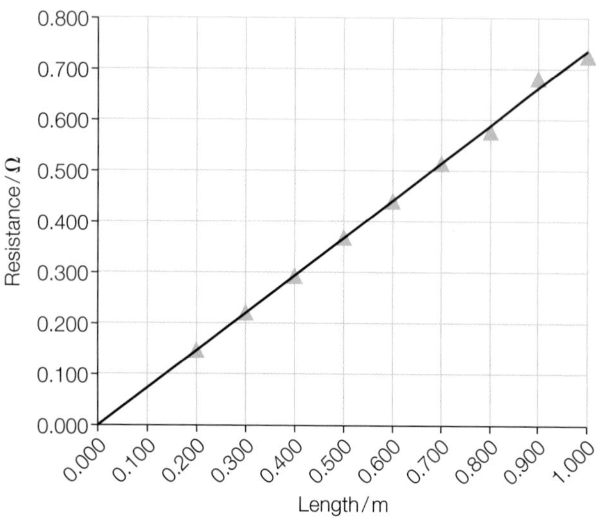

▲ **fig D** From the gradient of this graph, it is possible to calculate the resistivity of the aluminium wire in the investigation.

$$\text{gradient} = \frac{\Delta y}{\Delta x} = \frac{\Delta R}{\Delta l} = \frac{0.730 - 0.000}{1.000 - 0.000} = 0.730$$

$$R = \frac{\rho l}{A}$$

As we have R on the y-axis and l on the x-axis, comparing the equation above with $y = mx + c$ shows that the gradient is equivalent to $\frac{\rho}{A}$, so we can find the resistivity by multiplying the gradient by the cross-sectional area.

$$\therefore \quad \rho = \frac{RA}{l}$$

$$\therefore \quad \rho = 0.730 \times A = 0.730 \times 3.8 \times 10^{-8} = 2.8 \times 10^{-8}\,\Omega\,\text{m}$$

Experimentally, we have found that the resistivity of aluminium is $2.8 \times 10^{-8}\,\Omega\,\text{m}$.

CHECKPOINT

SKILLS CREATIVITY, INNOVATION

1. Calculate the resistance of a piece of constantan wire if it is 80 cm long and has a diameter of 0.84 mm.
2. Explain three ways of reducing uncertainties in an experimental investigation to find the resistivity of a metal wire.
3. Conductivity is the opposite of resistivity. It is calculated as the reciprocal of the resistivity and is measured in Siemens per metre (S m^{-1}). Calculate and compare the conductivity values for polyethene and copper.
4. In a resistance experiment, a gold ring was connected into a circuit as a resistor. The connections touched on opposite points on the circular ring. Its diameter is 2 cm, and the metal's cross-section is a rectangle 3 mm by 0.5 mm. A voltmeter connected across the ring measured 4.2 mV, whilst the current through it was measured at 18 A.

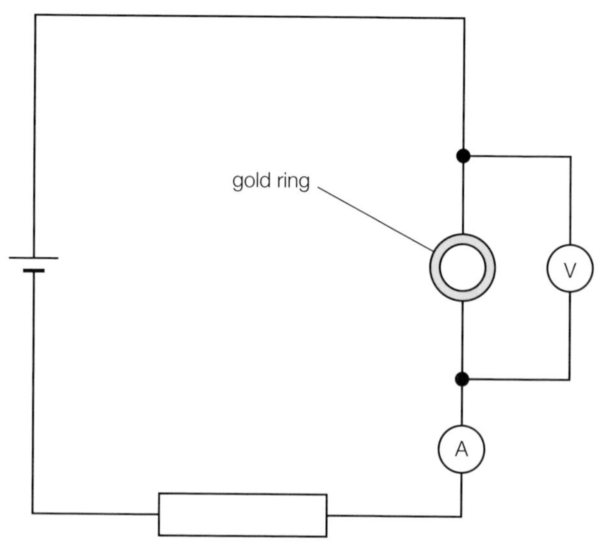

▲ **fig E**

(a) Calculate the resistivity of gold.

(b) Describe a practical difficulty with undertaking an experiment of this kind.

5. Estimate the volume of metal in the wire in a fuse.

SUBJECT VOCABULARY

resistivity for a material, the same value as the resistance between opposite faces of a cubic metre of the material

5 CONDUCTION AND RESISTANCE

SPECIFICATION REFERENCE 2.4.73

LEARNING OBJECTIVES

- Explain conduction in metals.
- Explain electrical resistance.
- Calculate the drift velocity of conduction electrons in metals.

In order to conduct electricity, solids need to have electrons that are delocalised from the solid's atoms, so that they can move through the solid causing an overall movement of charge – a current.

CONDUCTION IN METALS

The structure of metals has a regular lattice of metal atoms. These are bonded together through the sharing of electrons, which act as if they were associated with more than one atom. Many of the atoms also have an outer electron that is not needed for bonding between the atoms. These *free electrons* have a random motion, which changes as they collide with atoms or other electrons, but on average the overall position of all the charge in the metal is stationary. However, if a source of emf is connected across the metal, the electric field it sets up in the metal will push the negative electrons towards the positive end of the field. The slow overall movement of the electrons is called their **drift velocity**.

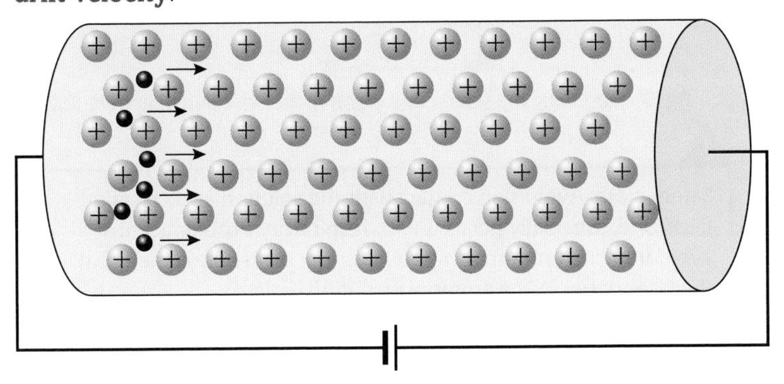

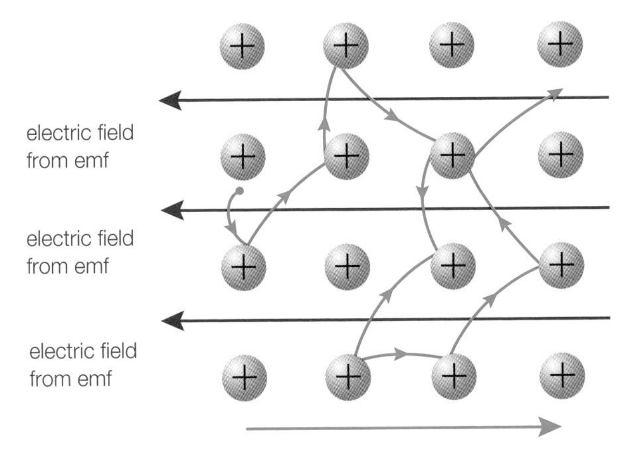

A movement in the direction of the positive side of the emf is superimposed on the random thermal motion.

▲ **fig A** Conduction in metals happens as the free electrons add an overall movement along the direction of the voltage across the conductor (towards the positive anode) to their random collisions and vibrations.

EXAM HINT

In exam answers, draw the direction of the electron movement changing sharply for each collision.

THE TRANSPORT EQUATION

DID YOU KNOW?

For a metal, the random thermal motion of the free electrons will be at speeds of thousands of kilometres per second. However, the drift velocity during conduction is usually only millimetres per second.

The value of the electric current in a metal can be calculated from the fundamental movement of the electrons if we remember the definition that $I = \frac{\Delta Q}{\Delta t}$

If we consider the cylinder shaded on the diagram in **fig B** as the length of the wire that the charges move through in a time Δt, then we need to calculate how much charge flows through it in that time. There are n electrons per cubic metre of this metal, and the wire has a cross-sectional area, A. Their movement is at a drift velocity, v, and the distance this takes them along the wire in that time is Δx. So: $\Delta x = v\Delta t$.

The total charge will be the number of electrons multiplied by the charge on each, e. The number of electrons will be their density, n, multiplied by the volume of the cylinder, V, that they travel through in Δt.

$$\therefore \quad \Delta Q = n \times V \times e = n \times A\Delta x \times e = n \times Av\Delta t \times e$$

$$\therefore \quad I = \frac{\Delta Q}{\Delta t} = \frac{nAv\Delta te}{\Delta t}$$

$$\therefore \quad I = nAve$$

This is called the **transport equation**.

EXAMPLE OF A TRANSPORT EQUATION

What is the drift velocity of the electrons in a copper wire which has a diameter of 0.22 mm and carries a current of 0.50 A? The number density of conduction electrons in copper, $n = 8.5 \times 10^{28}\,\text{m}^{-3}$.

wire radius = 0.11 mm = 1.1×10^{-4} m

cross-sectional area, $A = \pi r^2 = \pi \times (1.1 \times 10^{-4})^2 = 3.8 \times 10^{-8}\,\text{m}^2$

$$I = nAve$$

$$\therefore \quad v = \frac{I}{nAe}$$

$$= \frac{0.50}{8.5 \times 10^{28} \times 3.8 \times 10^{-8} \times 1.6 \times 10^{-19}}$$

$$\therefore \quad v = 9.67 \times 10^{-4}\,\text{ms}^{-1}$$

$$v = 0.97\,\text{mms}^{-1}$$

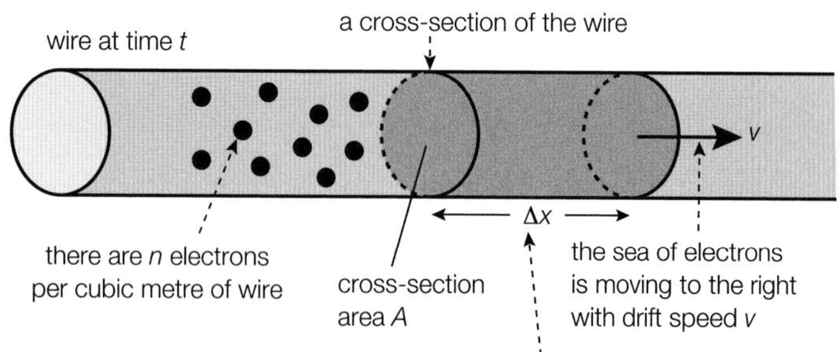

fig B A section of metal wire with dimensions to show how to make calculations using the transport equation.

PRACTICAL SKILLS

Investigating conduction velocity of coloured ions

In this experiment, you can observe the movement of coloured ions as charge carrier particles on a piece of filter paper soaked in ammonium hydroxide solution. A crystal of copper sulfate and one of potassium manganate(VII) will each dissolve, producing positive blue copper ions and negative purple ions of manganate(VII). Connecting a pd of 30 V across the wet filter paper will cause the ions to flow slowly across it, in opposite directions, and their velocity can be measured using a ruler and stopclock. Expect a velocity of about 1 mm per minute.

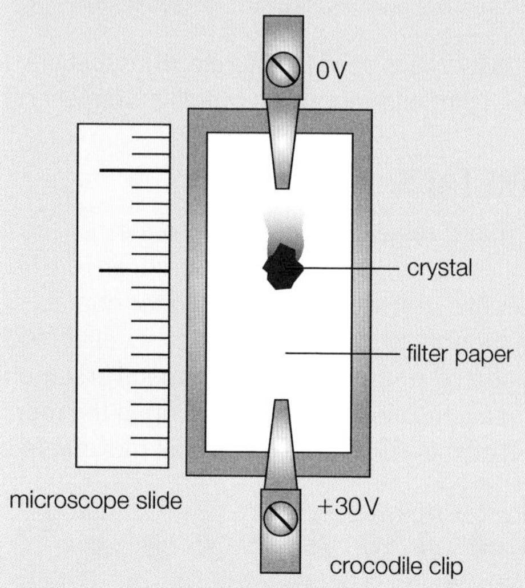

fig C Observing the slow speed of ion movement during current flow.

Safety Note: Avoid skin contact with the solution and crystals. Everyone, especially persons with respiratory problems, must avoid inhaling ammonia vapour. Use eye protection when soaking the filter paper in the ammonia solution.

RESISTANCE

We have seen that whilst the natural speed of electrons within a metal is very high, their progress through a metal's lattice of fixed atoms is very slow. They are constantly colliding with those atoms making their motion like taking 100 steps forward and 99 steps back. This is electrical resistance. The frequency of collisions is determined in part by the temperature, as this affects the vibration of the fixed atoms. The higher the metal's temperature, the more its atoms vibrate, and this means more collisions, further slowing the drift velocity of the electrons.

However, a higher current will cause more collisions, as more electrons move faster through the metal structure. This makes the metal atoms vibrate more – effectively increasing the temperature.

Looking back to the $I–V$ graph for a filament lamp, **fig C** in **Section 4A.3**, we can now explain how increased current leads to higher temperature, which leads to greater resistance, causing the curve to flatten out when higher voltages try to drive higher currents.

RESISTIVITY

Table A in **Section 4A.4** showed the resistivity of various materials, good conductors, semiconductors and insulators (which are discussed in detail in **Section 4A.6**). It also showed how the resistivity varies with temperature for metals and semiconductors. The resistivity of semiconductors was seen to fall as the temperature rises. **Table A** below shows how the density of conduction electrons, n in the transport equation, explains why the resistivity of metals is so much lower than for semiconductors. The negative temperature coefficients of resistivity for semiconductors show that their resistivity falls as the temperature goes up. The reason for this is that the value for n in the transport equation will be higher at higher temperatures (see also **Section 4A.6**). So the fact that the resistivity goes down is just a consequence of the fact that these materials, like silicon, *conduct better* at higher temperatures.

MATERIAL	n / m^{-3} AT 20 °C	n / m^{-3} AT 50 °C
copper	8.42×10^{28}	8.27×10^{28}
aluminium	18.2×10^{28}	17.9×10^{28}
silicon	8.49×10^{15}	7.42×10^{17}
germanium	1.56×10^{19}	6.82×10^{20}

table A The number of charge carriers for some common materials, at room temperature and at a higher temperature.

A slight decrease in n for metals at higher temperatures is due to their thermal expansion, rather than any change in the number of available conduction electrons. It is not as significant as the increase in collisions between metal atoms and conduction electrons caused by increased thermal vibrations.

CHECKPOINT

1. What is the 'drift velocity' of conduction electrons in a metal?
2. For an aluminium wire at room temperature with a diameter of 0.40 mm, calculate the electrons' drift velocity when a current of 2.2 A passes through the wire.
3. Explain why the resistance of a wire is greater if the wire is longer.
4. Develop a model to explain conduction and resistance in metals. Your explanation should include reference to at least the following ideas:
   - movement of charge carriers
   - fixed lattice ions
   - collisions between ions and charge carriers
   - increased vibrations of fixed lattice ions with temperature
   - an evaluation of the strengths and weaknesses of your model.
5. Estimate the number of free conduction electrons in a coin.

EXAM HINT

Take care not to confuse resistance with resistivity.

SUBJECT VOCABULARY

drift velocity the slow overall movement of the charges in a current
transport equation $I = nAvq$. This defines electric current, I, from a fundamental basis. It is the product of charge carriers, n; the charge on those carriers, q; the cross-sectional area of the conductor, A; and the drift velocity of the charge carriers in that conductor, v

SPECIFICATION REFERENCE 2.4.70 2.4.79 2.4.80

LEARNING OBJECTIVES

- Explain conduction in semiconductors.
- Explain electrical insulation.

CONDUCTION IN SEMICONDUCTORS

Semiconductors are generally solid materials that only have small numbers of delocalised electrons that are free to conduct. A typical example is silicon, one of the most common elements on Earth.

Free atoms have a series of discrete energy levels in which we can find their electrons, depending on the energy the electrons have received (see **Section 3D.4**). If the electron receives enough energy, it will leave the atom altogether, leaving behind an ion.

In solid materials, where there are many, many atoms close together, the allowed energy levels become much wider, forming energy bands. The electrons can have a large range of energies and still be within the same band. There are energy amounts which are forbidden for the electrons, but it is a very different situation from the highly limited energy levels of the isolated atom's electrons. As these energy bands are created by the collective grouping of the solid's atoms, the bands are attributed to the semiconductor as a whole rather than to individual atoms. There is an energy level called the **valence band**. Electrons with this amount of energy remain tied to atoms and do not form part of any electric current. Those that gain energy to jump up to the **conduction band** become delocalised and can move through the semiconductor as part of a current.

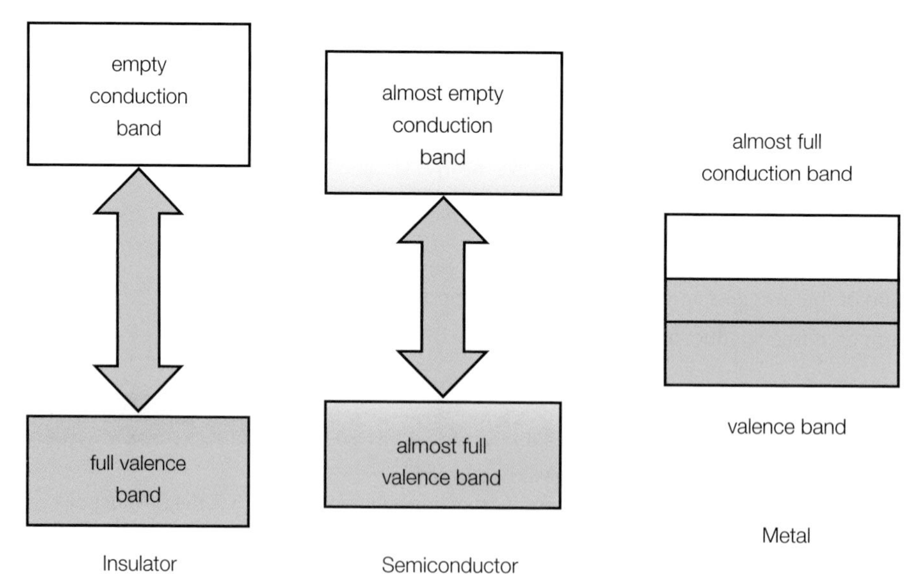

fig A The gaps between energy bands within materials explain why they are conductors or insulators. If there is a large energy gap, electrons will need to gain a lot of energy to leave their atoms and conduct a current.

The number of delocalised electrons in a semiconductor is low compared with metals, and so the current they will carry is therefore lower than metals for the same applied voltage. At higher temperatures they have more conduction electrons, as more electrons are elevated into the conduction band. There will be a temperature-related reduction in current due to increased collisions with fixed atoms, but the increase in available conduction electrons is a much bigger effect. Overall, a semiconductor will carry more current as the temperature goes up – its resistivity effectively drops as the temperature rises.

CONDUCTION 'HOLES'

When an electron enters the conduction band and moves away, this leaves the atom with a positive charge. The empty space the electron has left is referred to as a (positive) hole. If an electron from another atom moves to fill the hole, leaving its original atom with a hole, the hole has then moved. As the electrons will be attracted to jump in the positive direction of an applied voltage, the hole will slowly appear to move in the opposite direction. A positive hole moving towards the negative cathode is like another charge carrier, adding to the current flow in a semiconductor. It is like the current flow in electrolysis, where positively and negatively charged particles are moving in opposite directions at the same time.

LEARNING TIP

Note that conduction by holes does not mean there is overall movement of the positive metal ions. The holes are an absence of an electron, and it is this space that moves as electrons actually jump in the opposite direction.

I–V CHARACTERISTICS OF A SEMICONDUCTOR DIODE

A diode is made by joining different types of semiconductors, which normally creates an energy barrier at the junction between them. This blocks the movement of charge carriers (holes and electrons) across the barrier. This barrier can be overcome in the forward direction if a small forward voltage is applied. In the reverse direction, only very few charge carriers can pass through at low voltages. They account for a tiny 'leakage current'. Once the reverse voltage becomes large enough, it can overcome the large reverse energy barrier and force the conduction process in the opposite direction.

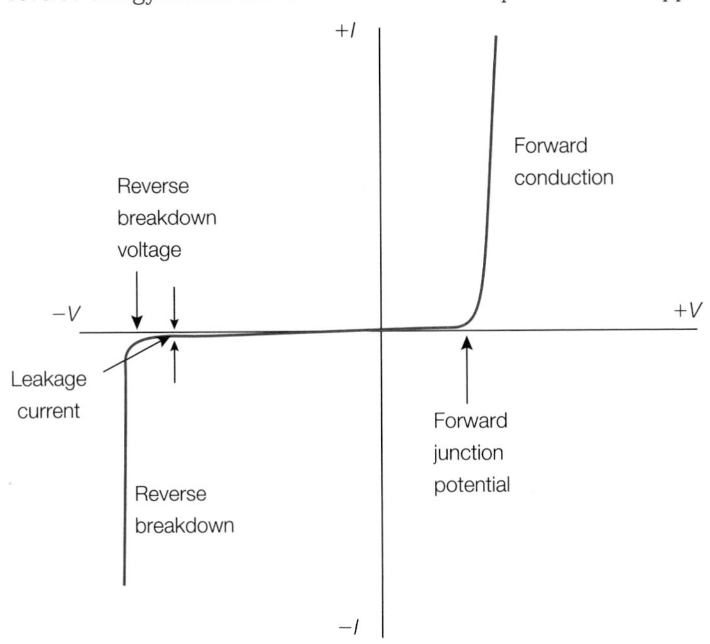

▲ **fig B** The *I–V* characteristics for a semiconductor diode in detail.

LIGHT DEPENDENT RESISTORS (LDRS) AND THERMISTORS

Light dependent resistors have the property that their resistance depends on the light level around them. In brighter conditions, the LDR will have a lower resistance. LDRs are made from semiconductor material, and light landing on the material can boost electrons from the valence energy band up to the conduction band, increasing the number of conduction electrons. The effect of this is to make the LDR conduct better – it has a lower resistance.

Thermistors work in exactly the same way, except that their resistance depends on thermal energy from the surroundings. The most common type of thermistors are referred to as negative temperature coefficient thermistors. These use thermal energy to boost their electrons into the conduction energy band, meaning their resistance falls as the temperature rises.

INSULATORS

Electrical insulators can be thought of as materials in which the energy gap between the valence band and the conduction band is so large that there are virtually zero electrons available for conduction. There will therefore be no conduction holes either. A very large input of energy is required in order to make the material conduct. Often this results in melting, or other damage, before the material becomes electrically conducting. For example, glass is normally an excellent electrical insulator, but a very high electric field, more than 10 million volts per metre, created across glass can cause it to become conducting.

SUPERCONDUCTIVITY

Resistance increases with higher temperatures, because the higher level of internal energy in the material causes more vibration of the fixed ions. These ions collide more often with charge carriers to reduce their speed of movement through the material. Reducing the temperature therefore reduces resistance, allowing greater current flow.

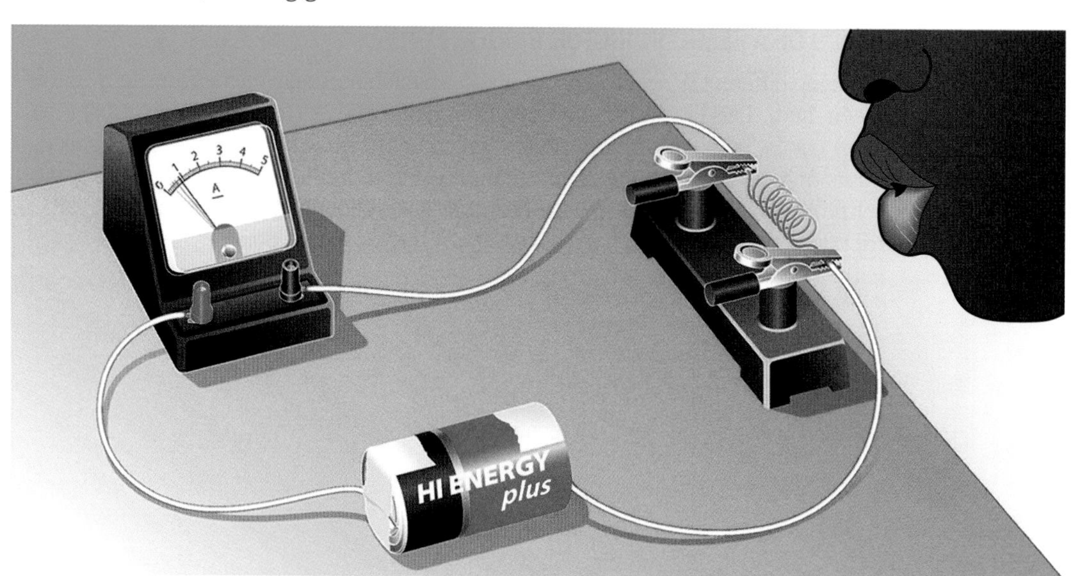

▲ **fig C** If we cool a conductor, the current through it is seen to increase because its resistance drops.

Cooling to ever lower temperatures continues this trend, until something quite unexpected happens. Below a certain **critical temperature**, the resistance suddenly drops to zero. This is called **superconductivity**. The critical temperature varies with material, but for most metals it will be below −243 °C. Complex ceramic superconductors have been created that have temperatures at which they conduct without any resistance as high as −135 °C.

Superconductors are especially useful in applications where a large current is needed, as a large current would normally waste too much energy or damage the surroundings with the heat dissipated. For example, the strong magnets needed in a particle accelerator will often be superconducting electromagnets, cooled to very low temperatures to maintain their superconductivity. The electromagnets in the Large Hadron Collider operate at about 1.9 K.

DID YOU KNOW?

Low temperatures are often expressed on the Kelvin temperature scale, where 0 Kelvin (K) is referred to as Absolute Zero.

0 K = −273.15 °C

0 °C = 273.15 K

We usually ignore the 0.15 ° and convert from Celsius to Kelvin by adding 273 °. Absolute Zero is the lowest temperature possible, as this represents a temperature where the particles have zero internal energy.

▲ **fig D** Superconductors can levitate magnets, as they do not accept penetration by magnetic fields. This is the Meissner Effect.

CHECKPOINT

1. Explain why a piece of silicon semiconductor will have a lower resistance at a higher temperature.
2. 'The critical temperature for superconductivity for lead is 7.2 K'. Explain what this statement means.
3. Some types of glass have been manufactured that do conduct electricity. Suggest and explain what properties impurities might have in these glasses to make them less insulating than standard glass.

SKILLS CRITICAL THINKING

WORKED EXAMPLE

semiconductors materials with a lower resistivity than insulators, but higher than conductors. They usually only have small numbers of delocalised electrons that are free to conduct

valence band a range of energy amounts that electrons in a solid material can have which keeps them close to one particular atom

conduction band a range of energy amounts that electrons in a solid material can have which delocalises them to move more freely through the solid

critical temperature the temperature below which a material's resistivity instantly drops to zero

superconductivity the electrical property of a material having zero resistivity

4A THINKING BIGGER

SHOCKING STUFF

SKILLS CRITICAL THINKING, PROBLEM SOLVING, ANALYSIS, REASONING/ARGUMENTATION, INTERPRETATION, ADAPTIVE LEARNING, INITIATIVE, PRODUCTIVITY, COMMUNICATION

A defibrillator is a medical device which is designed to deliver an electric shock to a patient's heart. It will restart it if the heart has stopped, or reset it if the heart it is beating with a dangerously unpredictable rhythm.

In this activity, we will consider how the electrical and other characteristics of one type of defibrillator help it to perform its function.

PRODUCT MANUALS

Extract 1:

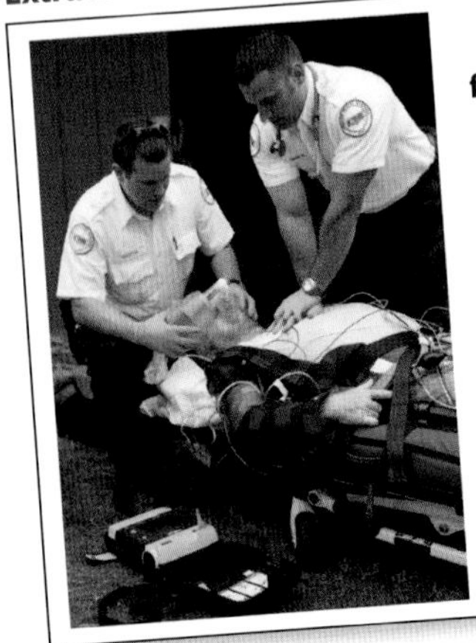

fig A The LIFEPAK 1000 can help paramedics with resuscitation by restarting the patient's heart using an electric shock through the heart.

Extract 2:

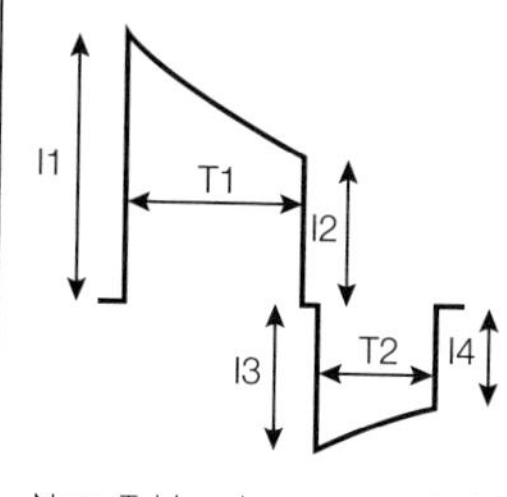

Note: Table values are nominal for a 200-joule shock.

Patient Impedance (Ω)	I1 (A)	I2 (A)	I3 (A)	I4 (A)	T1 (ms)	T2 (ms)
25	50.3	20.1	19.7	10.7	5.9	3.9
50	28.2	14.6	14.5	9.3	7.5	5.0
75	19.8	11.7	11.7	8.2	8.7	5.8
100	15.5	10.0	9.9	7.3	9.7	6.5
125	12.9	8.7	8.7	6.6	10.4	7.0
150	11.1	7.8	7.7	6.2	11.1	7.4
175	9.8	7.1	7.1	5.7	11.7	7.8

fig B Graph of the current against time during discharge. The defibrillator discharge has a large shock in one direction (T1) followed by a smaller shock in the reverse direction (T2) each of which follows a trapezium shape on the graph. The table shows the current values and times during the discharge, for various different resistance values of the patient.

Extract 3:

cprMAX Technology

The 1000 features our exclusive cprMAX technology, which gives you the flexibility to choose CPR settings that best accommodate your patient and CPR protocol requirements. The pre-shock CPR option allows adjustment of the CPR interval prior to the first shock, making the 1000 the only device that minimizes pre-shock pauses by allowing providers to continue compressions while the AED charges.

Recently published clinical data shows a relationship between increased compression fraction and survival to hospital discharge, and the 2010 AHA Guidelines place a strong emphasis on high-quality CPR. With the LIFEPAK 1000, you have more control over the CPR you provide in lifesaving settings than ever before.

LIFEPAK TOUGH™

Built for the harshest environments, the LIFEPAK 1000 is the toughest, most durable AED from Physio-Control. The device itself withstands rigorous drop-testing from any angle, and is enclosed in a highly protective case with bumpers. In addition, the 1000 has received an IP55 rating – the highest for any AED – signifying maximum protection from external elements.

360 Joules

Like every LIFEPAK defibrillator from Physio-Control, the 1000 can escalate energy up to 360 J. Studies show that for difficult-to-defibrillate patients, repeating 200 J shocks yields significantly lower VF termination rates. And the 2010 AHA Guidelines note that rescuers may consider using escalating energy up to 360 J if initial shocks at a lower dose aren't working.

Extract 4:

ENVIRONMENTAL

One Hour Operating Temperature (from room temperature to temperature extreme, one hour duration): –20 to 60 °C (–4 to +140 °F).
Operating Temperature: 0 ° to 50 °C (32 ° to 122 °F).
Storage Temperature: –30 ° to 60 °C (–22 ° to 144 °F) with battery and electrodes (maximum exposure limited to 7 days).
Atmospheric Pressure: 575 hPa to 1060 hPa (4572 to –382 meters; 15,000 to –1253 feet).
Relative Humidity: 5 to 95% (non-condensing).
Dust/Water Resistance: IP55 with battery and REDI-PAK™ electrodes installed (IEC 60529/EN 60529).
Bump: 15 g, 1000 bumps (IEC 600–68–2–29).
Shock: 40 g peak, 15–23 ms, 45 Hz cross over frequency.
Drop: 1 meter drop on each corner, edge and surface (MIL-STD-810F, 516.5, Procedure IV).

Extract 5:

physical characteristics

Height: 8.7 cm (3.4 in).
Width: 23.4 cm (9.2 in).
Depth: 27.7 cm (10.9 in).
Weight: 3.2 kg (7.1 lbs) with one set of REDI-PAK electrodes and one nonrechargeable battery.

Extract 6:

Picture the waiting room of a hospital or a corridor or a cafeteria. Now picture a visitor, patient or staff member struck down by sudden cardiac emergency. Who will respond first? It could be an ALS-trained member of the code team, but it's just as likely to be a nurse or a receptionist trained in the use of the 1000. The LIFEPAK 1000 defibrillator is ready for all these possibilities, combining the simplicity of one-push defibrillation with clear guidance, both onscreen and from audio prompts.

From the brochures *LP1000_Operations_Manual_3205213–002* and *LP1000_Brochure_3303851_A_LR by Physio Control, www.physio-control.com*

SCIENCE COMMUNICATION

1 The extracts opposite come from a manufacturer's sales brochure, and also from the product user manual. Consider the extracts and comment on the type of writing that is used in each case. Then answer the following questions.

(a) Identify which extracts are from the sales brochure, and which are from the operations manual. Justify your choices.

(b) Discuss the science presented in each booklet, in relation to the intended audience.

INTERPRETATION NOTE

Think about what you would expect to be the different purposes of the two booklets.

PHYSICS IN DETAIL

Now we will look at the physics in detail. Some of these questions link to topics elsewhere in this book, so you may need to combine concepts from different areas of physics to work out the answers.

2 (a) Using information from the discharge current waveform graph and table, calculate the average current for each of the two parts of the discharge. Then use this average current to calculate the charge transferred in the times given. Do these calculations for the 50 Ω, 100 Ω and 150 Ω entries in the table, and compare the variations in charge.

(b) For the same 50 Ω, 100 Ω and 150 Ω entries in the table, use your charge values to calculate the voltage needed to cause each shock to deliver exactly 200 J.

(c) Compare your answers to part (b) with the voltage values that Ohm's law gives you for the peak current in each case.

3 One of the Environmental specifications claims that the unit can survive a one metre drop onto its corner. Calculate the force that the unit would experience if it falls one metre and is stopped by the ground in 20 ms.

THINKING BIGGER TIP

Impedance is a term used to mean 'resistance' when charge flow is in changing directions.

ACTIVITY

The manufacturers of the LIFEPAK 1000 would like a brochure that its customers can leave in hospital waiting rooms for patients to read. Use the information above to design this leaflet.

4A EXAM PRACTICE

1 Which of the following is a unit equivalent to the volt?

A A h

B $C\,s^{-1}$

C $J\,C^{-1}$

D $J\,s^{-1}$ [1]

(Total for Question 1 = 1 mark)

2 An ammeter reads 1.42 A. How many electrons pass through the ammeter in one hour?

A 5.1×10^{3}

B 8.9×10^{18}

C 5.3×10^{20}

D 3.2×10^{22} [1]

(Total for Question 2 = 1 mark)

3 Which of the following is a correct explanation for the shape of a current/voltage graph for a filament bulb?

A Below a certain temperature, the resistivity of the filament material suddenly reduces to zero.

B Above a certain temperature, the resistivity of the filament material suddenly reduces to zero.

C Increased voltage increases the current, which raises the temperature, increasing the resistance.

D As the voltage increases, the current increases, increasing the temperature of the filament, causing a release of more conduction electrons. [1]

(Total for Question 3 = 1 mark)

4 Two wires are made of the same material with resistivity, ρ. Wire X has resistance R. Wire Y is twice as long and has twice the diameter of wire X.

Which row in the table correctly gives the resistances for both wires?

	Wire X resistance	Wire Y resistance
A	R	$\frac{R}{2}$
B	R	R
C	R	$2R$
D	R	$4R$

[1]

(Total for Question 4 = 1 mark)

5 Explain, in terms of energy, the difference between potential difference (pd) and electromotive force (emf). [2]

(Total for Question 5 = 2 marks)

6 The planet Jupiter has a moon Io. Volcanic activity on Io releases clouds of electrons which travel at high speeds towards Jupiter. During a 15 s time period, 2.6×10^{26} electrons reach Jupiter from Io.
Calculate the current. [3]

(Total for Question 6 = 3 marks)

7 An integrated circuit uses strips of gold as connectors and strips of silicon as resistors.
A strip of gold of cross-sectional area $3.0 \times 10^{-6}\,m^2$ carries a current of 8.0 mA. The charge carrier density n is $6.0 \times 10^{28}\,m^{-3}$.

(a) Show that the carrier drift velocity v for gold is approximately $3 \times 10^{-7}\,m\,s^{-1}$ [2]

(b) An approximate value of v for a sample of silicon of the same dimensions, carrying the same current, would be $0.2\,m\,s^{-1}$.
Compare this value with the one for gold and account for the difference in the values. [2]

(c) State and explain what happens to the resistance of a sample of silicon as its temperature increases. [2]

(Total for Question 7 = 6 marks)

8 (a) Explain the difference between resistance and resistivity. [2]

(b) The resistivity of copper is $1.7 \times 10^{-8}\,\Omega\,m$. A copper wire is 0.50 m long and has a cross sectional area of $1.0 \times 10^{-6}\,m^2$. Calculate its resistance. [2]

(Total for Question 8 = 4 marks)

9 When tidying a prep room, a teacher discovers a tray of resistance wires that have lost their labels. He decides to ask his students to carry out experiments to determine the material that each wire is made of by measuring the resistivity of the wires.

(a) Explain why the teacher asks the students to measure the resistivity and not the resistance of the wires. [2]

(b) Describe a method to determine accurately the resistivity of one of the metal wires.
Your description should include:

- the circuit diagram you would use
- the quantities you would measure
- the graph you would plot
- how you would determine the resistivity. [9]

(Total for Question 9 = 11 marks)

10 (a) State Ohm's law. [2]

(b) Using the axes below sketch graphs to show how resistance varies with potential difference for a fixed resistor and a 1.5 V filament lamp. [2]

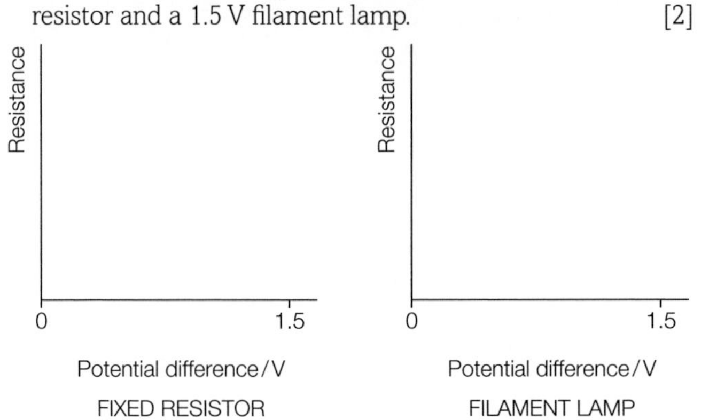

(c) The filament of a lamp is made of metal. Explain why the lamp does not demonstrate Ohm's law. [2]

(Total for Question 10 = 6 marks)

11 The graph shows the current–potential difference characteristic for an electrical component.

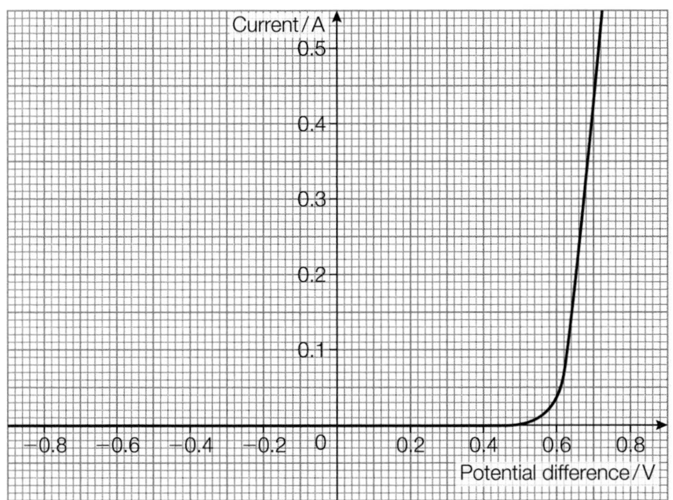

(a) State the name of the component. [1]

(b) State the resistance of the component when the potential difference is −0.7 V. [1]

(c) Calculate the resistance of the component when the potential difference is +0.7 V. [2]

(d) State a practical use for this component. [1]

(Total for Question 11 = 5 marks)

12 A pen which has ink that conducts electricity is used to draw a circle.

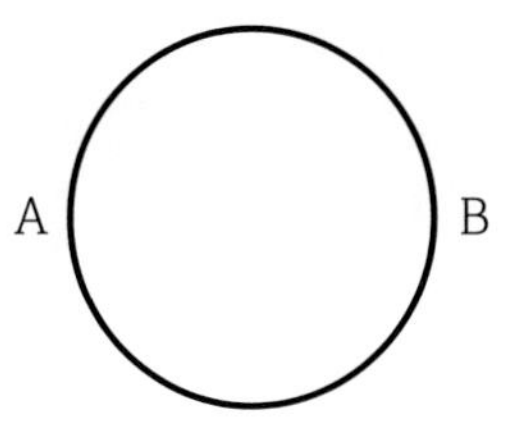

The diameter from point A to point B across the circle is 25.0 cm

(a) Describe an experimental method to measure the resistance of the ink circle between points A and B. [3]

(b) How would the resistance between points A and B change if the circle was redrawn with the ink line twice the width, but with the same size circle and the same depth of ink on the paper? [2]

(Total for Question 12 = 5 marks)

TOPIC 4 ELECTRIC CIRCUITS

CHAPTER 4B COMPLETE ELECTRICAL CIRCUITS

A smartphone uses over a billion tiny electrical circuits in order to perform a huge range of functions. However, it is basically powered by a DC battery that will supply a continuous voltage of approximately 4 volts. From this single electrical quantity, the billion or more circuits can each be powered separately to perform their individual function.

For the complex machines we use every day, the individual components need particular currents and voltages in order to function properly. In order to ensure that they receive appropriate currents and voltages, we must connect them together in circuits of the correct design and with other components. The design of the overall circuit, and designs of parts of it, rely on the calculations of the electrical properties at each point in the circuit.

In this chapter, you will learn how the fundamental electrical quantities change around different parts of a circuit, and how they can be altered by the design of the circuit.

MATHS SKILLS FOR THIS CHAPTER

- **Units of measurement** (*e.g. the ohm, Ω*)
- **Changing the subject of an equation** (*e.g. finding a particular branch current from electric current rules*)
- **Substituting numerical values into algebraic equations** (*e.g. calculating the electrical energy transferred by a component*)
- **Solving algebraic equations** (*e.g. the potential divider equation*)
- **Using ratios, fractions and percentages** (*e.g. the efficiency equation*)
- **Determining the slope and intercept of a linear graph** (*e.g. finding the emf and internal resistance from experimental results*)

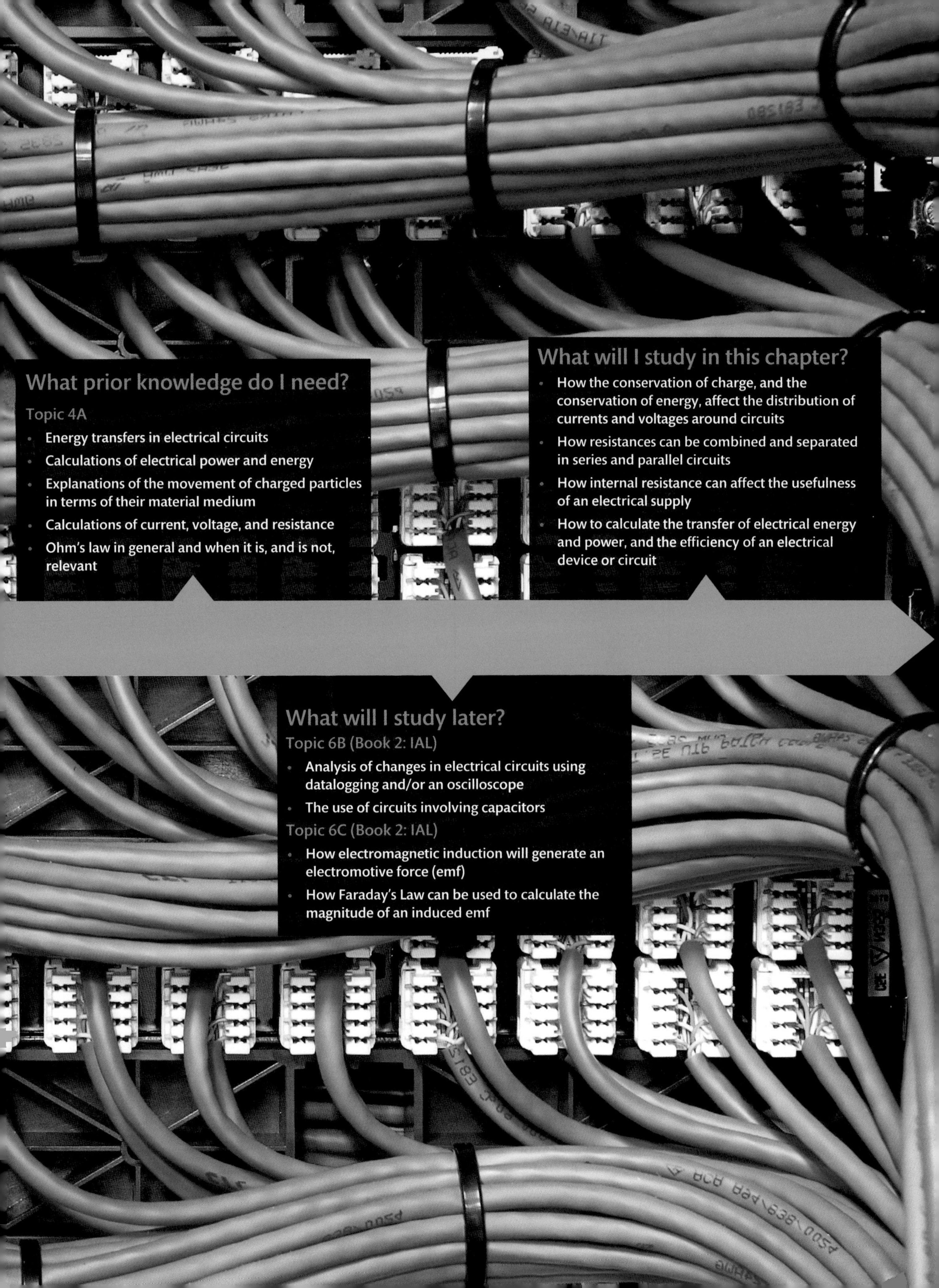

What prior knowledge do I need?

Topic 4A

- Energy transfers in electrical circuits
- Calculations of electrical power and energy
- Explanations of the movement of charged particles in terms of their material medium
- Calculations of current, voltage, and resistance
- Ohm's law in general and when it is, and is not, relevant

What will I study in this chapter?

- How the conservation of charge, and the conservation of energy, affect the distribution of currents and voltages around circuits
- How resistances can be combined and separated in series and parallel circuits
- How internal resistance can affect the usefulness of an electrical supply
- How to calculate the transfer of electrical energy and power, and the efficiency of an electrical device or circuit

What will I study later?

Topic 6B (Book 2: IAL)

- Analysis of changes in electrical circuits using datalogging and/or an oscilloscope
- The use of circuits involving capacitors

Topic 6C (Book 2: IAL)

- How electromagnetic induction will generate an electromotive force (emf)
- How Faraday's Law can be used to calculate the magnitude of an induced emf

4B 1 SERIES AND PARALLEL CIRCUITS

SPECIFICATION REFERENCE 2.4.67(a) 2.4.67(b) 2.4.68

LEARNING OBJECTIVES

- Calculate currents, voltages and resistances in series and parallel circuits.
- Derive the equations for combining resistances in series and parallel.

The most important rules governing measurements in electrical circuits are the conservation of charge and conservation of energy. These determine how currents, voltages and resistances change around any particular circuit.

CURRENT

The total amount of charge within a circuit cannot increase or decrease when the circuit is functioning.

LEARNING TIP

Currents are measures of charge flow through a component, so they must be measured in the same line as the component, i.e. the ammeter must always be placed in series next to the component.

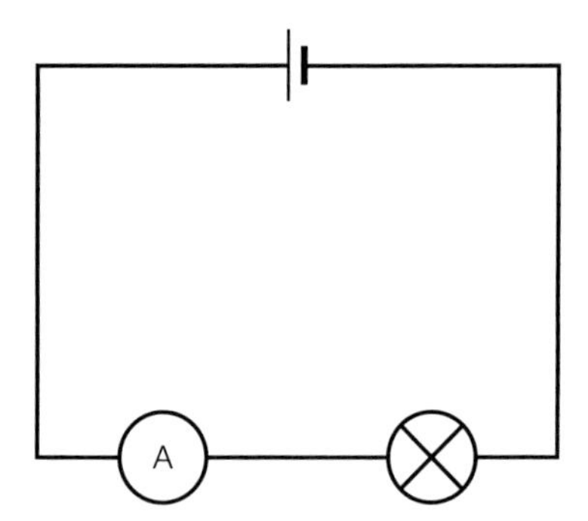

▲ **fig A** Ammeters must always be in series with the component they are measuring.

CURRENTS IN SERIES CIRCUITS

Any group of components that follow in series in a circuit, with no junctions in the circuit, must have the same current through them all.

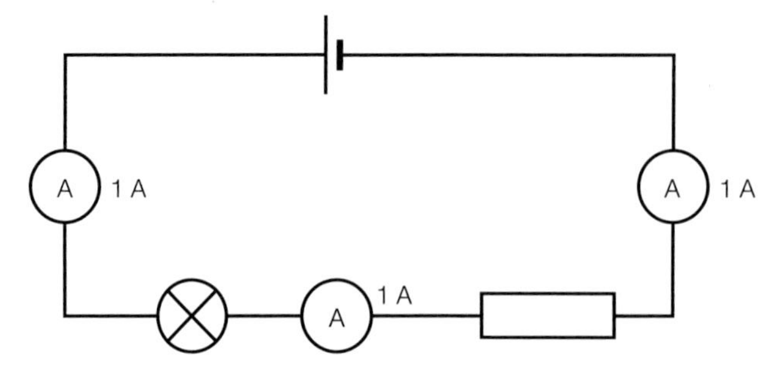

▲ **fig B** The current is constant throughout a series loop in a circuit.

CURRENTS IN PARALLEL CIRCUITS

Whenever a current encounters a junction in a circuit, the charges can only go one way or the other, so the current must split. The proportions that travel along each possible path will be in inverse proportion to the resistance along that path. If the path has high resistance, the current is less likely to go that way. However, the total along the branches must add up to the original total current.

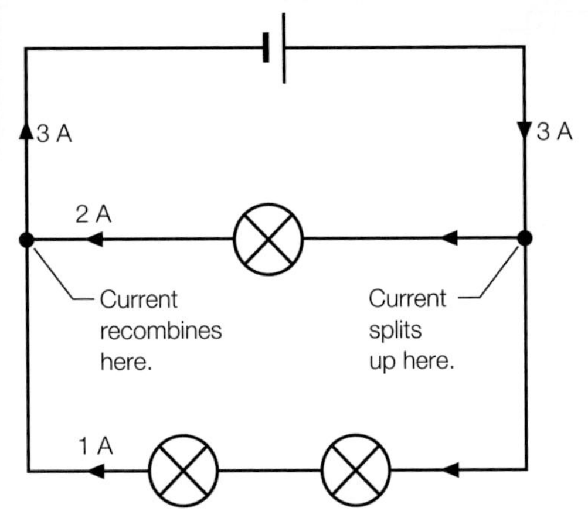

▲ **fig C** Current splits at circuit junctions, but the total current will remain the same, in order to conserve charge.

VOLTAGES AROUND CIRCUITS

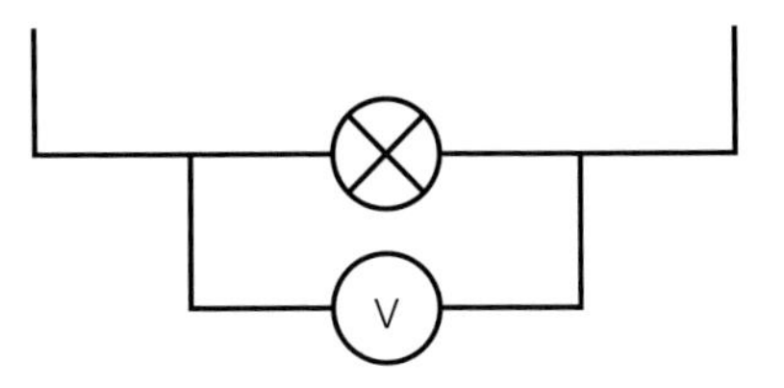

▲ **fig D** Voltmeters must always be in parallel across the component they are measuring.

> **LEARNING TIP**
>
> Voltages are measures of energy change as charge passes through a component. As such, they must be measured from one side to the other, i.e. the voltmeter must always be placed in parallel across the component.

VOLTAGES IN SERIES CIRCUITS

Any group of emfs that follow in series in a circuit, with no junctions in the circuit, will have a total emf that is the sum of their individual values, that accounts for the direction of their positive and negative sides. For example, two 1.5 V cells in a TV remote control will be in series, so that they supply an emf of 3.0 V. This is also true for any string of potential differences in series.

1.5 V 4.5 V

3.0 V 3.0 V

▲ **fig E** Emfs in series will add up. Take care to account for their direction, as cells in opposite directions oppose each other and cancel out the emf they would supply.

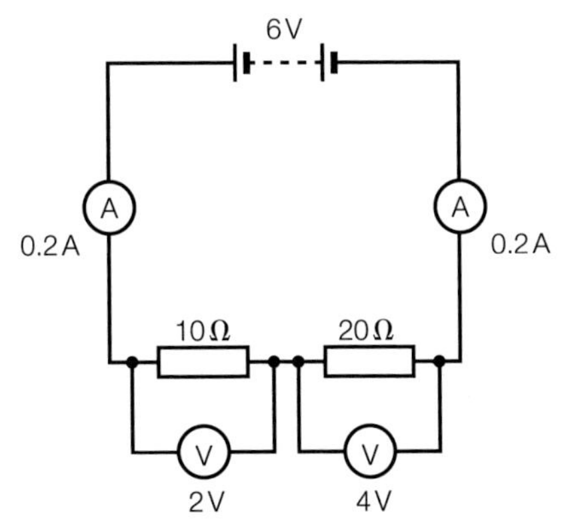

▲ **fig F** Potential differences in series will add up. What would be the total pd across the two resistors in this diagram?

> **WORKED EXAMPLE**
>
> In the circuit diagram of **fig F**, the two voltmeters measure the potential differences in the circuit.
> The total pd across both resistors will be the sum of their individual pds, so the total is 2 V + 4 V = 6 V.

VOLTAGES IN PARALLEL CIRCUITS

If the total voltage across any branch of a circuit is known, then the voltage across any other branch in parallel with it will be identical.

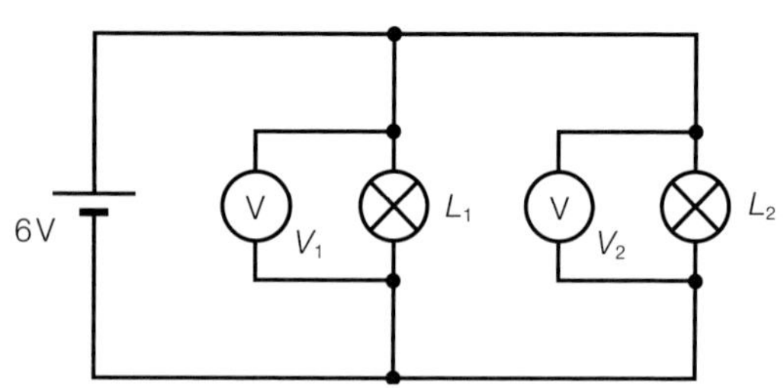

▲ **fig G** Both voltmeters here will read 6 V: the voltage is the same across all parallel branches within a circuit.

RESISTANCE

Resistance can be calculated from Ohm's law as $\frac{V}{I}$. This can be done for any individual component or a whole branch of a circuit, as long as the voltage and current are known. This also means that if we follow the rules for current and voltage variations around series and parallel circuits, we come up with simple rules for the total resistance in series and parallel combinations.

RESISTORS IN SERIES

Any group of resistances that follow in series in a circuit, with no junctions in the circuit, will have a total resistance that is the sum of their individual values. In **fig F**, we could use the total pd across both resistors and the current through them to calculate their total resistance:

$$\text{total } R = \frac{\text{total } V}{I} = \frac{6}{0.2} = 30\,\Omega$$

You will see that this matches with our rule that

$$R_{total} = R_1 + R_2 = 10 + 20 = 30\,\Omega$$

LEARNING TIP

Ammeters in series, and voltmeters in parallel, do not affect resistance and can be ignored.

DERIVATION

Resistors in series will have a total pd across them that is the sum of their individual pds. The current through them will be the same for each one. Take an example of three resistors in series:

potential difference: $V_{total} = V_1 + V_2 + V_3$

and $V = IR$

$\therefore \quad IR_{total} = IR_1 + IR_2 + IR_3$

$\therefore \quad R_{total} = R_1 + R_2 + R_3$

RESISTORS IN PARALLEL

The total resistance of a group of resistors in parallel can be calculated from the equation:

$$\frac{1}{R_{total}} = \frac{1}{R_1} + \frac{1}{R_2} + \frac{1}{R_3} + \ldots$$

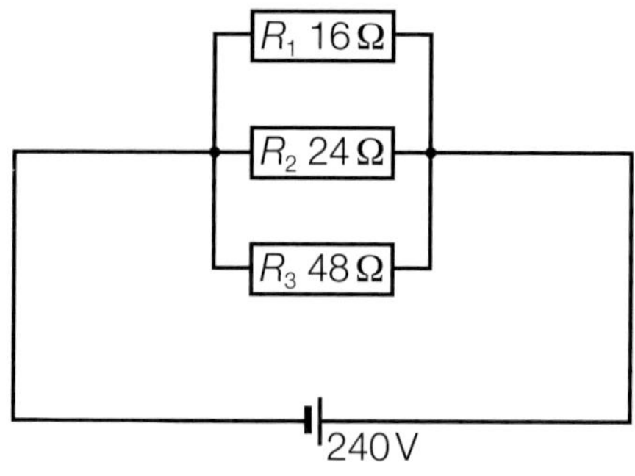

▲ **fig H** Resistors in parallel follow a reciprocal sum rule to find their total resistance.

WORKED EXAMPLE

In the circuit diagram of **fig H**, what is the overall resistance of the block of three resistors?

$$\frac{1}{R_{total}} = \frac{1}{R_1} + \frac{1}{R_2} + \frac{1}{R_3}$$

$$\therefore \quad \frac{1}{R_{total}} = \frac{1}{16} + \frac{1}{24} + \frac{1}{48}$$

$$\therefore \quad \frac{1}{R_{total}} = \frac{3}{48} + \frac{2}{48} + \frac{1}{48}$$

$$\therefore \quad \frac{1}{R_{total}} = \frac{6}{48}$$

$$\therefore \quad \frac{1}{R_{total}} = \frac{1}{8}$$

$$R_{total} = 8\,\Omega$$

EXAM HINT

Note that the steps and layout of the solution in this worked example are suitable for questions on calculating the overall resistance of resistors in parallel in the exam.

LEARNING TIP

Any combination of resistors in parallel will have a resistance that is smaller than the smallest individual resistance in the group.

DERIVATION

Resistors in parallel will have a total current through them that is the sum of their individual currents. The pd across them will be the same for each branch. Take an example of three resistors in parallel:

current: $I_{total} = I_1 + I_2 + I_3$

and $I = \frac{V}{R}$

$$\therefore \quad \frac{V}{R_{total}} = \frac{V}{R_1} + \frac{V}{R_2} + \frac{V}{R_3}$$

$$\therefore \quad \frac{1}{R_{total}} = \frac{1}{R_1} + \frac{1}{R_2} + \frac{1}{R_3}$$

RESISTORS BRANCH COMBINATIONS

If a circuit has a mixture of series and parallel combinations, we must use the rules we have for each group of resistors. Then, we should use the rules again to combine these sub-totals.

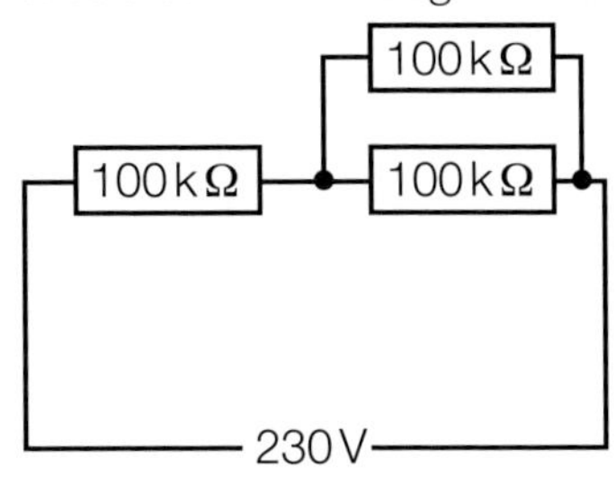

▲ **fig I** Calculate the total resistance of resistor groups separately, before using the appropriate rule to combine the sub-totals together.

WORKED EXAMPLE

In **fig I**, we must calculate the total resistance of the parallel pair first, and then add this total to the remaining 100 kΩ that is in series with the pair. The junction separating the single resistor from the parallel pair is the clue as to how to separate the calculations.

Parallel pair:

$$\frac{1}{R_{para}} = \frac{1}{100} + \frac{1}{100} = \frac{2}{100} = \frac{1}{50}$$

$$\therefore \quad R_{para} = 50\,\text{k}\Omega$$

Total resistance:

$$R_{total} = 100\,\text{k}\Omega + R_{para} = 100 + 50$$

$$R_{total} = 150\,\text{k}\Omega$$

EXAM HINT

Note that the steps and layout of the solution in this worked example are suitable for questions on calculating the overall resistance of resistors in parallel and series in the exam.

QUANTITY	SERIES CIRCUIT RULES	PARALLEL CIRCUIT RULES
current	current is the same throughout	$I_{total} = I_1 + I_2$
voltage	$V_{total} = V_1 + V_2$	voltage is the same on each branch
resistance	$R_{total} = R_1 + R_2$	$\frac{1}{R_{total}} = \frac{1}{R_1} + \frac{1}{R_2}$

table A Summary of the electrical rules in series and parallel circuits.

PRACTICAL SKILLS

Investigating circuit rules

Make a complex circuit with various components in series and parallel combinations. By using only an ammeter and voltmeter, your partner should be able to prove the circuit rules explained in this section. If the resistances are large, the ammeter may need to be a multimeter set to read milliamp currents.

Join various resistors in various combinations of series and parallel connections. Your partner should calculate what they expect the overall resistance of the combination to be. Check whether they have calculated correctly by measuring the overall resistance of the resistor combination using an ohmmeter.

Safety Note: Short circuits and overloading components will cause overheating and burns on skin contact.

CHECKPOINT

1. How does **fig B** illustrate the conservation of charge in a circuit?

2. If all four cell combinations in **fig E** were connected to each other in series, what would be the total emf?

3. If R_3 has a resistance of 24 Ω, how much current flows in **fig J**?

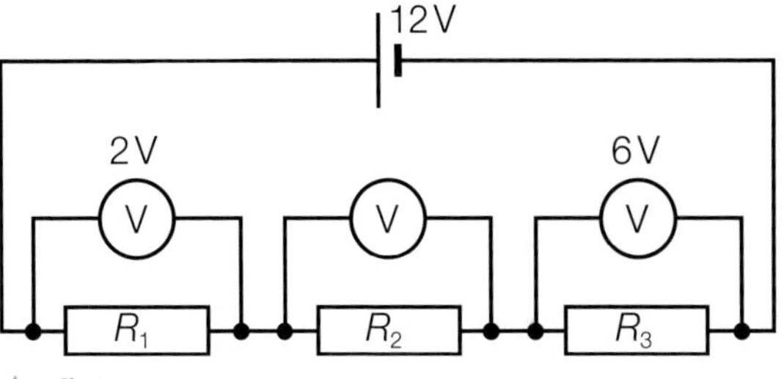

▲ **fig J**

4. Use Ohm's law on the resistors in **fig K** to verify that the voltage across parallel branches is always the same.

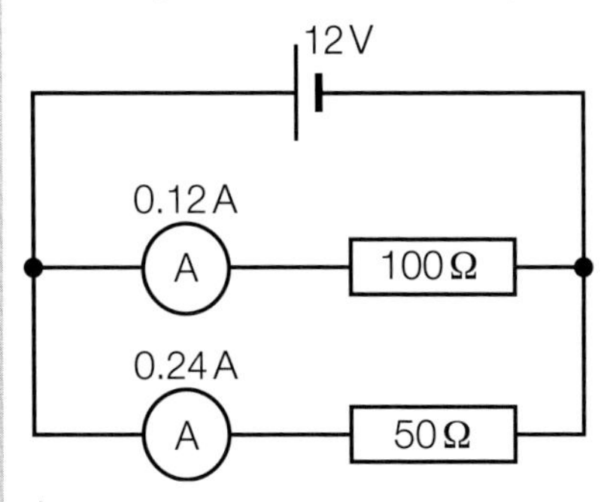

▲ **fig K**

5. What is the total current that flows in **fig I**?

6. (a) What is the total current flow through the cell in **fig H**?

(b) How much current flows in the 48 Ω resistor in **fig H**?

4B 2 ELECTRICAL CIRCUIT RULES

SPECIFICATION REFERENCE 2.4.67(a) 2.4.67(b) 2.4.68 2.4.75

LEARNING OBJECTIVES

- Understand how the distribution of current in a circuit is a consequence of charge conservation.
- Understand how the distribution of potential differences in a circuit is a consequence of energy conservation.
- Make calculations based on current and voltage circuit rules.

Gustav Kirchhoff was a Prussian physicist working in the middle of the nineteenth century. He made important contributions to several areas of physics, and there are also laws named after him in spectroscopy and thermochemistry. The electrical circuit rules presented here sometimes carry his name.

▲ **fig A** Gustav Kirchhoff presented the conservation of charge and of electrical energy within a circuit as two laws of physics.

ELECTRIC CURRENT RULE

Electric current rule: the algebraic sum of the currents entering a junction is equal to zero:

$$\Sigma I = 0$$

In order to conserve electric charge, the sum of all the currents arriving at any point (such as a junction) in a circuit is equal to the sum of all the currents leaving that point.

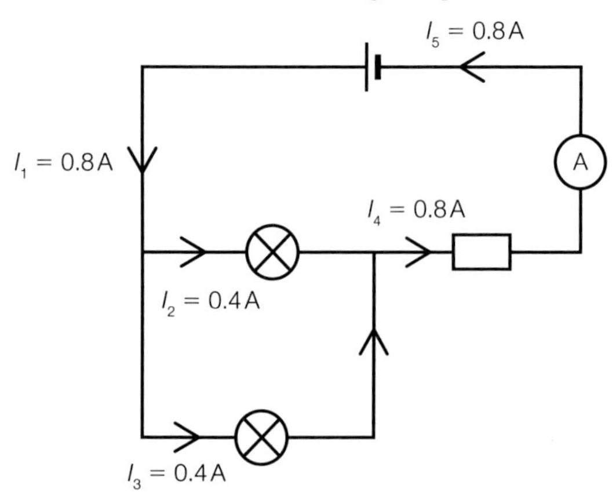

▲ **fig B** Current rule example circuit.

In the circuit diagram of **fig B**, we could think about the current entering and leaving the resistor. This is 0.8 A entering and 0.8 A leaving. If we added these together, accounting for the direction:

$$I_{in} + I_{out} = I_4 + -I_5 = 0.8 + -0.8 = 0$$

This simple example confirms the rule at work in the resistor.

Similarly, consider the currents at the point where the circuit splits to send current through the two bulbs:

$$I_1 + -I_2 + -I_3 = 0.8 + -0.4 + -0.4 = 0$$

The same overall result would appear at any point in the circuit – charge must always be conserved.

ELECTRICAL VOLTAGES RULE

In order to conserve electrical energy around any closed loop, **Electrical voltages rule:** the sum of emfs is equal to the sum of the pds around that loop.

Note that a loop being considered might not be the whole of a circuit – the only requirement is that the loop is complete. Note also that the potential differences referred to here are defined as being the products of component currents and resistances, as per Ohm's law. So, if we isolate a particular closed loop, it may have potential differences with positive values, and with negative values when the current flows in the opposite direction.

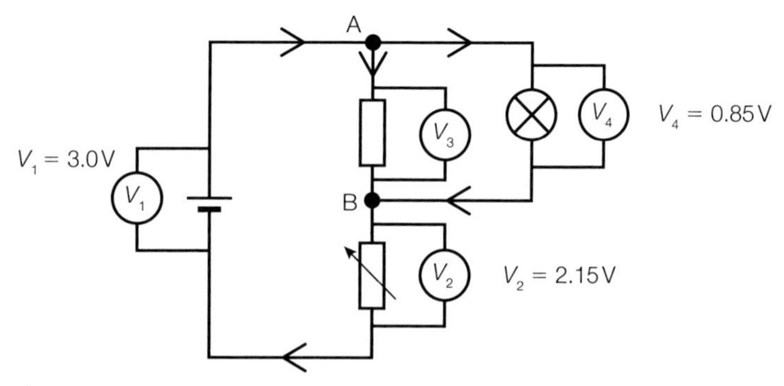

▲ **fig C** Voltages rule example circuit.

In **fig C**, consider the voltages around the closed loop that consists of the cell, variable resistor and fixed resistor. We know V_3 must be 0.85 V, as all parallel branches must have the same voltage and it is parallel to V_4. If you move clockwise around the loop, and start at the cell, the cell is the only emf, so the sum of all emfs is 3.0 V. Following the same route to sum the potential differences, we ignore the cell as it is an emf, not a pd. This means the sum of the pds will be:

$$\Sigma V = V_3 + V_4 = 0.85 + 2.15 = 3.0\,V$$

As the current is flowing in the same direction through the cell and both variable resistor and fixed resistor, these voltages are all in the same direction, and so:

$$\Sigma \varepsilon = \Sigma V$$

Next, consider just the closed loop containing the fixed resistor and the lamp. If $\Sigma\varepsilon = \Sigma V$, and there are no emfs, the pds must total zero. Starting at point A and moving clockwise around the loop, the current moves through the lamp, which is measured at $V_4 = 0.85$ V. Continuing to point B and then back up to complete the loop at A means going through the fixed resistor against the flow of the current. This means the pd will be negative, although also 0.85 V. So in this consideration, $V_3 = -0.85$ V.

$$\Sigma V = V_3 + V_4 = -0.85 + 0.85 = 0$$

$$\therefore \quad \Sigma\varepsilon = \Sigma V$$

VOLTAGES RULE WITH CALCULATED PDS

Potential differences are the product of currents passing through resistances, thus transferring electrical energy. Each pd could be calculated from Ohm's law as $V = IR$. So this becomes:

$$\Sigma\varepsilon = \Sigma IR$$

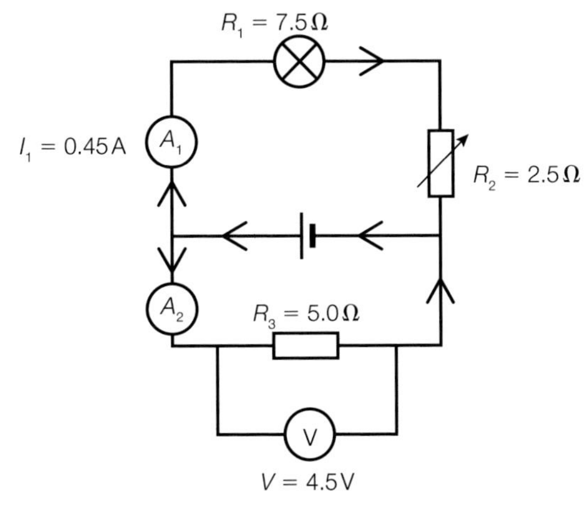

fig D Confirming $\Sigma\varepsilon = \Sigma IR$.

In **fig D**, we can confirm the voltages rule around the upper closed loop, through the cell, ammeter A_1, lamp and variable resistor. The only emf is the cell, and this will have an emf of 4.5 V, as shown on the voltmeter at the bottom of the diagram.

$$\Sigma\varepsilon = 4.5\text{ V}$$

Ammeters have zero resistance, and so A_1 contributes nothing to the sum of pds. The current must be the same through the lamp and the variable resistor, so the sum of the potential differences will be calculated from:

$$\Sigma IR = I_1R_1 + I_1R_2 = (0.45 \times 7.5) + (0.45 \times 2.5) = 3.375 + 1.125$$

$$\Sigma IR = 4.5\text{ V}$$

$$\therefore \quad \Sigma\varepsilon = \Sigma IR$$

CHECKPOINT

SKILLS PROBLEM SOLVING

1. In **fig D**, use Ohm's law to find the current through ammeter A_2.
2. The current through the cell in **fig D** is 1.35 A. Use your answer from question 1 to show that the circuit junction between the two ammeters in **fig D** conserves current.
3. Demonstrate by calculation that the lower loop (cell and resistor) in the circuit in **fig D** conserves energy.
4. Find the emf of the cell in **fig E**.

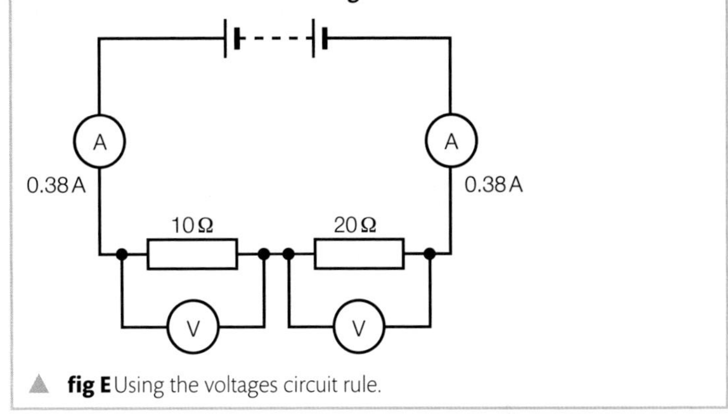

fig E Using the voltages circuit rule.

SUBJECT VOCABULARY

electric current rule the algebraic sum of the currents entering a junction is equal to zero:

$$\Sigma I = 0$$

voltages circuit rule around a closed loop, the algebraic sum of the emfs is equal to the algebraic sum of the pds:

$$\Sigma\varepsilon = \Sigma IR$$

4B 3 POTENTIAL DIVIDERS

SPECIFICATION REFERENCE 2.4.74 2.4.75 2.4.76

LEARNING OBJECTIVES

- Describe and explain potential divider circuits.
- Make calculations in potential divider circuits.
- Explain uses for potential divider circuits.

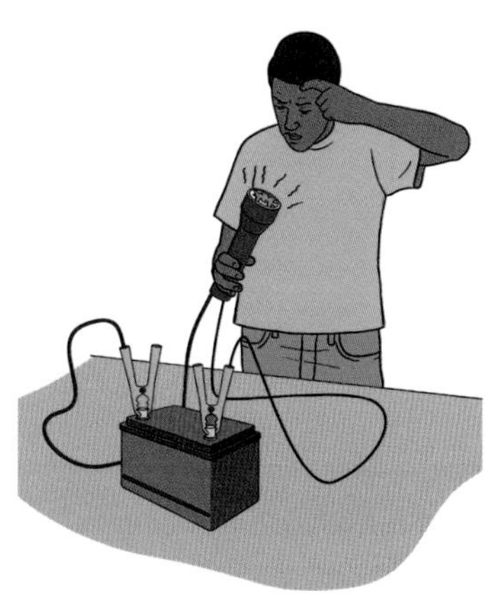

▲ **fig A** Fixed emfs can be problematic when components require a lower voltage.

We have seen that all the voltage supplied by emfs in a circuit loop must be used by components as potential differences at other points around the circuit loop. The way that the voltage splits up is in proportion to the resistances of the components in the circuit loop, as shown in **fig B**. The voltages across the two resistors must add up to 6 V. The 20 Ω resistor has twice the resistance, so takes twice the voltage of the 10 Ω resistor. This is a **potential divider** circuit.

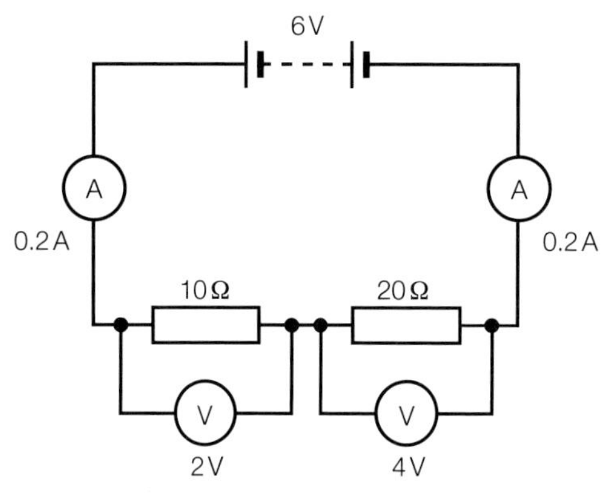

▲ **fig B** All of the 6 volts supplied by the battery must be used up by the resistors. The total pd of 6 V is split into 2 V and 4 V in proportion to their resistances. Ammeters have zero resistance, so transfer zero energy and have zero pd.

Ohm's law explains why the voltage is split in proportion to the resistance. Consider the resistances and voltages in **fig B** in more general terms, as R_1 (for the 10 Ω resistor) and R_2, and their corresponding pds V_1 and V_2. This will make the calculation valid for all values of resistances, and calculating the current through each one gives:

$$I_1 = \frac{V_1}{R_1} \quad \text{and} \quad I_2 = \frac{V_2}{R_2}$$

However, they both have the same current passing through them, so $I_1 = I_2$.

$$\therefore \quad \frac{V_1}{R_1} = \frac{V_2}{R_2}$$

$$\therefore \quad \frac{V_1}{V_2} = \frac{R_1}{R_2}$$

So the voltage is split in the same proportion as the ratio of the resistances for any values of resistance.

This means that for a known emf supply voltage, we can use carefully chosen resistances to share the voltage and provide a specific value of pd on one component. If we then also remember that all parallel branches in a circuit must have the same voltage, then any branch we set up in parallel with that specific pd will also have the same voltage. So, we can set up a circuit to provide an exactly chosen value of voltage to the parallel branch. Using variable resistors, this set up can then be used to provide whatever voltage we choose. This can be particularly useful if our emf source is a fixed value, like a battery, which can only supply, say, 6 V.

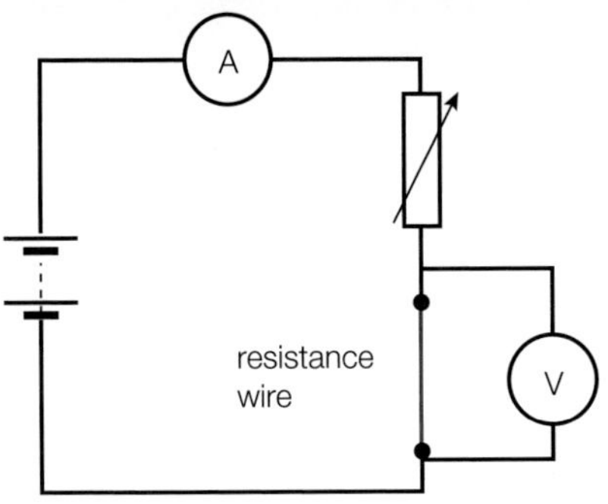

▲ **fig C** Adjusting the variable resistor alters the proportion of the voltage it takes, and so the pd that is left for the resistance wire can be varied to any value we choose.

In **Section 4A.3**, we saw graphs of current against potential difference for various components. In order to undertake the investigation to produce these data, a potential divider circuit could be used. **Fig C** shows a piece of resistance wire for which the current and voltage can be measured. If the variable resistor is set to a high resistance, then it will have a high potential difference, and only a small voltage will be left for the test wire. By decreasing the resistance of the variable resistor, the pd across the test wire will increase. We can then adjust the variable resistor to set the values of pd on the test wire to those we need for the I–V plot.

THE POTENTIAL DIVIDER EQUATION

As the pd in a potential divider circuit is split in proportion to the resistances of the components, we can calculate the pd across them mathematically.

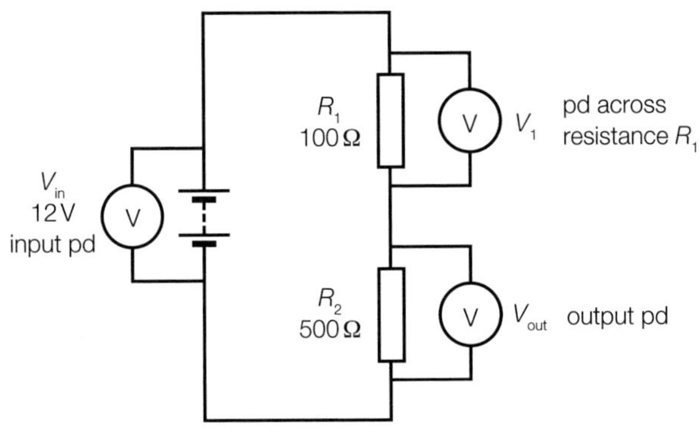

▲ **fig D** Generalised potential divider circuit.

In the circuit diagram of **fig D**, we have the general terminology for the values used in calculations of a potential divider circuit: V_{out}, R_1, R_2 and V_{in}. Using these, we can calculate the pd across the component of interest, which is usually drawn as the second resistance, R_2, using the potential divider equation:

$$V_{out} = V_{in} \times \frac{R_2}{(R_1 + R_2)}$$

WORKED EXAMPLE

Using the values in **fig D**, we can find the pd across resistor R_2:

$$V_{out} = V_{in} \times \frac{R_2}{(R_1 + R_2)}$$

$$V_{out} = 12 \times \frac{500}{(100 + 500)} = 12 \times \frac{5}{6}$$

$$V_{out} = 10\text{ V}$$

EXAM HINT

Note that the steps and layout of the solution in this worked example are suitable for questions on finding the pd across a resistor in the exam.

THE POTENTIOMETER

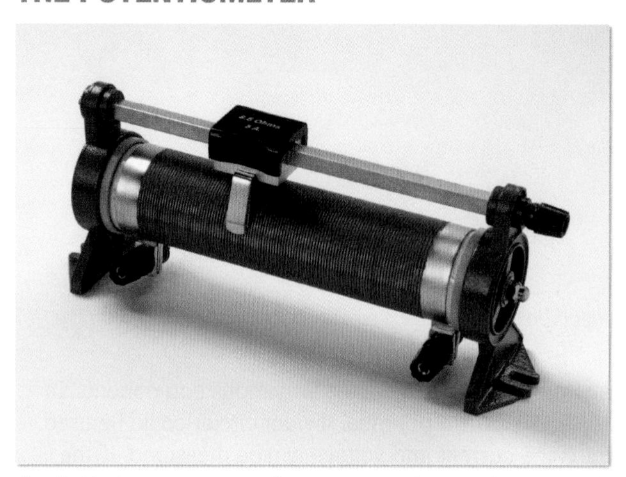

▲ **fig E** A potentiometer allows us to supply any voltage up to the maximum provided by our emf source.

The idea of using a potential divider to supply a variable voltage as required has been incorporated into an electrical component called the **potentiometer**. A moveable contact can slide over a fixed length of resistance wire which adjusts the length that is on either side of this 'wiper'. The two lengths of wire are the two resistors in our potential divider circuit. Adjusting their relative lengths alters their relative resistances and the voltage across each. If we want to power a separate circuit with a variable voltage, we connect it across one side of the potentiometer. This is shown in **fig F**, where we would connect the right-hand side connections across R_2 to our main circuit. The blocks marked R_1 and R_2 are a continuous piece of resistance wire, which will be split into two resistances by the position of the wiper. Adjusting the wiper up and down allows us to have any desired voltage up to the maximum of V_{supply}.

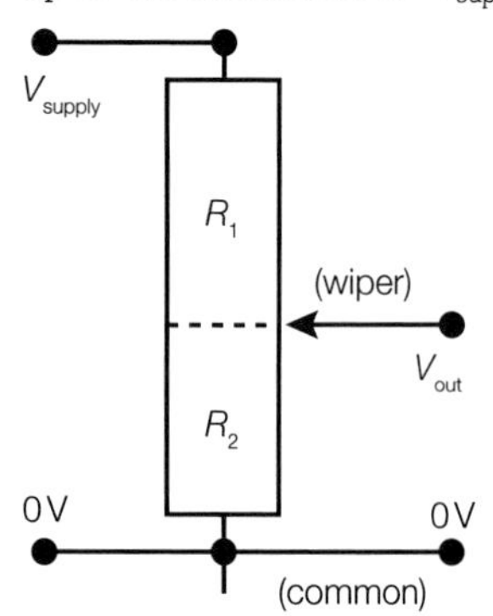

▲ **fig F** Adjusting the position of the slider (wiper) on the potentiometer changes its position on the resistance wire, so that the proportions of resistance on either side change, altering the potential differences on either side of the wiper.

POTENTIAL DIVIDERS IN SENSOR CIRCUITS

If one of the resistors in a potential divider is a sensor component (such as thermistor or LDR) then we can use its changing resistance to control an external circuit.

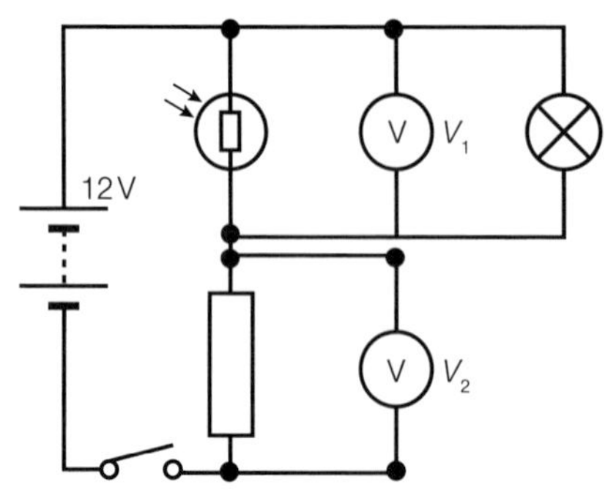

▲ **fig G** A sense and control circuit which makes the light brighter as the light level in the room drops.

In **fig G**, the lamp is in parallel with the LDR, so will be powered by the same voltage as that across the LDR. The resistance of a LDR increases if the ambient light level drops (darker surroundings make its resistance go up). In bright conditions, the LDR has a very low resistance, so it takes a very small proportion of the 12 V supplied by the battery. This will make the lamp very dim, probably so much that it will appear to be off; it is not needed if the surroundings are bright. As conditions become darker, the LDR resistance goes up, increasing the proportion of the 12 V supplied that drop across it. This increases the voltage across the bulb, and makes it brighter. This circuit increases the brightness of the lamp as the surroundings become darker.

Many electronic components have a switch-on voltage of 5 V. These could be used with a sensor in a potential divider circuit, and would only switch on when the sensor's resistance reached a level where its pd was 5 V. Altering the resistor in this circuit would then control the sensor level at which the electronic component was switched on.

CHECKPOINT

1. A potential divider circuit supplied by an emf of 8 V has two resistors, $R_1 = 125\,\Omega$ and $R_2 = 135\,\Omega$. What will be the output voltage across R_2?
2. A potentiometer, as in **fig F**, is supplied by an emf of 15 V and has a 40 cm length of resistance wire. What voltage would V_{out} be if the wiper was placed 10 cm from the top?
3. Draw a potential divider circuit, like that in **fig G**, which will drive a fan motor in a kitchen faster as it becomes hotter. (A standard thermistor has a higher resistance at lower temperatures.)
4. **Fig H** shows a length of resistance wire, with a contact that can rotate around a pivot to contact with more or less of the resistance wire. Explain how this represents a potential divider circuit and what will happen to the brightness of the lamp as the contact is rotated clockwise from the position shown.

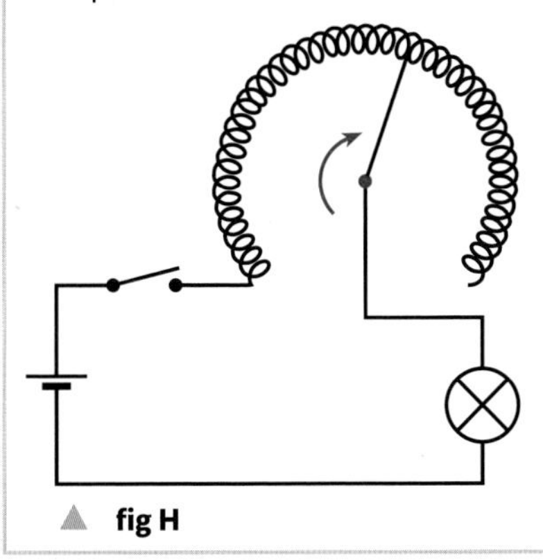

▲ **fig H**

SUBJECT VOCABULARY

potential divider a circuit designed to provide specific voltage values by splitting an emf across two resistors

potentiometer a version of the potential divider in which a single resistance wire is used in two parts to form the two resistances. A sliding connection on the wire can be adjusted to alter the comparative resistances and thus alter the output pd from the potentiometer

4 EMF AND INTERNAL RESISTANCE

SPECIFICATION REFERENCE

2.4.77 2.4.78 CP8

CP8 LAB BOOK PAGE 32

LEARNING OBJECTIVES

- Explain what is meant by internal resistance.
- Describe an investigation to determine an emf's internal resistance.
- Make calculations of internal resistance.

All components have some resistance. The best ammeters have such a small resistance that the effect on their current measurements can be ignored, but it is not actually zero. Sources of electromotive force (emf), such as batteries, will also have a small, but non-zero, resistance. The resistance of an emf source is called its **internal resistance**. In most situations, the internal resistance of an emf source will not affect the performance of a circuit. It may not even be noticeable. However, as the power lost – energy wasted as heat – through a component increases with the current through it (see **Section 4B.5**), there can be a significant power loss when large currents are used. The net effect of this energy transfer by the emf source itself will be for it to supply a smaller voltage to the rest of the circuit, which could affect the circuit's performance. A large current through the internal resistance could also damage the emf source as a result of ohmic (resistance) heating.

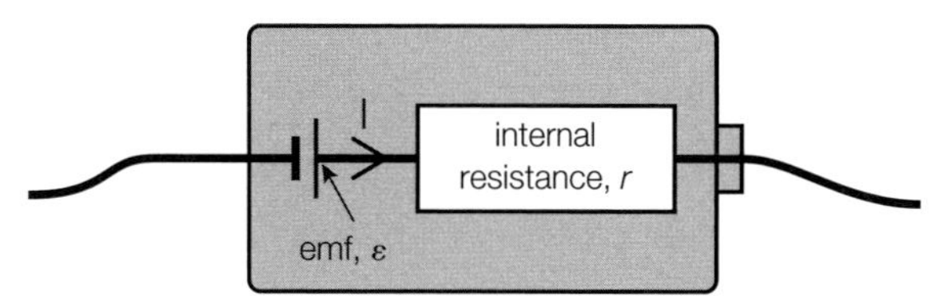

fig A Any emf source could be considered to be a pure emf with a small resistive potential difference acting in series with the emf.

The effect of having an internal resistance is that an emf will never be able to fully supply its maximum voltage. There will always be a small drop in voltage over the internal resistance and this drop will be bigger with a higher current. The pd over the internal resistance, r, can be found from Ohm's law:

$$V_{internal} = Ir$$

This pd is sometimes referred to as 'lost volts', as the circuit appears to have fewer volts in use by the load than the emf should be supplying. A measurement of voltage across the terminals of a cell powering a circuit would not measure the emf; it would measure the emf minus the lost volts. This 'terminal pd' would then be:

$$V_{terminal} = \varepsilon - Ir$$

As all parallel branches have the same voltage across them, the terminal pd would also equal the pd across the load resistance.

$$V_{terminal} = V_{load} = \varepsilon - Ir$$

EXAM HINT

The internal resistance is an internal part of the emf, and cannot be accessed separately. Do not draw a voltmeter measuring across r, as this is impossible.

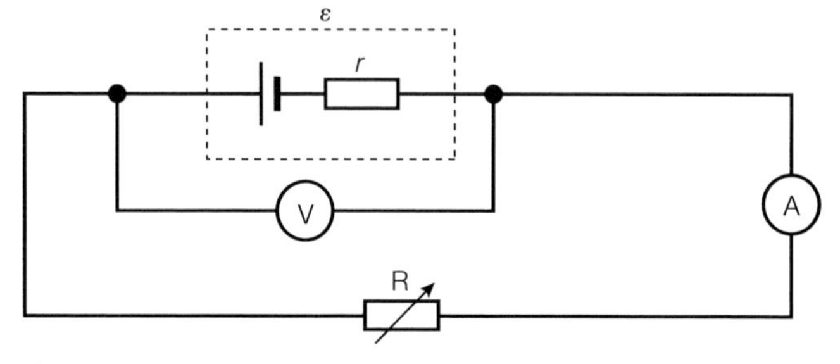

fig B Circuit for demonstrating the effect of internal resistance.

By considering energy conservation, we can draw up an equation for the circuit that includes the internal resistance.

$$\Sigma\varepsilon = \Sigma V$$

$$\therefore \quad \varepsilon = V_{load} + V_{internal}$$

$$\therefore \quad \varepsilon = V_{load} + Ir = IR + Ir$$

The circuit shown in **fig B** demonstrates what the mathematics means practically. The voltmeter will be measuring the voltage actually supplied by the cell, so that will be its pure emf minus the pd across the internal resistance, or lost volts. However, as the ammeter should have zero pd, the voltmeter will also be reading the value of the potential difference across the variable resistor, *R*.

$$\varepsilon - V_r = V_R$$

Or, in terms of the ammeter reading, *I*, and the voltmeter reading, *V*:

$$\varepsilon - Ir = V$$

$$\therefore \quad \varepsilon = V + Ir$$

which is, again, a statement of energy conservation in this circuit.

PRACTICAL SKILLS CP8

Investigating internal resistance

We can find the internal resistance for a source of emf such as a cell in an experiment. Using a circuit like that shown in **fig B**, we adjust the variable resistor to change the current through the circuit, and at each current value we record the voltmeter reading.

The equation we have previously seen for the quantities in this circuit is:

$$\varepsilon - Ir = V$$

or:

$$V = -rI + \varepsilon$$

Comparing this with the equation for a straight-line graph $y = mx + c$ shows us that a graph plotted with V on the y-axis and I on the x-axis will have a gradient equal in magnitude to the internal resistance, r. The y-intercept, c, here represents the voltmeter reading when no current flows. If there is no current in the circuit, the internal resistance can take no energy from the circuit, so there will be zero lost volts and the voltmeter will read the full value of the emf. So, the y-intercept is the value of the emf, ε.

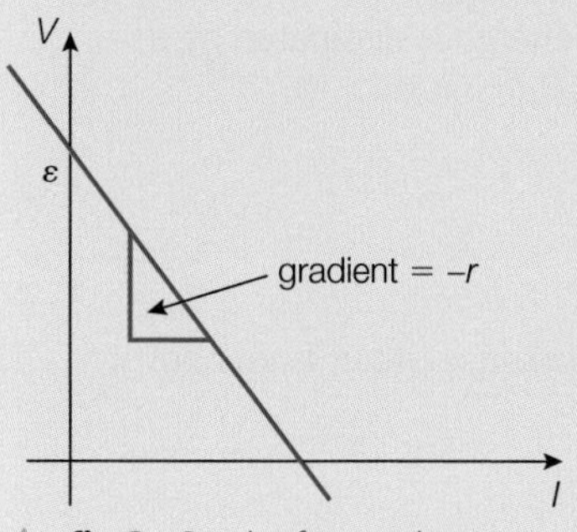

fig C Graph of internal resistance investigation results.

CHECKPOINT

SKILLS REASONING/ARGUMENTATION, ANALYSIS

1. What is the internal resistance of a 6 V battery if it supplies 5.8 V to a lamp when the current flowing is 0.40 A?

2. What is the internal resistance of a 1.5 V dry cell, that delivers 1.3 V to a lamp which has a resistance of 5.0 Ω?

3. The starter motor for a car engine needs a very high current, for example 100 A.
   (a) Explain why a car battery should have a very low internal resistance.
   (b) Explain why the car's headlights are observed to dim if they are on when the car's starter motor is used.

4. A student investigating the emf and internal resistance of a dry cell used the circuit in **fig B**, with the cell in the place marked by the dashed lines. He collected the results in **table A**.

   Plot a graph to find both the internal resistance and the emf of the cell that the student tested.

AMMETER READING, *I* / A	VOLTMETER READING, *V* / V
0.10	1.45
0.20	1.42
0.30	1.39
0.40	1.31
0.80	1.18
1.10	1.09

table A Current and voltage readings from a student experiment to find internal resistance of a dry cell.

SUBJECT VOCABULARY

internal resistance the resistance of an emf source

5 POWER IN ELECTRIC CIRCUITS

SPECIFICATION REFERENCE 1.3.22 2.4.69

LEARNING OBJECTIVES

- Explain work done by electric circuits.
- Make calculations of work and power in electric circuits.
- Explain efficiency and make calculations of it within electric circuits.

The measurement of voltages in electric circuits has shown us that these deal with the transfer of energy by components. As we saw in **Section 1B.2**, energy must always be conserved, and electric circuits are no exception to this. When components transfer electrical energy to other useful stores, they can be considered to be doing work, as this is defined as a transfer of energy.

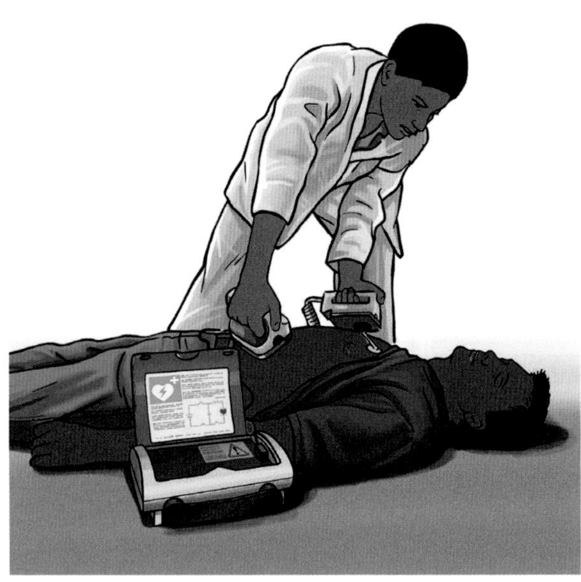

fig A As part of this circuit, the heart will transfer some electrical energy to kinetic energy. The circuit is doing work.

ELECTRICAL WORK

Work done has the symbol *W*, and as the equation defining potential difference includes a term for the amount of energy transferred, *E*, these two will be the same.

$$W = E$$

The equation for pd, in rearranged form, gives us:

$$W = V \times Q$$

The definition of current, in rearranged form, is:

$$Q = I \times t$$

Combining these will give us the work done by a component in an electric circuit:

$$W = V \times I \times t$$

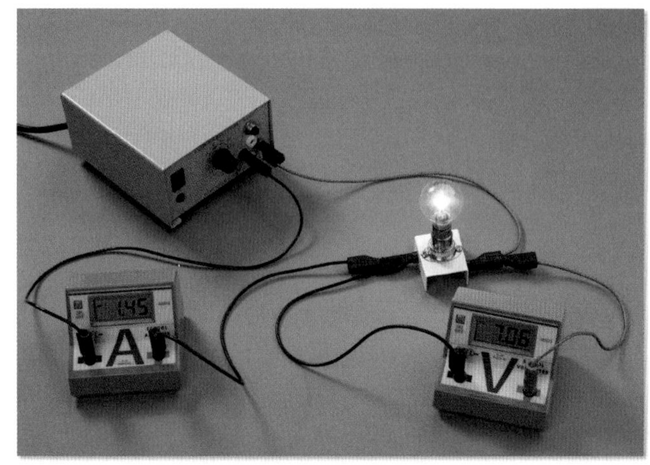

fig B Electrical work done by a light bulb.

WORKED EXAMPLE

In **fig B**, if the bulb lights up for 4.85 seconds, how much work has the bulb done?

$$W = VIt$$

$$W = 7.06 \times 1.45 \times 4.85$$

$$W = 49.6\,\text{J}$$

EXAM HINT

Note that the steps and layout of the solution in this worked example are suitable for questions on finding the work done in the exam.

ELECTRICAL POWER

Power, *P*, is the rate of transfer of energy, or the rate of doing work. In an electrical circuit, the energy is dissipated by a component. The mathematical definition is:

$$P = \frac{E}{t}$$

OR

$$P = \frac{W}{t}$$

Incorporating the equation for work in a circuit from above:

$$P = \frac{VIt}{t}$$

$$P = VI$$

LEARNING TIP

Electrical work and power equations for all components:

$$W = VIt$$

$$P = VI$$

WORKED EXAMPLE

In **fig B**, what is the power of the bulb?

$P = VI$

$P = 7.06 \times 1.45$

$P = 10.2\,\text{W}$

EXAM HINT

Note that the steps and layout of the solution in this worked example are suitable for questions on calculating electrical power in the exam.

POWER DISSIPATED BY RESISTORS

We can also write variations of the electrical power equation, substituting in terms from Ohm's law so that we can calculate the power dissipated by resistors in a circuit.

$$P = VI$$

and

$$V = IR$$

If we only know current and resistance:

$$P = IR \times I = I^2R$$

Or, if we only know pd and resistance:

$$P = V \times \frac{V}{R} = \frac{V^2}{R}$$

LEARNING TIP

Resistance power dissipation equations:

$$P = I^2R$$

$$P = \frac{V^2}{R}$$

EFFICIENCY

▲ **fig C** Modern electrical appliances come with a label indicating their efficiency, so that consumers can choose good value for running costs when making purchases.

From the equations above, we can easily calculate the work and power of electrical devices. However, this calculation isn't meaningful if we do not also consider the purpose of the device. If most of the energy is not actually transferred to a store that is useful to us, then the cost of that electrical energy will be wasted. We saw in **Section 1B.2** that the ability of a device to transfer energy usefully is called **efficiency**.

You will remember that efficiency is defined mathematically – and this applies to electrical energy transfers exactly as it did to mechanical energy transfers – as:

$$\text{efficiency} = \frac{\text{useful energy output}}{\text{total energy input}}$$

$$\text{efficiency} = \frac{\text{useful power output}}{\text{total power input}}$$

The answer will be a decimal between zero and one. It is common to convert this to a percentage value (multiply the decimal by 100).

EXAM HINT

Remember, work done means the same as energy transferred.

EXAM HINT

Note that the steps and layout of the solution in this worked example are suitable for questions on calculating efficiency in the exam.

WORKED EXAMPLE

In **fig B**, what is the efficiency of the bulb if it transfers 3.57 joules of light energy per second?

$$\text{efficiency} = \frac{\text{useful power}}{\text{total power input}}$$

$$\text{efficiency} = \frac{3.57}{10.2}$$

$$\text{efficiency} = 0.35 = 35\,\%$$

PRACTICAL SKILLS

Investigating efficiency

We can find the efficiency of an electric motor by measuring the electrical energy it transfers, and by comparing this with its useful energy output in lifting a mass against its weight to give the mass gravitational potential energy.

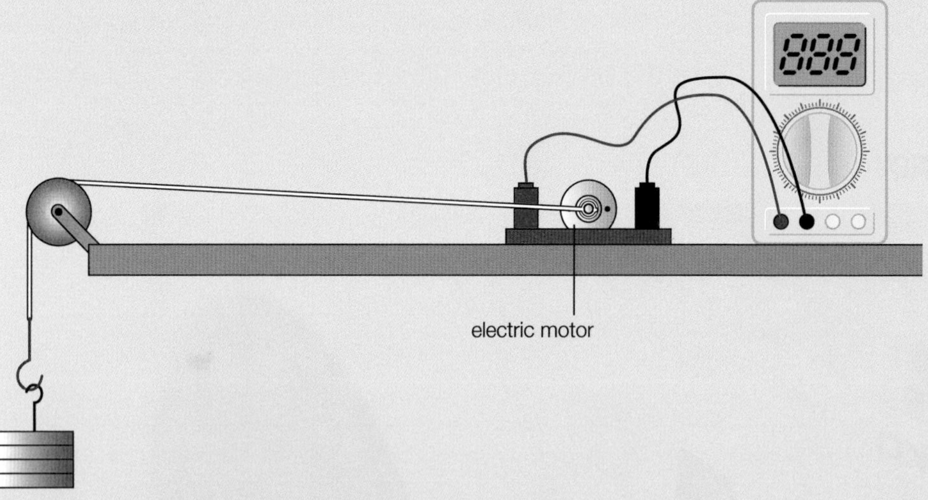

fig D Finding the efficiency of an electric motor lifting a weight.

You switch on the motor and find the time (t) it takes to lift the known mass (m) through a fixed height (Δh). During the lifting process, you measure the average pd (V) across the motor and current through it (I) using an ammeter and a voltmeter. You use these measurements to calculate the total energy input and the useful energy output for use in the efficiency equation.

useful energy output = gravitational potential energy gained = $mg\Delta h$

total energy input = electrical energy used = VIt

$$\text{motor efficiency} = \frac{\text{useful energy output}}{\text{total energy input}} = \frac{mg\Delta h}{VIt}$$

Safety Note: Secure the motor to a firm base. Only use low voltage motors and small masses.

CHECKPOINT

1. How much work is done by the resistor in the circuit in **fig E** if the circuit is switched on for 11 seconds?

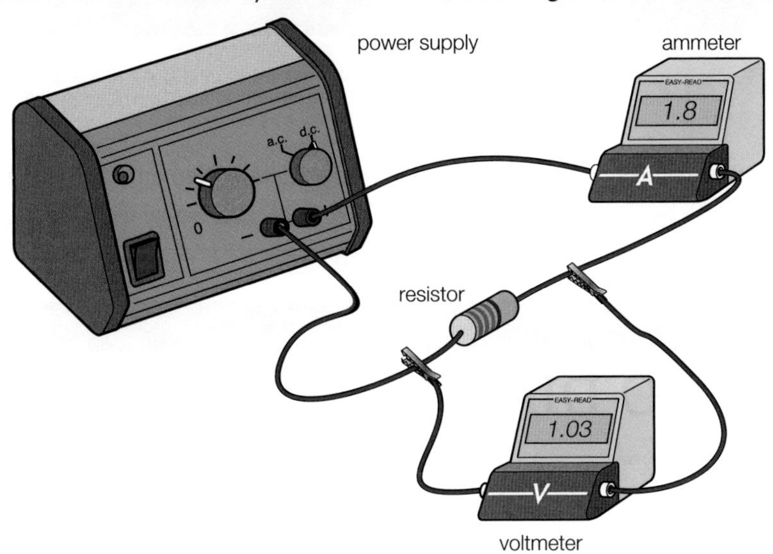

▲ **fig E** A resistor circuit.

2. What is the power dissipated by the resistor in **fig E**?

3. What power is dissipated by a 470 Ω resistor when:

(a) a current of 1.76 A passes through it?

(b) it is measured with a pd of 9.0 V?

4. (a) What is the efficiency of the electric heater in **fig F**?

(b) Why would your answer to part (a) be the same if you consider the heater switched on for one hour, and if you consider it switched on for two hours?

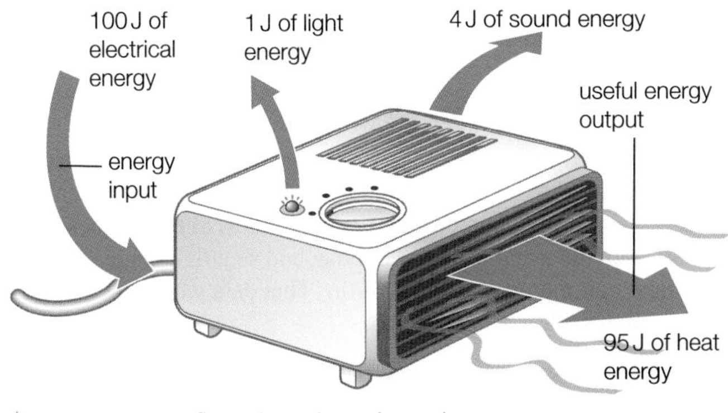

▲ **fig F** Energy flows through an electric heater.

SUBJECT VOCABULARY

efficiency the ability of a machine to transfer energy usefully:

$$\text{efficiency} = \frac{\text{useful work done}}{\text{total energy input}}$$

$$\text{efficiency} = \frac{\text{useful energy output}}{\text{total energy input}}$$

4B THINKING BIGGER

CURIOUS VOLTAGE DROP

SKILLS CRITICAL THINKING, PROBLEM SOLVING, ANALYSIS, ADAPTIVE LEARNING, PRODUCTIVITY, COMMUNICATION

One of the most important components for robotic spacecraft is their electrical power supply. The electrical system on NASA's Mars explorer, Curiosity, is designed to function over a large range of voltages, in case of supply disruption.

ONLINE NEWS STORY

ROVER TEAM WORKING TO DIAGNOSE ELECTRICAL ISSUE

This picture of NASA's Mars rover Curiosity combines 66 exposures taken by the rover's Mars Hand Lens Imager (MAHLI) during the 177th Martian day, or sol, of Curiosity's work on Mars.

Science observations by NASA's Mars rover Curiosity have been stopped for a few days while engineers run tests to check possible causes of a voltage change that was noticed on Nov. 17.

'The vehicle is safe and stable and fully capable of operating in its present condition, but we are taking the precaution of investigating what may be a soft short,' said Mars Science Laboratory Project Manager Jim Erickson at NASA's Jet Propulsion Laboratory, Pasadena, Calif.

A 'soft' short is a leak through something that's partially conductive of electricity, rather than a hard short such as one electrical wire contacting another.

The team noticed a change in the voltage difference between the chassis and the 32-volt power bus that distributes electricity to systems throughout the rover. Data indicating the change were received on Sunday, during Curiosity's 456th Martian day. The level had been about 11 volts since landing day, and is now about 4 volts. The rover's electrical system is designed with the flexibility to work properly throughout that range and more with a design feature called 'floating bus.'

A soft short can cause such a voltage change. Curiosity had already experienced one soft short on the day it landed on Mars. That one was linked to explosive-release devices used for deployments shortly before and after the landing. It lowered the bus-to-chassis voltage from about 16 volts to about 11 volts but has not affected rover operations since then.

The analysis work to find the cause of the voltage change has gained an advantage from an automated response by the rover's onboard software when it detected the voltage change on Nov. 17. The rover increased the rate at which it recorded electrical variables, to eight times per second from the usual once per minute, and sent that engineering data in its next communication with Earth. 'That data was quite helpful,' Zimmerman said.

The electrical issue did not cause the rover to enter a safe-mode status, in which most activities automatically stop until further instructions are given. Also, there is no indication the issue is related to a computer reboot that caused a 'safe-mode' earlier this month.

NASA's Mars Science Laboratory Project is using Curiosity inside Gale Crater to assess ancient habitable environments and major changes in Martian environmental conditions. JPL, a division of the California Institute of Technology in Pasadena, built the rover and manages the project for NASA's Science Mission Directorate in Washington.

From a news story on the website of NASA, www.jpl.nasa.gov/news

SCIENCE COMMUNICATION

1 The extract opposite consists of an article from the News section of the NASA website. Consider the extract and comment on the type of writing that is used. Try and answer the following questions:
  (a) Discuss the level of scientific detail included in the article, particularly considering the intended audience.
  (b) Explain **three** different reasons why NASA would post this news article.

INTERPRETATION NOTE

Think about what NASA might gain from this being made public information. NASA is the American space agency – the National Aeronautics and Space Administration.

PHYSICS IN DETAIL

Now we will look at the physics in detail. Some of these questions link to topics elsewhere in this book, so you may need to combine concepts from different areas of physics to work out the answers.

2 Curiosity rover is powered by a radioisotope electricity generating unit, which has a power of 110 watts when functioning correctly. What would be the current through the main 32 volt power bus that distributes all of the electrical power?

3 The power bus uses potential dividers to provide the appropriate voltage to each separate electrical circuit from its basic 32 V supply.
  (a) The article refers to an earlier voltage drop incident from the standard of 16 V down to 11 V. Suggest **two** resistance values that could be used in a potential divider, from the main power bus supply, to give an output voltage equal to 16 V.
  (b) If a 'soft short' had caused that drop down to 11 V, suggest what the resistance of the short circuiting object would be, based on your answer to part (a).
  (c) What further resistance would another soft short object need to further reduce the voltage from 11 V down to 4 V?

4 NASA claim that the rover can function perfectly normally across these variations in supply voltage.
  (a) Why would it be important that the plutonium cell generating emf has as low a resistance as possible, especially in the event of a short circuit reducing the overall resistance?
  (b) Explain how an item of debris touching electrical circuitry could reduce the resistance of that circuit.

THINKING BIGGER TIP

Choose any values of R_1 and R_2 for your potential divider that will share the voltage to give the correct output values. Use your initial suggested values as a basis for answering all parts of the question.

THINKING BIGGER TIP

The 'flying bus' system uses a number of different available resistances to change how the voltage is divided, depending on the emf available. Some of the components will be made from metal, and some will be made from semi-conductors. The temperature on the surface of Mars ranges from 20 °C to −150 °C, with −55 °C being a typical value.

ACTIVITY

Once the problem with Curiosity's voltage supply was determined and rectified, NASA had to produce another press release explaining this. Prepare a half page of text to be included in the second press release which explains how the materials used in the electric circuitry on the rover will be affected by the cold temperatures on Mars. Focus your explanation on the electrical properties of the materials.

4B EXAM PRACTICE

1 Which of the following is an expression for calculating the power in an electrical circuit?

A $E \times t$

B $V \times I$

C $V \times I \times t$

D $Q \times I \times t$ [1]

(Total for Question 1 = 1 mark)

2 Three 12 Ω resistors are connected together. Which value below could **not** be the total resistance of the combination?

A 2 Ω

B 4 Ω

C 8 Ω

D 18 Ω [1]

(Total for Question 2 = 1 mark)

3 A 6 V emf cell lights a filament bulb. Current and potential difference readings on the filament bulb are: $V = 5.8$ V, $I = 0.44$ A. What is the internal resistance of the cell?

A 0.20 Ω

B 0.45 Ω

C 7.8 Ω

D 13.6 Ω (1)

(Total for Question 3 = 1 mark)

4 A student drew this graph from an investigation of the pd across and current through a cell, which was connected in series to a variable resistor.

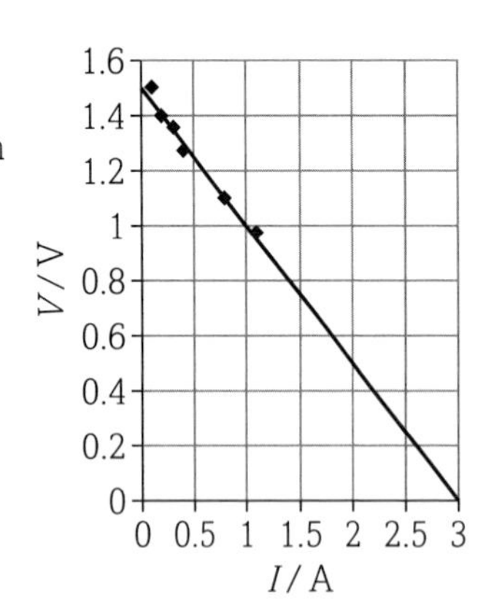

Which row in the table correctly shows the emf and internal resistance for this cell?

	Emf / V	Internal resistance / V
A	0	1.5
B	0.5	3.0
C	1.5	0.5
D	3.0	1.5

(1)

(Total for Question 4 = 1 mark)

5 Which of these is a correct definition of electromotive force (emf)?

A A supply voltage equal to the ratio of energy transferred per unit charge passing.

B The maximum voltage a component in a circuit may receive.

C A measure of the energy transferred by a component per unit charge passing through it.

D A measure of the energy transferred by a component per unit time it is connected to a power supply. [1]

(Total for Question 5 = 1 mark)

6 (a) Two lamps A and B are connected in series with a battery.

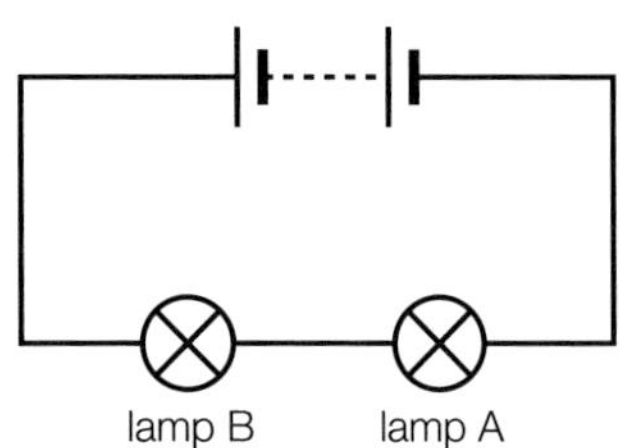

Lamp B glows more brightly than lamp A. Complete each of the sentences by choosing one of the phrases in the box.

equal to	greater than	less than

The current in lamp A is .... the current in lamp B.

The pd across lamp A is .... the pd across lamp B.

The resistance of lamp A is .... the resistance of lamp B. [3]

(b) The same bulbs are now connected in parallel with the battery.

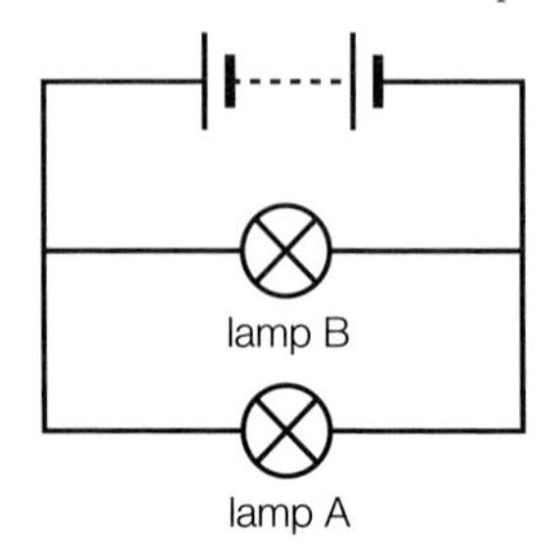

State which bulb will be brighter and explain your answer. [3]

(Total for Question 6 = 6 marks)

7 A student is asked to explain which of two filament lamps will be brighter when they are connected in parallel across a power supply. He is told that the resistance of lamp A is greater than the resistance of lamp B.

The student wrote the following explanation that contains some mistakes:

'The current is the same in both lamps. Because lamp A has a higher resistance, it is harder for the electrons to move through this lamp so they will lose more energy. Lamp A will therefore be brighter than lamp B.'

Write a correct explanation. [5]

(Total for Question 7 = 5 marks)

8 A circuit is set up as shown in the diagram. The battery has negligible internal resistance.

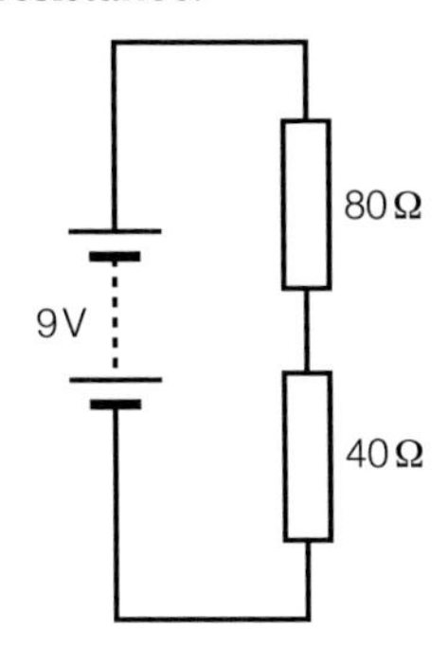

(a) Calculate the potential difference across the 40 Ω resistor. [2]

(b) A thermistor is connected in parallel with the 40 Ω resistor as shown.

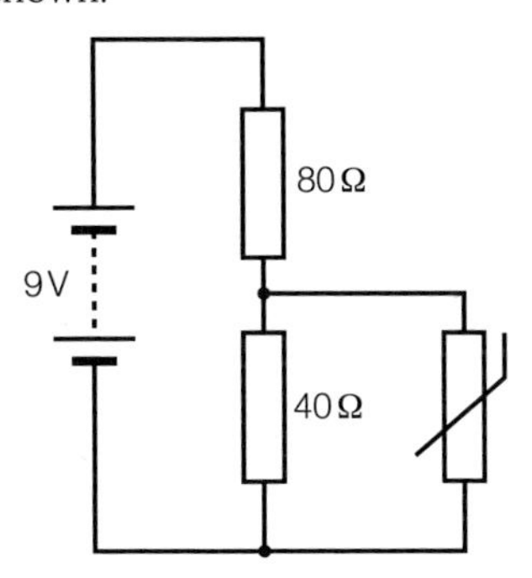

The thermistor is initially at a temperature of 100° C and its resistance is 20 Ω. As the thermistor cools down, its resistance increases.

Explain what happens to the current through the battery as the temperature of the thermistor decreases. [3]

(Total for Question 8 = 5 marks)

9 (a) A kettle is rated at 1 kW, 220 V.
Calculate the working resistance of the kettle. [2]

(b) When connected to a 220 V supplied, it takes 3 minutes for the water in the kettle to reach boiling point.
Calculate how much energy has been supplied. [2]

(c) Different countries supply mains electricity at different voltages. Many hotels now offer a choice of voltage supplies as shown in the photograph.

(i) By mistake, the kettle is connected to the 110 V supply. Assuming that the working resistance of the kettle does not change, calculate the time it would take for the same amount of water to reach boiling point. [3]

(ii) Explain what might happen if a kettle designed to operate at 110 V is connected to a 220 V supply. [2]

(Total for Question 9 = 9 marks)

10 Mobile phones have a rechargeable battery which is recharged by means of a mains adaptor. One such adaptor has an input power of 4.8 W at a voltage of 230 V.

(a) Calculate the input current to the adaptor when it is in use. [2]

(b) The adaptor's output is labelled as 5 V 0.1 A 0.5 V A.

(i) Show that the unit V A is equivalent to the watt. [1]

(ii) Calculate the efficiency of the adaptor. [2]

(iii) Suggest a reason why the efficiency is less than 100%. [1]

(Total for Question 10 = 6 marks)

11 Electrically heated gloves are used by skiers and climbers to provide extra warmth for their hands.

Each glove has a heating element of resistance 3.6 Ω. Two cells each of e.m.f. 1.5 V and internal resistance 0.2 Ω are used to operate each heating element.

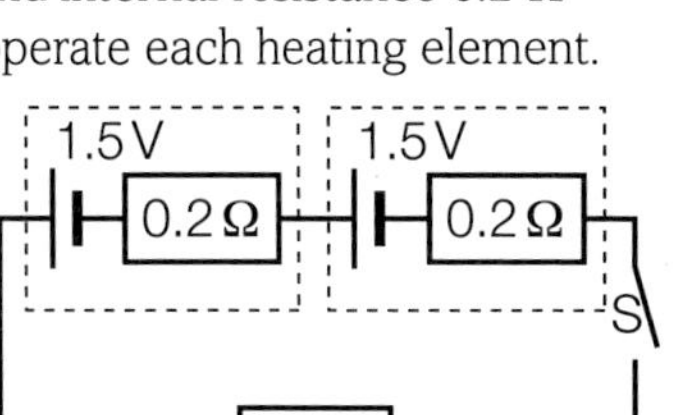

(a) When the switch is closed:

(i) Calculate the total resistance in the circuit. [1]

(ii) Calculate the current in the heating element. [2]

(iii) Calculate the power output from the heating element. [2]

(b) When in use, the internal resistance of each cell gradually increases.

State and explain the effect this will have on the power output of the heating element. [3]

(Total for Question 11 = 8 marks)

12 Explain how the current and voltage rules for electrical circuits are expressions of the conservation of charge and the conservation of energy. [6]

(Total for Question 12 = 6 marks)

MATHS SKILLS

In order to be able to develop your skills, knowledge and understanding in Physics, you will need to have developed your mathematical skills in a number of key areas. This section gives more explanation and examples of some key mathematical concepts you need to understand. Further examples relevant to your International AS/A level Physics studies are given throughout the book. In particular, the Working as a Physicist section explores important ideas about units and estimation.

ARITHMETIC AND NUMERICAL COMPUTATION

USING STANDARD FORM

Dealing with very large or small numbers can be difficult. To make them easier to handle, you can write them in the format $a \times 10^b$. This is called standard form.

To change a number from decimal form to standard form:

- Count the number of positions you need to move the decimal point by until it is directly to the right of the first number which is not zero.
- This number is the index number that tells you how many multiples of 10 you need. If the original number was a decimal, your index number must be negative.

Here are some examples:

DECIMAL NOTATION	STANDARD FORM NOTATION
0.000 000 012	1.2×10^{-8}
15	1.5×10^{1}
1000	1.0×10^{3}
3 700 000	3.7×10^{6}

WORKED EXAMPLE

Calculate the momentum of a bullet of mass 5.2 grams fired at a speed of $825\ \text{m s}^{-1}$ in kg m s^{-1}.
The momentum of this bullet would be:

$$p = m \times v$$
$$p = 5.2 \times 10^{-3} \times 825$$
$$p = 4290 \times 10^{-3}$$
$$p = 4.29\ \text{kg m s}^{-1}$$

ALGEBRA

CHANGING THE SUBJECT OF AN EQUATION

It can be very helpful to rearrange an equation to express the variable that you are interested in in terms of the variables it is related to. Always remember that any operation that you apply to one side of the equation must also be applied to the other side.

WORKED EXAMPLE

Consider the following equation which relates v = velocity, u = initial velocity, a = acceleration and s = displacement:

$$v^2 = u^2 + 2as$$

If we wished to rearrange this equation to make u the subject, we would first subtract $2as$ from each side to obtain:

$$v^2 - 2as = u^2$$

Now to obtain the formula in terms of u, we have to 'undo' the square term by doing the opposite of squaring for each side. In essence, we have to take the square root of each side:

$$\sqrt{u^2} = \sqrt{v^2 - 2as}$$
$$u = \sqrt{v^2 - 2as}$$

HANDLING DATA

USING SIGNIFICANT FIGURES

Often when you do a calculation, your answer will have many more figures than you need. Using an appropriate number of significant figures will help you to interpret results in a meaningful way.

Remember the 'rules' for significant figures:

- The first significant figure is the first figure which is not zero.
- Digits 1–9 are always significant.
- Zeros which come after the first significant figure are significant unless the number has already been rounded.

Here are some examples:

EXACT NUMBER	TO ONE S.F.	TO TWO S.F	TO THREE S.F
45 678	50 000	46 000	45 700
45 000	50 000	45 000	45 000
0.002 755	0.003	0.002 8	0.002 76

GRAPHS

UNDERSTAND THAT $y = mx + c$ REPRESENTS A LINEAR RELATIONSHIP

Two variables are in a linear relationship if they increase at a constant rate in relation to one another. If you plotted a graph with one variable on the *x*-axis and the other variable on the *y*-axis, you would get a straight line. Any linear relationship can be represented by the equation $y = mx + c$ where the gradient of the line is m and the value at which the line crosses the *y*-axis is c. An example of a linear relationship is the relationship between degrees Celsius and degrees Fahrenheit, which can be represented by the equation $F = \frac{9}{5}C + 32$ where C is temperature in degrees Celsius and F is temperature in degrees Fahrenheit.

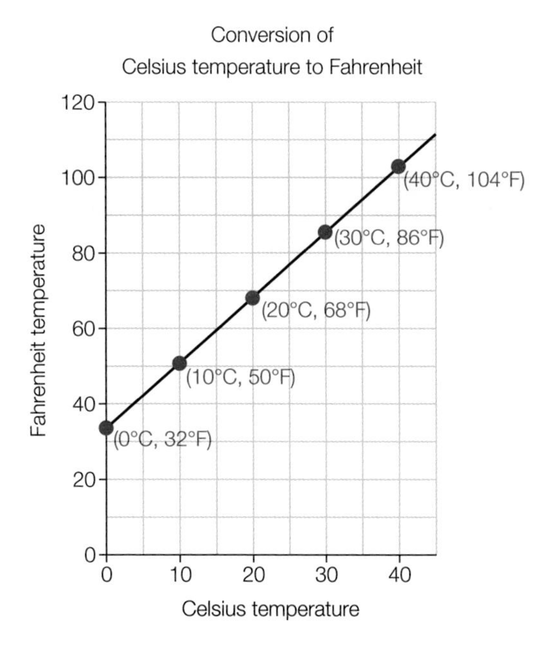

DRAW AND USE THE SLOPE OF A TANGENT TO A CURVE AS A MEASURE OF A RATE OF CHANGE

Sir Isaac Newton was fascinated by rates of change. He drew tangents to curves at various points to find the rates of change of graphs as part of his journey towards discovering the calculus – an amazing branch of mathematics. He argued that the gradient of a curve at a given point is exactly equal to the gradient of the tangent of a curve at that point.

Technique:

1 Use a ruler to draw a tangent to the curve.
2 Calculate the gradient of the tangent using the technique given for a linear relationship. This is equal to the gradient of the curve at the point of the tangent.
3 State the unit for your answer.

GEOMETRY AND TRIGONOMETRY

USING SIN, COS AND TAN WITH RIGHT-ANGLED TRIANGLES

For a right-angled triangle, Pythagoras' Theorem can be used to find the length of a side given the length of two other sides. For problems that involve angles as well as side lengths, we can use the ratios sine, cosine and tangent. These ratios are called trigonometric ratios.

$$\sin \theta = \frac{\text{opposite}}{\text{hypotenuse}}$$

$$\cos \theta = \frac{\text{adjacent}}{\text{hypotenuse}}$$

$$\tan \theta = \frac{\text{opposite}}{\text{adjacent}}$$

hypotenuse *opposite* q *adjacent*

To help you to remember these rules, you can learn a mnemonic for the letters 'SOH CAH TOA'. For example:

Some **O**ld **H**orses

Can **A**lways **H**ear

Their **O**wners **A**pproaching

FINDING THE LENGTH OF AN UNKNOWN SIDE

Follow this four-step procedure:

WORKED EXAMPLE

Calculate the length of side x.

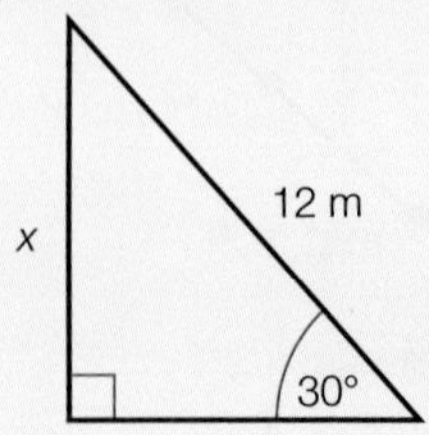

1 Summary of data

First, write down all of the data involved:

Angle = 30°
Hypotenuse = 12 m
Opposite = x

2 Selection

Next, select the appropriate trigonometric ratio using 'SOH CAH TOA'. It can be helpful to place a small tick above any variable involved.

In this case we see:

✓✓ ✓ ✓
SOH CAH TOA

3 Substitution

Write out the trigonometric ratio that you have chosen. You may need to rearrange it to express the value you are interested in.

$$\sin A = \frac{\text{opposite}}{\text{hypotenuse}}$$

$$\text{opposite} = \sin A \times \text{hypotenuse}$$

$$\text{opposite} = \sin 30 \times 12\text{ m}$$

4 Calculate and check

Calculate your answer using the required function on your scientific calculator. Then perform three checks: check that your answer is sensible, check whether it needs to be rounded and check whether you have used the correct units.

$x = 0.5 \times 12\text{ m} = 6\text{ m}$

FINDING AN UNKNOWN ANGLE

Follow the same four-step procedure. The only difference is that you will need to use the inverse function on your calculator.

WORKED EXAMPLE

Calculate the angle x.

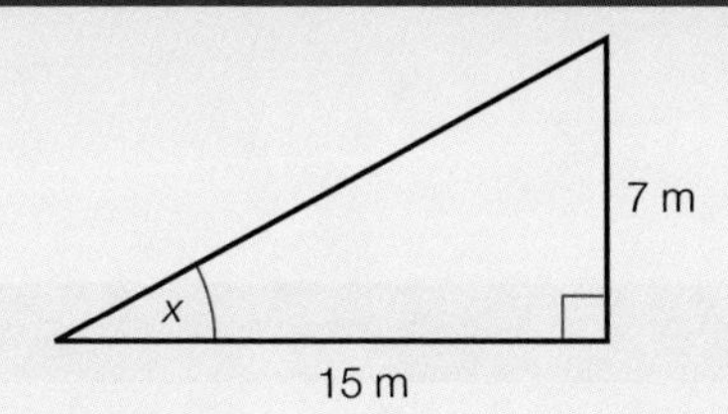

1 Summary of data

Opposite = 7 m

Adjacent = 15 m

Angle = x

2 Selection

✓ ✓ ✓✓
SOH CAH TOA

The correct ratio to use is tangent.

3 Substitution

$$\tan \theta = \frac{\text{opposite}}{\text{adjacent}}$$

$$\theta = \tan^{-1}\left(\frac{\text{opposite}}{\text{adjacent}}\right)$$

$$\theta = \tan^{-1}\left(\frac{7}{5}\right)$$

4 Calculate and check

$x = 25.016\,89$

$x = 25.0°$ (1 decimal place)

APPLYING YOUR SKILLS

You will often find that you need to use more than one maths technique to answer a question. In this section, we will look at four example questions and consider which maths skills are required and how to apply them.

WORKED EXAMPLE

A tree stands in a field. The tree is 50 m north of the gate to the field and is 20 m east of the gate.

(a) Calculate the distance of the tree from the gate.

(b) Calculate the angle of the direction to the tree from north.

You recognise a right angle triangle. You need to use Pythagoras.

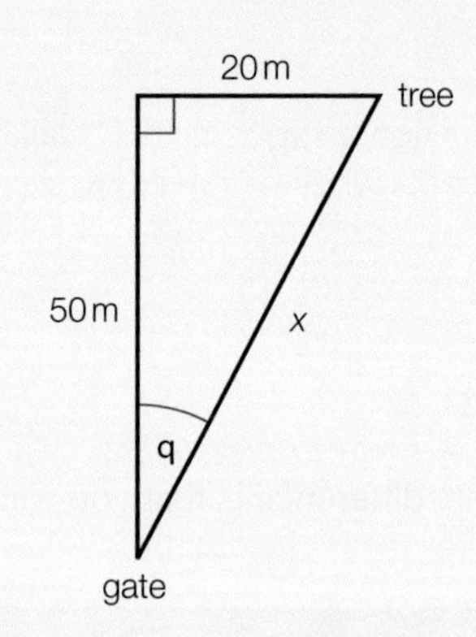

(a) $x^2 = 50^2 + 20^2$

$x^2 = 2500 + 400 = 2900$

$x = \sqrt{2900}$

$x = 54$ m (2s.f.)

(b) $\tan \theta = \frac{20}{50}$

so $\theta = 22°$

WORKED EXAMPLE

A boat sails due east (090°) at 3 m s⁻¹. The tide comes from the south-east (135°) at 1 m s⁻¹.

Draw a vector diagram to show the movement of the boat over the ground and calculate the speed.

You need to draw a diagram that represents the information in the question. If you draw a scale diagram, choose a simple scale like 1 cm : 10 m.

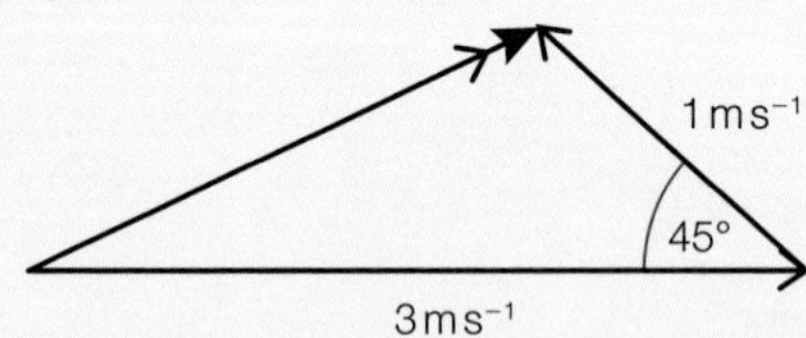

Now we must split both of the vectors up into two components, one going north and one going east.

North The boat speed vector 3 m s^{-1} has no component going north. The tide speed vector of 1 m s^{-1} has a component 1 m s^{-1} × sin 45 = **0.707 m s^{-1} going north**.

East The boat speed vector 3 m s^{-1} is going due east. The tide speed vector of 1 m s^{-1} has a component 1 m s^{-1} × cos 45 = −0.707 m s^{-1} going east; the minus sign means that the component is to the west and so we must subtract it since it reduces the boat speed to the east. So the **boat speed to the east** is (3 − 0.707) = 2.293 or **2.29 m s^{-1}** using 3 significant figures.

Now we must combine the north and east components to find the net speed. You recognise a right angle triangle. You need to use Pythagoras.

$(\text{speed})^2 = 0.707^2 + 2.29^2 = 0.500 + 5.24 = 5.74$

so speed $= \sqrt{5.74} = 2.40$ m s^{-1}.

WORKED EXAMPLE

A student measures the resistivity of some metal in the form of a wire. He starts by looking up the definition of resistivity which is $\rho = \frac{R \times A}{l}$ where R is the resistance of a length l of a wire of cross-sectional area A.

He measures the resistance of different lengths of the wire and plots a graph of his readings.

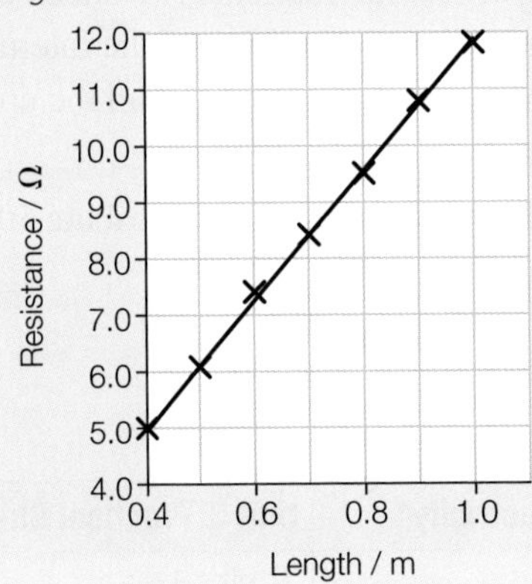

(a) *Explain whether the readings should give a line of best fit that passes through the origin.*

(b) *He measures the diameter of the wire as 0.234 mm. Determine the cross-sectional area A.*

(c) *Take measurements from the graph to show that the gradient m of the line of best fit is about $m = 11.7\,\Omega\,m^{-1}$ and hence calculate a value for the resistivity of the metal of the wire.*

(a) You need to recall the equation of a straight line $y = mx + c$ where the data is plotted on the x and y axes and m is the gradient and c is the intercept on the y-axis.

You must then compare this with the equation for resistivity where he has measured R and l.

Since $R = \rho \times \frac{l}{A}$ we can see that if R is plotted on the y-axis and l on the x-axis then the gradient of the line will be $\frac{\rho}{A}$ which is a constant so the line will be straight. There are no other terms in the resistivity equation so the value for c will be zero and the line should pass through the origin.

Note that the origin is not marked on this graph, it starts at (0.4, 4.0)

(b) You need to remember the formula for the area of a circle $A = \pi \times r^2$

Note you can only measure the diameter of a piece of wire so you must divide by 2 to get the radius r.

So $A = \left(0.234 \times \frac{10^{-2}}{2}\right)^2 = 4.30 \times 10^{-8}\,m^2$

(c) Take measurements from the graph to show that the gradient m of the line of best fit is about $m = 11.7\,\Omega\,m^{-1}$ and hence calculate a value for the resistivity of the metal of the wire.

Gradient = rise over run, so take measurements from the line of best fit – not the plots

$$\text{Gradient} = \frac{11.9 - 4.9}{1.0 - 0.4} = \frac{7}{0.6} = 11.7 = \frac{\rho}{A}$$

So $\rho = 11.7\,\Omega\,m^{-1} \times 4.30 \times 10^{-8}\,m^2 = 5.03 \times 10^{-8}\,\Omega\,m$

WORKED EXAMPLE

A ball is dropped from a height of 1.50 m above the ground.

(a) *Show that it hits the ground at about $5.4\,m\,s^{-1}$.*

(b) *The ball is now thrown horizontally at a height of 1.5 m with a horizontal velocity of $20\,m\,s^{-1}$. Calculate the angle at which it hits the ground.*

(a) You need a kinematics equation. You know $u = 0\,m\,s^{-1}$, $g = 9.81\,m\,s^{-2}$ and $s = 1.50\,m$. You need to find v.

The equation you need is $v^2 = u^2 + 2as$

So $v^2 = 0 + 2 \times 9.81 \times 1.50 = 29.4$ so $v = \sqrt{29.4} = 5.42\,m\,s^{-1}$

(b) Remember for projectiles that the horizontal velocity remains constant. While the ball is falling it is travelling horizontally at a constant velocity of $20\,m\,s^{-1}$

It hits the ground horizontally at $20\,m\,s^{-1}$ and vertically at $5.42\,m\,s^{-1}$.

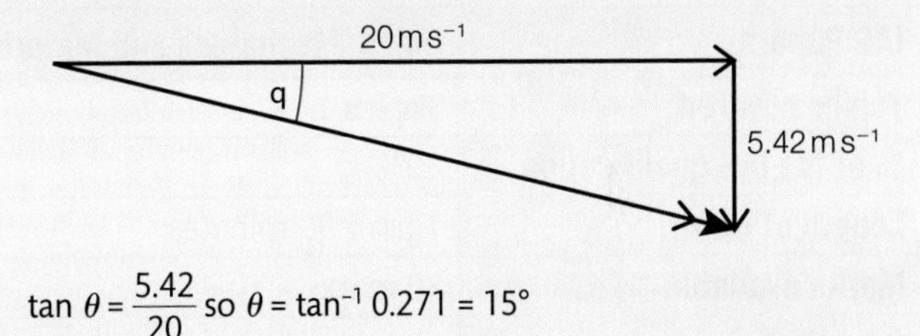

$\tan\theta = \frac{5.42}{20}$ so $\theta = \tan^{-1} 0.271 = 15°$

PREPARING FOR YOUR EXAMS

IAS AND IAL OVERVIEW

The Pearson Edexcel International Advanced Subsidiary in Physics and the Pearson Edexcel International Advanced Level in Physics are modular qualifications. The International Advanced Subsidiary can be claimed on completion of the International Advanced Subsidiary (IAS) units. The International Advanced Level (IAL) can be claimed on completion of all the units (IAS and IA2 units).

- International AS students will sit three exam papers. The IAS qualification can either be standalone or contribute 50% of the marks for the International Advanced Level.
- International A level students will sit six exam papers, the three IAS papers and three IAL papers.

The tables below give details of the exam papers for each qualification.

IAS Papers	Unit 1: Mechanics and Materials	Unit 2: Waves and Electricity*	Unit 3: Practical Skills in Physics 1
Topics covered	Topics 1–2	Topics 3–4	Topics 1–4
% of the IAS qualification	40%	40%	20%
Length of exam	1 hour 30 minutes	1 hour 30 minutes	1 hour 20 minutes
Marks available	80 marks	80 marks	50 marks
Question types	multiple-choice short open open-response calculation extended writing	multiple-choice short open open-response calculation extended writing	short open open-response calculation extended writing
Mathematics	For Unit 1 and Unit 2, a minimum of 32 marks will be awarded for mathematics at Level 2 or above. For Unit 3, a minimum of 20 marks will be awarded for mathematics at Level 2 or above.		

* This paper will contain some synoptic questions which require knowledge and understanding from Unit 1.

IAL Papers	Unit 4: Further Mechanics, Fields and Particles**	Unit 5: Thermodynamics, Radiation, Oscillations and Cosmology†	Unit 6: Practical Skills in Physics 2
Topics covered	Topics 5–7	Topics 8–11	Topics 5–11
% of the IAL qualification	20%	20%	10%
Length of exam	1 hour 45 minutes	1 hour 45 minutes	1 hours 20 minutes
Marks available	90 marks	90 marks	50 marks
Question types	multiple-choice short open open-response calculation extended writing	multiple-choice short open open-response calculation extended writing	short open open-response calculation extended writing synoptic
Mathematics	For Unit 4 and Unit 5, a minimum of 36 marks will be awarded for mathematics at Level 2 or above. For Unit 6, a minimum of 20 marks will be awarded for mathematics at Level 2 or above.		

** This paper will contain some synoptic questions which require knowledge and understanding from Units 1 and 2.

† This paper will contain some synoptic questions which require knowledge and understanding from Units 1, 2 and 4.

EXAM STRATEGY

ARRIVE EQUIPPED

Make sure you have all of the correct equipment needed for your exam. As a minimum you should take:

- pen (black ink or ball-point pen)
- pencil (HB)
- rule (ideally 30 cm)
- rubber (make sure it's clean and doesn't smudge the pencil marks or rip the paper)
- calculator (scientific).

MAKE SURE YOUR ANSWERS ARE LEGIBLE

Your handwriting does not have to be perfect but the examiner must be able to read it! When you're in a hurry it's easy to write key words that are difficult to understand.

PLAN YOUR TIME

Note how many marks are available on the paper and how many minutes you have to complete it. This will give you an idea of how long to spend on each question. Be sure to leave some time at the end of the exam for checking answers. A rough guide of a minute a mark is a good start, but short answers and multiple choice questions may be quicker. Longer answers might require more time.

UNDERSTAND THE QUESTION

Always read the question carefully and spend a few moments working out what you are being asked to do. The command word used will give you an indication of what is required in your answer.

Be scientific and accurate, even when writing longer answers. Use the technical terms you've been taught.

Always show your working for any calculations. Marks may be available for individual steps, not just for the final answer. Also, even if you make a calculation error, you may be awarded marks for applying the correct technique.

PLAN YOUR ANSWER

In questions marked with an *, marks will be awarded for your ability to structure your answer logically showing how the points that you make are related or follow on from each other where appropriate. Read the question fully and carefully (at least twice!) before beginning your answer.

MAKE THE MOST OF GRAPHS AND DIAGRAMS

Diagrams and sketch graphs can earn marks – often more easily and quickly than written explanations – but they will only earn marks if they are carefully drawn.

- If you are asked to read a graph, pay attention to the labels and numbers on the x and y axes. Remember that each axis is a number line.
- If asked to draw or sketch a graph, always ensure you use a sensible scale and label both axes with quantities and units. If plotting a graph, use a pencil and draw small crosses or dots for the points.
- Diagrams must always be neat, clear and fully labelled.

CHECK YOUR ANSWERS

For open-response and extended writing questions, check the number of marks that are available. If three marks are available, have you made three distinct points?

For calculations, read through each stage of your working. Substituting your final answer into the original question can be a simple way of checking that the final answer is correct. Another simple strategy is to consider whether the answer seems sensible. Pay particular attention to using the correct units.

SAMPLE EXAM ANSWERS

QUESTION TYPE: MULTIPLE CHOICE

The unit of electric current is the ampere. One ampere is equivalent to:

A $0.1\,C\,s^{-1}$ ☐ **B** $1\,C\,s^{-1}$ ☐

C $0.1\,s\,C^{-1}$ ☐ **D** $0.1\,C\,s$ ☐ [1]

In all types of Physics exam questions, using the correct unit in your answer is vital. Make sure you are totally familiar with all the units you need, as well as the order of magnitude prefixes, like 'mega-'.

Question analysis

- Multiple choice questions look easy until you try to answer them. Very often they require some working out and thinking. A good approach is to work out the answer as if it was an open response question, and then find your answer among the choices.
- In multiple choice questions you are given the correct answer along with three incorrect answers (called distractors). You need to select the correct answer and put a cross in the box of the letter next to it.
- If you change your mind, put a line through the box () and then mark your new answer with a cross (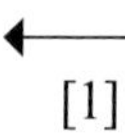).

Multiple choice questions always have one mark and the answer is given! For this reason students often make the mistake of thinking that they are the easiest questions on the paper. Unfortunately, this is not the case. These questions often require several answers to be worked out and an error in one of them will lead to the wrong answer being selected. The three incorrect answers supplied (distractors) will feature the answers that students arrive at if they make typical or common errors. The trick is to answer the question before you look at any of the answers.

Average student answer

D $0.1\,C\,s$ 

One ampere (1 A) is the movement of one coulomb (1 C) of charge per second (1 s). Therefore the correct answer is B.

COMMENTARY

This is an incorrect answer because:

- The student did not know the definition of an ampere and chose an answer with incorrect units.

If you have any time left at the end of the paper go back and check your answer to each part of a multiple choice question so that a slip like this does not cost you a mark.

QUESTION TYPE: SHORT OPEN

Write the equation for calculating the total resistance of three resistors in parallel. [1]

The command word write is used when you are being asked to write down an equation. Since only one mark is available, your equation needs to be completely correct to gain the mark.

Question analysis

- Generally one piece of information is required for each mark given in the question. There is one mark available for this question and so one piece of information is required.
- Clarity and brevity are the keys to success on short open questions. For one mark, it is not always necessary to write complete sentences.

Average student answer

$$\frac{1}{R^1} + \frac{1}{R_2} + \frac{1}{R_3}$$

While this answer is the correct idea, it would not get the mark as the student has not written an equation as the question asked.

COMMENTARY

This is an incorrect answer because:

- The student did not give a complete equation. This is easy to see because there is no equals sign!

QUESTION TYPE: OPEN RESPONSE

Describe a circuit that could be used to control the heating in a greenhouse to come on when the ambient temperature drops below a certain value. [3]

The command word in this question is describe. This means that you need to give an account of something. You do not need to include a justification or reason.

Question analysis

- With any question worth three or more marks, think about your answer and the points that you need to make before you write anything down. Keep your answer concise, and the information you write down relevant to the question. You will not gain marks for writing down physics that is not relevant to the question (even if correct) but it will cost you time.
- Remember that you can use bullet points or diagrams in y our answer.

Average student answer

A potential divider circuit with a thermistor in it could do this job. This controls the heating so it was on when too cold.

In many answers, a clear (labelled) diagram can easily gain marks. Here it is essential.

COMMENTARY

This is a weak answer because:

- The student has suggested including a thermistor but has not described what it would do.
- The student's answer restates information from the question. This will never score marks so is a poor use of time and space.

QUESTION TYPE: EXTENDED WRITING

The maximum speed limit on motorways has not been raised since the 1960s. However, developments in engineering mean that modern cars have much higher top speeds than cars in the past and can stop safely in shorter distances.

Discuss the risks and benefits of raising the maximum speed limits on motorways. [4]

It is reasonable to assume that there will be equal numbers of marks available for each side of the argument so you should balance the viewpoints you give accordingly. However, you should also remember that marks will also be available for giving an overall conclusion so you should be careful not to omit that.

Question analysis

- All extended writing answers need to discuss the physics behind the scenario. To gain full marks, your answer must be complete and coherent.
- It is vital to plan out your answer before you write it down. There is always space given on an exam paper to do this so just jot down the points that you want to make before you answer the question in the space provided. This will help to ensure that your answer is coherent and logical and that you don't end up contradicting yourself. However, once you have written your answer go back and cross these notes out so that it is clear they do not form part of the answer.

Average student answer

It would be more dangerous because with more momentum, people wouldn't be able to stop in time. Weaving in and out of the lanes would be more common as people try to get past each other at higher speeds, which would also be more dangerous. Shouldn't be done.

There is no need to make judgements in physics questions. Explaining the risks and benefits is all that is required. Including your own opinion is likely to cause you to forget to include points from the other side of the argument, just like this student has done.

COMMENTARY

This is a weak answer because:

- The phrase 'more dangerous' does not explain a specific event that will be dangerous and more likely at higher speeds.
- The student needs to explain the physics behind why these dangerous events become more likely at higher speeds.
- This student's mention of momentum does not follow up with why momentum increase means cars take longer to stop, and so it would not score a mark.
- There must be at least one benefit explained in order to score full marks.

QUESTION TYPE: CALCULATION

An electric motor takes 45.0 s to lift a mass of 800 kg through a vertical height of 14.0 m. The potential difference across the motor is 230 V and the current is 13.0 A.

Calculate the efficiency of the motor. [3]

The command word here is calculate. This means that you need to obtain a numerical answer to the question, showing relevant working. If the answer has a unit, this must be included.

Question analysis

- The important thing with calculations is that you must show your working clearly and fully. The correct answer on the line will gain all the available marks, however, an incorrect answer can gain all but one of the available marks if your working is shown and is correct.
- Show the calculation that you are performing at each stage and not just the result. When you have finished, look at your result and see if it is sensible.

Average student answer

GPE = 800 kg × 9.81 m s–2 × 14 m = 109 900 J

P = 230 V × 13.0 A = 2990 W

E = P × t = 2990 W × 45 s = 134 600 J

efficiency = 109 900 J/134 600 J = 0.816

Answers using any standard unit will be acceptable (providing correctly converted). Here, giving the final answer as 81.6% would also score full marks.

COMMENTARY

This is a strong answer because:

- The answer has been laid out very clearly so that even if an error had been made then a mark could have been awarded for part of the calculation and errors carried forward lose no further marks.

COMMAND WORDS

The following table lists the command words used across the IAS/IAL Science qualifications in the external assessments. You should make sure you understand what is required when these words are used in questions in the exam.

COMMAND WORD	THIS TYPE OF QUESTION WILL REQUIRE STUDENTS TO:
ADD/LABEL	Requires the addition or labelling to stimulus material given in the question, for example, labelling a diagram or adding units to a table.
ASSESS	Give careful consideration to all the factors or events that apply and identify which are the most important or relevant. Make a judgement on the importance of something, and come to a conclusion where needed.
CALCULATE	Obtain a numerical answer, showing relevant working. If the answer has a unit, this must be included.
COMMENT ON	Requires the synthesis of a number of factors from data/information to form a judgement. More than two factors need to be synthesised.
COMPARE AND CONTRAST	Looking for the similarities **and** differences of two (or more) things. Should not require the drawing of a conclusion. Answer must relate to both (or all) things mentioned in the question. The answer must include at least one similarity and one difference.
COMPLETE/RECORD	Requires the completion of a table/diagram/equation.
CRITICISE	Inspect a set of data, an experimental plan or a scientific statement and consider the elements. Look at the merits and/or faults of the information presented and back judgements made.
DEDUCE	Draw/reach conclusion(s) from the information provided.
DERIVE	Combine two or more equations or principles to develop a new equation.
DESCRIBE	To give an account of something. Statements in the response need to be developed as they are often linked but do not need to include a justification or reason.
DETERMINE	The answer must have an element which is quantitative from the stimulus provided, or must show how the answer can be reached quantitatively.
DEVISE	Plan or invent a procedure from existing principles/ideas.
DISCUSS	Identify the issue/situation/problem/argument that is being assessed within the question. Explore all aspects of an issue/situation/problem. Investigate the issue/situation/problem, etc. by reasoning or argument.

COMMAND WORD	THIS TYPE OF QUESTION WILL REQUIRE STUDENTS TO:
DRAW	Produce a diagram either using a ruler or freehand.
ESTIMATE	Give an approximate value for a physical quantity or measurement or uncertainty.
EVALUATE	Review information then bring it together to form a conclusion, drawing on evidence including strengths, weaknesses, alternative actions, relevant data or information. Come to a supported judgement of a subject's qualities and relation to its context.
EXPLAIN	An explanation requires a justification/exemplification of a point. The answer must contain some element of reasoning/justification; this can include mathematical explanations.
GIVE/STATE/NAME	All of these command words are really synonyms. They generally all require recall of one or more pieces of information.
GIVE A REASON/REASONS	When a statement has been made and the requirement is only to give the reasons why.
IDENTIFY	Usually requires some key information to be selected from a given stimulus/resource.
JUSTIFY	Give evidence to support (either the statement given in the question or an earlier answer).
PLOT	Produce a graph by marking points accurately on a grid from data that is provided and then drawing a line of best fit through these points. A suitable scale and appropriately labelled axes must be included if these are not provided in the question.
PREDICT	Give an expected result or outcome.
SHOW THAT	Prove that a numerical figure is as stated in the question. The answer must be to at least one more significant figure than the numerical figure in the question.
SKETCH	Produce a freehand drawing. For a graph, this would need a line and labelled axes with important features indicated; the axes are not scaled.
STATE WHAT IS MEANT BY	When the meaning of a term is expected but there are different ways of how these can be described.
SUGGEST	Use your knowledge and understanding in an unfamiliar context. May include material or ideas that have not been learnt directly from the specification.
WRITE	When the question asks for an equation.

GLOSSARY

acceleration the vector defined as the rate of change of velocity:

$$\text{acceleration (m s}^{-2}\text{)} = \frac{\text{change in velocity (m s}^{-1}\text{)}}{\text{time taken to change the velocity (s)}}$$

$$\boldsymbol{a} = \frac{\boldsymbol{v} - \boldsymbol{u}}{t} \text{ OR } \boldsymbol{a} = \frac{\Delta \boldsymbol{v}}{\Delta t}$$

ampere the unit of measurement for electric current: one amphere (1 A) is the movement of one coulomb (1 C) of change per second (1 s)

amplitude the magnitude of the maximum displacement reached by an oscillation in the wave

anchor an object, usually on a rope or chain, designed to give support

antinodes regions on a stationary wave where the amplitude of oscillation is at its maximum

Archimedes' principle the upthrust on an object is equal to the weight of fluid displaced

average speed speed for a whole journey, calculated by dividing the total distance for a journey by the total time for the journey:

$$\text{average speed (m s}^{-1}\text{)} = \frac{\text{total distance (m)}}{\text{total time (s)}}$$

bisects to divide something into two, usually equal, parts

catapult a device that can throw objects at a high speed

centre of gravity the point through which the weight of an object appears to act

charge a fundamental property of some particles. It is the cause of the electromagnetic force, and it is a basic aspect of describing electrical effects

coefficient of viscosity a numerical value given to a fluid to indicate how much it resists flow

coherence waves which must have the same frequency and a constant phase relationship. Coherent waves are needed to form a stable standing wave

compression an area in a longitudinal wave in which the particle oscillations put them closer to each other than their equilibrium state

compression a force acting within a material in a direction that would squash the material. Also, the decrease in size of a material sample under a compressive force

conduction band a range of energy amounts that electrons in a solid material can have which delocalises them to move more freely through the solid

conservation of energy the rule that requires that energy can never be created or destroyed

conservation of linear momentum the vector sum of the momenta of all objects in a system is the same before and after any interaction (collision) between the objects

constructive interference the superposition effect of two waves that are in phase, producing a larger amplitude resultant wave

coulomb, C the unit of measurement for charge: one coulomb is the quantity of charge that passes a point in a conductor per second when one ampere of current is flowing in the conductor. The amount of charge on a single electron in these units is -1.6×10^{-19} C

critical angle is the largest angle of incidence that a ray in a more optically dense medium can have and still emerge into less dense medium. Beyond this angle, the ray will be totally internally reflected

critical temperature the temperature below which a material's resistivity instantly drops to zero

Electric **current** the rate of flow of charge. Current can be calculated from the equation:

$$\text{current (A)} = \frac{\text{charge passing a point (C)}}{\text{time for that charge to pass (s)}}$$

$$I = \frac{\Delta Q}{\Delta t}$$

deformation the process of alteration of form or shape

density a measure of the mass per unit volume of a substance

destructive interference the superposition effect of two waves that are out of phase, producing a smaller amplitude resultant wave

diffraction when a wave passes close by an object or through a gap, the wave energy spreads out

displacement the position of a particular point on a wave, at a particular instant in time, measured from the mean (equilibrium) position

displacement the vector measurement of distance in a certain direction

displacement–time graph a graph showing the positions visited on a journey, with displacement on the *y*-axis and time on the *x*-axis

drift velocity the slow overall movement of the charges in a current

efficiency the ability of a machine to transfer energy usefully:

$$\text{efficiency} = \frac{\text{useful energy output}}{\text{total energy input}}$$

$$\text{efficiency} = \frac{\text{useful power output}}{\text{total power input}}$$

elastic limit the maximum extension or compression that a material can undergo and still return to its original dimensions when the force is removed

electric current rule the algebraic sum of the currents entering a junction is equal to zero:

$$\Sigma I = 0$$

electromotive force, or **emf**, a voltage as defined below, with the energy coming into the circuit

electronvolt the amount of energy an electron gains by passing through a voltage of 1 V

$$1 \text{ eV} = 1.6 \times 10^{-19} \text{ J}$$

$$1 \text{ mega electronvolt} = 1 \text{ MeV} = 1.6 \times 10^{-13} \text{ J}$$

energy the property of an object related to doing work. A change in the amount of energy of an object could be equated to work being done, even if this is not mechanical – a change in the thermal energy of a sample of gas, for example

equilibrium the situation for a body where there is zero resultant force and zero resultant moment. It will have zero acceleration

excitation an energy state for a system that is higher energy than the ground state, for example, in an atom, if an electron is in a higher energy level than the ground state, the atom is said to be 'excited'

explosion a situation in which a stationary object (or system of joined objects) separates into component parts, which move off at different velocities. Momentum must be conserved in explosions

extension an increase in size of a material sample caused by a tension force

fluid any substance that can flow

free-body force diagram diagram showing an object isolated, and all the forces that act on it are drawn in at the points where they act, using arrows to represent the forces

frequency the number of complete wave cycles per second:

$$\text{frequency (Hz)} = \frac{1}{\text{time period (s)}}$$

$$f = \frac{1}{T}$$

gradient the slope of a line or surface

gravitational potential energy (E_{grav}) the energy an object stores by virtue of its position in a gravitational field:

$$\text{gpe (J)} = \text{mass (kg)} \times \text{gravitational field strength (N kg}^{-1}) \times \text{height } (m)$$

$$E_{grav} = mgh \text{ OR } \Delta E_{grav} = mg\Delta h$$

ground state the lowest energy level for a system, for example, when all the electrons in an atom are in the lowest energy levels they can occupy, the atom is said to be in its ground state

hydrometer an instrument used to determine the density of a fluid

hysteresis where the extension under a certain load will be different depending on its history of past loads and extensions

instantaneous speed the speed at any particular instant in time on a journey, which can be found from the gradient of the tangent to a distance–time graph at that time

interference the superposition outcome of a combination of waves. An interference pattern will only be observed under certain conditions, such as the waves being coherent

internal resistance the resistance of an emf source

ionisation energy the minimum energy required by an electron in an atom's ground state in order to remove the electron completely from the atom

kinematics the study of the description of the motion of objects

kinetic energy (E_k) the energy an object stores by virtue of its movement:

$$\text{kinetic energy (J)} = \tfrac{1}{2} \times \text{mass (kg)} \times (\text{speed})^2 \text{ m}^2\text{ s}^{-2}$$

$$E_k = \tfrac{1}{2} \times m \times v^2$$

laminar flow/streamline flow a fluid moves with uniform lines in which the velocity is constant over time

limit of proportionality the maximum extension (or strain) that an object (or sample) can have, which is still proportional to the load (or stress) applied

line spectrum a series of individual lines of colour showing the frequencies present in a light source

longitudinal wave a wave in which the oscillations occur parallel to the direction of movement of the wave energy

maximum (plural **maxima**) in a diffraction or interference pattern, the bright spots

minimum (plural **minima**) in a diffraction or interference pattern, the dark spots

momentum (kg m s^{-1}) = mass (kg) × velocity (m s^{-1})

$$\boldsymbol{p} = m \times \boldsymbol{v}$$

monochromatic containing or using only one colour. Light of a single wavelength

Newton's first law of motion an object will remain at rest, or in a state of uniform motion, until acted upon by a resultant force

Newton's second law of motion if an object's mass is constant, the resultant force needed to cause an acceleration is given by the equation:

$$\sum \boldsymbol{F} = m\boldsymbol{a}$$

Newton's third law of motion for every action, there is an equal and opposite reaction

nodes regions on a stationary wave where the amplitude of oscillation is zero

Ohm's law the current through a component is directly proportional to the voltage across it, providing the temperature remains the same. The equation for this is often expressed as:

$$\text{voltage } (V) = \text{current } (A) \times \text{resistance } (\Omega)$$

$$V = I \times R$$

period (also **time period**) the time taken for one complete oscillation at one point on the wave:

$$\text{time period (s)} = \frac{1}{\text{frequency (Hz)}}$$

$$T = \frac{1}{f}$$

phase the stage a given point on a wave is through a complete cycle, measured in angle units, rad

photoelectrons electrons released from a metal surface as a result of its exposure to electromagnetic radiation

photons 'packets' of electromagnetic radiation energy where the amount of energy $E = hf$, which is Planck's constant multiplied by the frequency of the radiation: the quantum unit that is being considered when electromagnetic radiation is understood using a particle model

polarisation the orientation of the plane of oscillation of a transverse wave. If the wave is (plane) polarised, all its oscillations occur in one single plane

potential difference or **pd** the correct term for the voltage of a component that is using electrical energy in a circuit and transferring this energy to other stores

potential divider a circuit designed to provide specific voltage values by splitting an emf across two resistors

potentiometer a version of the potential divider in which a single resistance wire is used in two parts to form the two resistances. A sliding connection on the wire can be adjusted to alter the comparative resistances and thus alter the output pd from the potentiometer

power the rate of energy transfer:

$$P = \frac{E}{t} = \frac{\Delta W}{t}$$

principle of moments a body will be in equilibrium if the sum of clockwise moments acting on it is equal to the sum of the anticlockwise moments

progressive wave a means for transferring energy via oscillations

projectile a moving object on which the only force of significance acting is gravity. The trajectory is thus pre-determined by its initial velocity

quantisation the concept that there is a minimum smallest amount by which a quantity can change: infinitesimal changes are not permitted in a quantum universe. The quantisation of a quantity is like the idea of the precision of an instrument measuring it

rarefaction an area in a longitudinal wave in which the particle oscillations put them further apart from each other than their equilibrium state

refraction a change in wave speed when the wave moves from one medium to another. There is a corresponding change in wave direction, governed by Snell's law

refractive index, n, the amount that a material changes the speed of waves when they pass through the material from a different material:

$$\text{refractive index} = \frac{\text{speed of light in vacuum}}{\text{speed of light in the medium}}$$

$$n = \frac{c}{v}$$

resistance the opposition to the flow of electrical current. It can be calculated from the equation:

$$\text{resistance } (\Omega) = \frac{\text{potential difference } (V)}{\text{current } (A)}$$

$$R = \frac{V}{I}$$

resistivity for a material, the same value as the resistance between opposite faces of a cubic metre of the material

resolution or **resolving vectors** the determination of a pair of vectors, at right angles to each other, that sums to give the single vector they were resolved from

resultant force the total force (vector sum) acting on a body when all the forces are added together accounting for their directions

scalar a quantity that has magnitude only

semiconductors materials with lower resistivity than insulators, but higher than conductors. They usually only have small numbers of delocalised electrons that are free to conduct

Snell's law a rule governing the change in direction of waves as they are refracted:

$$n_1 \sin\theta_1 = n_2 \sin\theta_2$$

The values of n_1 and n_2 are the refractive indices in each medium, and the values of θ_1 and θ_2 are the angles that the ray makes to the normal to the interface between the two media, at the point the ray meets that interface

sonometer an apparatus for experimenting with the frequency relationships of a string under tension, usually consisting of a horizontal wooden sounding box and a metal wire stretched along the top of the box

speed the rate of change of distance:

$$\text{speed } (\text{m s}^{-1}) = \frac{\text{distance (m)}}{\text{time (s)}}$$

$$v = \frac{d}{t}$$

spring constant the Hooke's law constant of proportionality, k, for a spring under tension

stationary or **standing wave** a wave which has oscillations in a fixed space, with regions of significant oscillation and regions with zero oscillation, which remain in the same locations at all times

strain a proportionate measure of the extension (or compression) of a sample:

$$\text{strain (no units)} = \frac{\text{extension (m)}}{\text{original length (m)}}$$

$$\varepsilon = \frac{\Delta x}{x}$$

streamlines lines of laminar flow in which the velocity is constant over time

stress a proportionate measure of the force on a sample:

$$\text{stress (pascals Pa, or N m}^{-2}) = \frac{\text{force (N)}}{\text{cross-sectional area (m}^2)}$$

$$\sigma = \frac{F}{A}$$

stopping voltage the minimum voltage needed to reduce the photoelectric current to zero, when illuminated with a particular frequency of light

superconductivity the electrical property of a material having zero resistivity

superposition when more than one wave is in the same location, the overall effect is the vector sum of their individual displacements at each point where they meet

tension a force acting within a material in a direction that would extend the material

terminal velocity the velocity of a falling object when its weight is balanced by the sum of the drag and upthrust acting on it

thermal connected with heat

threshold frequency the minimum frequency of electromagnetic radiation that can cause the emission of photoelectrons from the metal

total internal reflection (TIR) waves reflect back into the same medium at a boundary between two media. This requires two conditions to be met:

- the ray is attempting to emerge from the more dense medium
- the angle between the ray and the normal to the interface is greater than the critical angle

transport equation $I = nAvq$. This defines electric current, I, from a fundamental basis. It is the product of the number density of charge carriers, n; the charge on those carriers, q; the cross-sectional area of the conductor, A; and the drift velocity of the charge carriers in that conductor, v

transverse wave a wave in which the oscillations occur perpendicular to the direction of movement of the wave energy

turbulent flow fluid velocity in a particular place changes over time, often in an unpredictable manner

twin-beam oscilloscope an oscilloscope with two inputs. It displays each as a line on the screen, and both are shown at the same time to compare the inputs

uniform motion motion when there is no acceleration:

$$\text{velocity } (\text{m s}^{-1}) = \frac{\text{displacement (m)}}{\text{time (s)}}$$

$$\mathbf{v} = \frac{\mathbf{s}}{t}$$

upthrust an upwards force on an object caused by the object displacing fluid

valence band a range of energy amounts that electrons in a solid material can have which keeps them close to one particular atom

vector a quantity that must have both magnitude and direction

velocity the rate of change of displacement:

$$\text{velocity } (\text{m s}^{-1}) = \frac{\text{displacement (m)}}{\text{time (s)}}$$

$$\mathbf{v} = \frac{\mathbf{s}}{t} \text{ OR } \mathbf{v} = \frac{\Delta \mathbf{s}}{\Delta t}$$

velocity–time graph a graph showing the velocities on a journey, with velocity on the y-axis and time on the x-axis

viscosity how resistant a fluid is to flowing

voltage a measure of the amount of energy a component transfers per unit of charge passing through it. It can be calculated by the equation:

$$\text{voltage (V)} = \frac{\text{energy transferred (J)}}{\text{charge passing (C)}}$$

$$V = \frac{E}{Q}$$

voltages circuit rule around a closed loop, the algebraic sum of the emfs is equal to the algebraic sum of the pds:

$$\Sigma\varepsilon = \Sigma IR$$

wave a means for transferring energy via oscillations

wave equation:

$$\text{wave speed (m s}^{-1}\text{)} = \text{frequency (Hz)} \times \text{wavelength (m)}$$

$$v = f\lambda$$

wave speed the rate of movement of the wave (not the rate of movement within oscillations)

wavefronts lines connecting points on the wave that are at exactly the same phase position

wavelength the distance between a point on a wave and the same point on the next cycle of the wave

wave–particle duality the principle that the behaviour of electromagnetic radiation can be described in terms of both waves and photons

work done in a mechanical system. This is the product of a force and the distance moved in the direction of the force:

$$\text{work done (J)} = \text{force (N)} \times \text{distance moved in the direction of the force (m)}$$

$$\Delta W = \boldsymbol{F}\Delta \boldsymbol{s}$$

work function the minimum energy needed by an electron at the surface of a metal to escape from the metal

yield point a strain value beyond which a material undergoes a sudden and large plastic deformation

Young modulus the stiffness constant for a material, equal to the stress divided by its corresponding strain

INDEX

Picture Credits
The publisher would like to thank the following for their kind permission to reproduce their photographs:

(Key: b-bottom; c-centre; l-left; r-right; t-top)

123RF: David Holm 48t, Khunaspix 45tr; **Alamy Stock Photo:** Sciencephotos 62, Blickwinkel 63, Brian Jackson 100-100, Buzz Pictures 45br, Cal Sport Media 50b, Chris Hellier 70r, Chuck Myers/ZUMA Press 42r, Cultura RM 60, Daimages Photo Agency 42l, David A. Eastley 117, Derek Bayes/Lebrecht Music & Arts 20, dpa picture alliance archive 48l, dpa picture alliance 64c, Education Images/ Universal Images Group North America LLC 130-131, Hugh Threlfall 193c, Justin Kase z11z 11b, vi, Paul John Fearn 150t, Phil Crow 11t, vi, Pictorial Press Ltd 178, Richard Cooke 108l, Simon Belcher 187, Stephen Barnes/Sport 39, Terry Oakley 32, viic; **Corbis:** Aurelien Meunier/Icon 155, Markus Moellenberg 54l, Monalyn Gracia 84; **Fotolia:** Anthony Brown 2-3, Boggy 40, Dario Lo Presti 50t, Joggie Botma 67l; **Getty Images:** AFP 4tl, Alexander Gerst/ESA 2-3, Brian Sytnyk/Photographer's Choice 88-89, Deep Desert Photo/RooM 148-149, FelixR/ E+ 172-173, Fuse 167, Gary Gladstone 3b, Hanneke Luijting 36t, Jamie Terry/ EyeEm 74-75, Jonathan Wood 65, Matthew Lewis 15, Moazzam Ali Brohi 8-9, viit, Oliver Furrer/Photographer's Choice 34-35, Peter Cade 3t, Rob Monk/Procycling Magazine 54r, Stocktrek Images 46-47, Westend61 58-58, Zeljkosantrac 73tl; **GNU**: 121; **IBM Research**: 139cr; **Jeff La Favre:** 114; **Lascells Ltd:** 124; **Library of Congress:** Philosophiae naturalis principia mathematica/Isaac Newton 49; **Martyn F. Chillmaid:** 14 , 38 , 91t; **NASA:** JPL-Caltech/MSSS 190; **National Geophysical Data Center:** 96; **National Institute of Standards and Technology:** 4br; **OBO Goalkeeping, New Zealand:** 54b; **Pearson Education:** 86, 87, Coleman Yuen 109b, Martyn F. Chillmaid 150c, Trevor Clifford 52t, 52b, 57, 64tl, 83, 182, Tsz-shan Kwok ix, 120, 159; **PhotoDisc:** Doug Menuez 48r; **Physio Control:** 168; **Prof. (emeritus) Dr. Michael M. Raith:** 126l; **Reprinted courtesy of the Central Research Laboratory, Hitachi, Ltd., Japan:** Demonstration of single-electron buildup of an interference pattern: Tonomura, A.; Endo, J. ; Matsuda, T.; Kawasaki, T.; Ezawa,H. American Journal of Physics, Volume 57, Issue 2, pp.117–120 (1989) 139cl; **Rex Features:** LumiGram 129; **Robert Harding World Imagery:** Amanda Hall 64br, James Emmerson 64bl; **Science & Society Picture Library:** Science Museum 133; **Science Photo Library:** 67r, Andrew Lambert Photography 105, 108r, 139tl, Cordelia Molloy 193b, Edward Kinsman 103, 109c, 142t, 142b, Erich Schrempp 112, GIPhotostock 122, 132, 138, Jim Amos 91b, Peter Aprahamian/ Sharples Stress Engineers Ltd 125r, Richard Beech Photography 118-119, Sheila Terry 72tr; **Shutterstock:** Brian Kinney 36b, catwalker 6r, Drohn 126r, Gualtiero boffi 17, Ilona Ignatova 37t, imagedb.com viib, 30, Leremy 51, mTaira 48b, Naipung 110, Pete Niesen vi, 10, Salvador Tali 153, Sebastian Kaulitzk 139tr, Serhii Borodin 6l, spe 70l, Vadim Sadovski 2-3, vectorlib.com 44, VoodooDot 6l; **Toone Images:** 156; **TSG@MIT Physics:** 125l; **Veer/Corbis:** Jason Maehl 37b; **Wikipedia:** Fizped/ Zátonyi Sándor 123;
All other images © Pearson Education
We are grateful to the following for permission to reproduce copyright material:

Figures
Figure on page 14 adapted from Acceleration–time graph of a skydiver, 'Exploring Mathematics with Sage' by Paul Lucas, http://www.arachnoid.com/sage/ terminal_velocity.html, copyright © 2009, P. Lutus; Figure on page 17 adapted from 'balanced seesaw', www.animatedscience.co.uk, copyright © 2014 Animated Science. Reproduced by kind permission; Figures on page 54 adapted from http:// www.obo.co.nz/the-o-lab, OBO. Reproduced by permission; Figure on page 82 adapted from 'stress strain graph for an IMPRESS Alloy', Mechanical%20 Properties/Question_Mechanical_ Properties_23.html, copyright © ESA; Figures on page 84 from 'static elongation test' and 'fall test', copyright © Pit Schubert, Neville McMillan and Georg Sojer, 2009; and 'new equations', Dynamic Mountaineering Ropes, UIAA-101, EN892, www.theuiaa.org, copyright © UIAA. Reproduced with permission; Figures on page 84 adapted from Mountaineering: The Freedom of the Hills, 6th Edition, published by Mountaineers Books, Seattle and Quiller Publishing, Shrewsbury 148, copyright © 1997. Illustration and table reprinted with permission of the publisher; Figure on page 94 adapted from 'Appendix: Pressure waves vs. Displacement waves. Physics 2010, Sound Waves Experiment 6' http:// www.colorado.edu/physics/phys2010/phys2010LabMan2000/2010labhtml/Lab6/ EXP6LAB99.html, copyright © Regents of the University of Colorado; Figures on page 96 adapted from 'Seismic Waves, 'and ' Seismogram recorded in the UK' http://www.bgs.ac.uk/discoveringGeology/hazards /earthquakes/, copyright © NERC 2015; Figure on page 103 adapted from 'Physics 3B, Principle of superposition' http://www.greenwood.wa.edu.au/resources/Physics%203B%20 WestOne/content /001_mechanical_waves/page_15.html, SCIENCE1194, copyright © WestOne Services. Image courtesy of Dept Training & Workforce Development WA; Figure on page 103 adapted from 'Did the Draupner wave occur in a crossing sea?', Proceedings A, DOI: 10.1098/rspa.2011.0049, Figure 1 (T. A. A. Adcock, P. H. Taylor, S. Yan, Q. W. Ma, P. A. E. M. Janssen), June 2011, The Royal Society, copyright © 2015, The Royal Society; Figures on pages 105 and 106 adapted from 'Creating musical sounds'. Figure 6: Formation of a standing wave from two travelling waves travelling in opposite directions, http://projects.kmi.open.ac.uk/ role/moodle/mod/page/view.php?id=1139; and 'Creating musical sounds'. 5.4 Vibrating string: normal modes of vibration. Figure 10: The first four normal modes of vibration o f a string fixed at each end, http://projects.kmi.open.ac.uk/role/, ROLE. Reproduced by permission of The Open University; Figure on page 112 adapted from Friday, November 12, 2010, Wave Theory and Principles Interpretted. Part II, Source: http://fish-kc79.blogspot.co.uk/2010/11/wave-theory-and-principles-interpretted_12.html; Figure on page 112 adapted from 'Waves Tutorial 7 – Interference, Q4', http://www.antonineeducation.co.uk/Pages/Physics_2/Waves/ WAV_07/Waves _Page_7.html Reproduced with permission; Figures on page 112 from 'Tuning the Marimba Bar and Resonator', 'third torsional mode', http://lafavre. us/tuning-marimba,copyright © 2007 Jeffrey La Favre; Figure on page 122 adapted from 'The Digital Camera: Snell's Window (the view from a fish's eye)', 4 August 2011, http://uptrout.com/2011/08/04/the-digital-camera-snells-window-the-view-from-a-fishs-eye/, copyright © 2011 Timothy Schulz; Figure on page 125 adapted from 'Reflection can polarise waves' in Glare and Polarized Light in Machine Vision Applications by Chris Walker, 08/17/2012, http://chriswalkertechblog.blogspot. co.uk/2012/08/glare-and-polarized-light-in-machine.htm. Reproduced with kind permission; Figure on page 126 adapted from Guide to Thin Section Microscopy, Second Edition by Michael M. Raith, Peter Raase and Jürgen Reinhardt, Fig. 4-18, 2012, http://www.minsocam.org/msa/openaccess_publications/Thin_Sctn_ Mcrscpy_2_rdcd_eng.pdf, copyright © 2012 by M.M. Raith (University of Bonn) P. Raase (University of Kiel), J. Reinhardt (University of KwaZulu-Natal); Figure on page 132 from 'Huygens' postulate from Huygens' Construction', http://www.a-levelphysicstutor.com/wav-Huygens.php, copyright © 2011 A-level Physics Tutor. All rights reserved; Figure on page 137 adapted from 'KE vs frequency', from 'Resource Lesson Famous Discoveries: The Photoelectric Effect', http:// dev.physicslab.org/Document.aspx?doctype=3& filename=AtomicNuclear_ PhotoelectricEffect.xml, copyright © 1997–2014, Catharine H. Colwell. All rights reserved; Figure on page 144 from 'Amazing Map: Total Solar Panels To Power The United States', December 23, 2013, by Ken Jorgustin, http://modernsurvivalblog. com/alternative-energy/amazing-total-area-of-solar-panels-to-power-the-united-states/, map data copyright © Google; Figure on page 164 adapted from 'What is an electrical resistance?', http://www.artinaid.com/2013/04/what-is-an-electrical-resistance/.Reproduced by permission of Artinaid; Figure on page 165 'Diodes' adapted from http://www.learnabout-electronics.org/diodes_01.php, copyright © 2007–2013 Eric Coates MA BSc. (Hons) All rights reserved; Figure on page 168 adapted from 'Current waveform image and data table for Lifepak 1000', p.52, A-2, http://www.physiocontrol.com/uploadedFiles/Physio85/Contents/Emergency_ Medical_Care/Products/Operating_Instructions/LP1000_OI_3205213008.pdf, copyright © 2012 Physio-Control, Inc.